AF389202

Advances in Angle-Only Filtering and Tracking in Two and Three Dimensions

Advances in Angle-Only Filtering and Tracking in Two and Three Dimensions

Editors

Mahendra Mallick
Ratnasingham Tharmarasa

MDPI • Basel • Beijing • Wuhan • Barcelona • Belgrade • Manchester • Tokyo • Cluj • Tianjin

Editors
Mahendra Mallick   Ratnasingham Tharmarasa
Independent Consultant   McMaster University
USA   Canada

Editorial Office
MDPI
St. Alban-Anlage 66
4052 Basel, Switzerland

This is a reprint of articles from the Special Issue published online in the open access journal *Sensors* (ISSN 1424-8220) (available at: https://www.mdpi.com/journal/sensors/special_issues/ Angle-Only_Filtering_Tracking).

For citation purposes, cite each article independently as indicated on the article page online and as indicated below:

LastName, A.A.; LastName, B.B.; LastName, C.C. Article Title. *Journal Name* **Year**, *Volume Number*, Page Range.

ISBN 978-3-0365-6854-6 (Hbk)
ISBN 978-3-0365-6855-3 (PDF)

Contents

About the Editors

Mahendra Mallick

Mahendra Mallick received an M.S. degree in computer science from Johns Hopkins University, Baltimore, MD, USA, in 1987, and a Ph.D. degree in quantum solid-state theory from the State University of New York, Albany, NY, USA, in 1981. He is a co-editor and a co-author of the book entitled Integrated Tracking, Classification, and Sensor Management: Theory and Applications (New York, NY, Wiley/IEEE, 2012). His research interests include nonlinear filtering, multisensor multitarget tracking, multiple hypothesis tracking, random-finite-set-based multitarget tracking, satellite orbit and attitude determination, over the horizon radar tracking, and distributed fusion.

Dr. Mallick was the Associate Editor-in-Chief of the online journal of the International Society of Information Fusion (ISIF) in 2008–2009. He was the Lead Guest Editor of the special issue on "Multitarget Tracking" in the IEEE Journal of Selected Topics in Signal Processing in June 2013. He is currently an Associate Editor for target tracking and multisensor systems of the IEEE Transactions on Aerospace and Electronic Systems. He is the Lead Guest Editor of the Sensors 2021-2022 Special Issue on Advances in Angle-Only Filtering and Tracking in Two and Three Dimensions. He is a Senior Life Member of IEEE.

Ratnasingham Tharmarasa

Ratnasingham Tharmarasa received the B.Sc. Eng. degree in electronic and telecommunication engineering from University of Moratuwa, Sri Lanka in 2001, and the M.A.Sc and Ph.D. degrees in electrical engineering from McMaster University, Canada in 2003 and 2007, respectively.

Currently he is an assistant professor in the Electrical and Computer Engineering Department at McMaster University, Canada. From 2001 to 2002 he was an instructor in electronic and telecommunication engineering at the University of Moratuwa, Sri Lanka. During 2002-2007 he was a graduate student/research assistant in ECE department at the McMaster University, Canada. In 2008, he has worked as a researcher at DRS Technologies Canada Limited, Ottawa, Canada. From 2008 to 2019 he was a research associate in ECE department at the McMaster University, Canada. He is currently an Associate Editor for Elsevier Signal Processing. He is the Guest Editor of the Sensors 2021-2022 Special Issue on Advances in Angle-Only Filtering and Tracking in Two and Three Dimensions.

His main areas of expertise are target tracking, information fusion, sensor resource management and performance prediction with application to surveillance systems, autonomous vehicles and intelligent transportation. He has published more than 60 peer-reviewed journal articles and 65 conference papers in the above areas, in addition to three book chapters. He has also worked on the development of many real-world target tracking, sensor fusion and resource management systems.

Preface to "Advances in Angle-Only Filtering and Tracking in Two and Three Dimensions"

Two-dimensional bearing-only filtering (BOF) arises in many real-world tracking problems, including underwater tracking using a passive sonar, aircraft surveillance using a passive radar, robot navigation using a passive sonar, and undersea exploration of natural resources using sonar. Single-sensor BOF is also a challenging nonlinear filtering problem due to poor observability and the nonlinear measurement model. This filtering problem and the associated tracking problem have been studied extensively.

Angle-only filtering (AOF) in 3D is the counterpart of BOF in 2D. Real-world AOF problems include passive ranging using an infrared search and track (IRST) sensor, passive sonar, passive radar in the presence of jamming, ballistic missile and satellite tracking using a telescope, satellite-to-satellite passive tracking, and missile guidance using bearing-only seekers. The number of publications in the AOF and angle-only tracking in 3D is rather limited compared with the corresponding problems in 2D. This Special Issue contains twelve papers, which are grouped into three sections. The first section deals with nonlinear filters, estimators, and localization, and has six papers.

The first paper by Lebon et al. discusses target motion analysis (TMA) using cosine of conical angles acquired by a towed array. It estimates the trajectory of a target moving with constant velocity (CV) at an unknown constant depth. The sound emitted by the target can reach the antenna through the direct path as well as through bounces off the sea bottom and/or off the sea surface. The authors analyze the observability by first identifying the presence of ghost targets and then estimating the target trajectory efficiently.

The second paper by Bu et al. considers the pseudolinear Kalman filter (PLKF) for the bearings-only tracking problem. Current PLKF algorithms use bias compensation to improve the accuracy of the state estimator. The authors present a stable PLKF without bias compensation in the minimum mean square error (MMSE) framework and compare their results with the posterior Cramer–Rao Lower Bound (PCRLB).

The next paper by Li et al. studies the hybrid source localization problem using differential received signal strength (DRSS) and angle of arrival (AOA) measurements. The authors develop a closed-form pseudolinear estimator by incorporating the AOA measurements into a linearized form of DRSS equations. Finally, they propose a selected-hybrid-measurement-weighted instrumental variable (SHM-WIV) estimator that has superior bias reduction and mean-squared error performance.

In the fourth paper, Dogancay studies the optimal sensor placement for target localization using AOA measurements in 2D with a Gaussian prior. Optimal sensor locations are determined analytically for a single AOA sensor using the D- and A-optimality criteria and an approximation of the Bayesian Fisher information matrix (BFIM). Then, these results are extended to a target moving with nearly constant velocity (NCV). For the NCV trajectory of the target, it is observed that the two optimality criteria generate significantly different optimal sensor trajectories.

Luo et al. consider the robust bearings-only source localization in impulsive noise with symmetric distribution based on the Lp-norm minimization criterion. To reduce the bias due to the correlation between system matrices and noise vectors, the authors propose a Total Lp-norm Optimization (TLPO) algorithm by minimizing the errors in all elements of the system matrix and data vector based on the minimum dispersion criterion. They obtain an equivalent form of TLPO and propose two algorithms to solve the TLPO problem by using Iterative Generalized Eigenvalue

Decomposition (IGED) and Generalized Lagrange Multiplier (GLM), respectively.

Oliva et al. investigates the TMA with bearings-only measurements in 2D using an intelligence-aware batch processing algorithm. A constrained maximum likelihood estimation (MLE) is developed by extending the Cramér–Rao lower bound (CRLB) for the MLE problems with inequality constraints. The ownship motion is selected based on the Artificial Potential Fields technique that is typically used by mobile robots to reach a goal while avoiding obstacles. The MLE is performed by evolutionary ant colony optimization software.

Three papers in the next section investigate 3D AOF . The paper by Mitra et al. considers the 3D tracking of a ground target with terrain uncertainty and bias in angle measurements of an airborne sensor. They derive equations for the PCRLB by considering terrain uncertainty and sensor measurement bias in addition to the process and measurement noises. It is shown that the biased PCRLB provides a tighter lower bound when compared with the PCRLB while evaluating the position error. The authors propose an algorithm to select optimal targets of opportunity and optimal platform trajectory to estimate the bias using the biased PCRLB.

The second paper by Mallick et al. studies AOF of a maneuvering target in 3D using bearing and elevation measurements from a passive IRST sensor. The target moves with a nearly constant turn (NCT) in the XY-plane and nearly constant velocity (NCV) along the Z-axis. The NCT motion in the XY-plane cannot be discretized exactly. The continuous-time NCT model is discretized using the first and second-order Taylor approximations and the polar velocity and Cartesian-velocity-based states for the NCT model in a cubature Kalman filter (CKF). Numerical results for realistic scenarios show that the second-order Taylor approximation provides the best accuracy using the polar velocity or Cartesian-velocity-based models.

The third paper by Urooj et al. considers 2D and 3D AOF using the maximum correntropy (MC) Kalman filters, where the measurement noise is non-Gaussian. The authors formulate the UKF and new sigma point Kalman filter (NSKF) using the Gaussian kernel (GK) and Cauchy kernel (CK) in the MC framework. Consequently, they develop the MC-UKF-CK, MC-NSKF-GK and MC-NSKF-CK for the non-Gaussian AOF filtering problems.

Tracking and localization with multiple angle-only sensors are addressed in the third section. Wei et al. consider 2D bearings-only multisensor-multitarget tracking using multidimensional assignment (MDA). They propose a new coarse gating strategy to reduce the computational cost of the MDA algorithm. The two-stage multiple hypothesis tracking framework uses the proposed coarse gating strategy for multisensor-multitarget tracking.

Chen et al. investigate the multitarget localization problem using a sensor network with 3D angle-only sensors. The authors propose a data association technique based on minimum distance and analyze the target localization accuracy as a function of the angle measurement accuracy and platform self-positioning accuracy. Their numerical result shows that the proposed algorithm can achieve a required data association rate and a high positioning accuracy with a low computation cost.

The paper by Zhou et al. considers the position and velocity estimation of a target moving with constant velocity using a passive sensor network with communication links. The sensors move with constant velocity and measure the azimuth and elevation angles. The authors present centralized and distributed fusion algorithms based on least squares (LS) and analyze the communication and computation requirements of each algorithm.

Mahendra Mallick and Ratnasingham Tharmarasa

Editors

Article

TMA from Cosines of Conical Angles Acquired by a Towed Array

Antoine Lebon [1], Annie-Claude Perez [2], Claude Jauffret [2,*] and Dann Laneuville [3]

1   Naval Group, 83190 Ollioules, France; antoine-lebon@etud.univ-tln.fr
2   Université de Toulon, Aix Marseille Univ, CNRS, IM2NP, CS 60584, 83041 Toulon, CEDEX 9, France; annie-claude.perez@univ-tln.fr
3   Naval Group—Technocampus Océan, 44340 Bouguenais, France; dann.laneuville@naval-group.com
*   Correspondence: claude.jauffret@univ-tln.fr; Tel.: +33-494-142-414

Abstract: This paper deals with the estimation of the trajectory of a target in constant velocity motion at an unknown constant depth, from measurements of conical angles supplied by a linear array. Sound emitted by the target does not necessarily navigate along a direct path toward the antenna, but can bounce off the sea bottom and/or off the surface. Observability is thoroughly analyzed to identify the ghost targets before proposing an efficient way to estimate the trajectory of the target of interest and of the ghost targets when they exist.

Keywords: target motion analysis; observability; fisher information matrix; Cramér–Rao lower bound; conical angles; nonlinear estimation

Citation: Lebon, A.; Perez, A.-C.; Jauffret, C.; Laneuville, D. TMA from Cosines of Conical Angles Acquired by a Towed Array. *Sensors* **2021**, *21*, 4797. https://doi.org/10.3390/s21144797

Academic Editors: Mahendra Mallick and Ratnasingham Tharmarasa

Received: 18 June 2021
Accepted: 12 July 2021
Published: 14 July 2021

Publisher's Note: MDPI stays neutral with regard to jurisdictional claims in published maps and institutional affiliations.

1. Introduction

Bearings-only target motion analysis (BOTMA) is a problem that has been widely studied and various solutions have been proposed in the literature: batch [1–5] or recursive filter (such as extended Kalman filter [6–8], unscented Kalman filter [9], particle filter [10], modified instrumental variable [11–13]), or a mix of recursive and batch methods [14]. Citing all the papers dealing with this topic is now a hard task. Among the abundant literature, most papers share the same assumption: the target is moving in a straight line with a constant speed, while the passive observer is maneuvering adequately in order to ensure the observability of the target [15–17]. The bearings are the measurements.

In this paper, we are concerned with the same problem, except that the available measurements are the cosine of the relative bearings, also called conical angles because the target belongs to the cone of ambiguity whose revolution axis is the line along which the towed array is moving (see [18] p. 39). Implicitly, we consider a target moving in 3D at a constant and unknown depth in near field; in this case, the two more energetic rays are the direct and the reflected paths (bottom or surface). In most cases, the sound bounces off the sea bottom. Therefore, we extend our analysis to surface- and sea bottom-bounced rays.

Indeed, the array detects the cosine of the relative angle of the direction of arrival by means of a suitable spatial filtering method such as beamforming, or more sophisticated techniques (see [19]). In the near field, sound can propagate to the sensor array along the direct path and/or the bottom-reflected path, and/or the surface-reflected path. Most of the time, at most, two rays coming from the same target are detected [18,20].

Unlike Gong [21] and Blanc-Benon [22], who addressed the three-dimensional target motion analysis (TMA) from a sequence of time differences of arrival (TDOA) of a signal traveling by two different paths coupled with a sequence of azimuths, we assume in this paper that the available measurements are the cosines of the conical angles only. In [23], a similar problem was addressed, but observability was not studied. We will consider two situations: the first case is devoted to TMA when sound propagates along a non-direct path at each sampling time. This will be the topic of Section 3: we will conduct observability analysis and identify all the ghost targets, given a set of noise-free measurements. We will prove that an assumption of the target's depth makes the target's trajectory observable, but not estimable (in the sense that the asymptotic performance given by the Cramér–Rao lower bound—CRLB—of the estimator of the depth is out of the physical constraints, that is, the source is navigating between the surface and the sea bottom).

In the fourth section, we will consider scenarios in which the antenna changes its own route. We will prove that the trajectory of the target is almost certainly observable.

In the fifth section, we will assume that sound will propagate along the direct path and the bottom-reflected path. The two rays will be assumed as being detected. Observability analysis will reveal that only three ghost targets at most exist without maneuvering the antenna. We will check that, in this case, the depth is not "estimable". We will give a palliative, allowing us to propose an estimator which is operationally acceptable, the price being a small bias. Convincing simulations will be given at the end of this section, proving that, even when the duration of the scenario is short, the estimated trajectory is very close to the true one. A conclusion ends the paper.

2. Notation and Problem Formulation

We consider two underwater vehicles moving at their own constant depth. The first mobile is a surface vessel or a submarine towing a horizontal sensor array, and the second one is the target of interest. Given a Cartesian coordinate system, the acoustic center of the array is located at time t at $\begin{pmatrix} x_O(t) & y_O(t) & z_O \end{pmatrix}^T$. At the same time, the target is at $\begin{pmatrix} x_T(t) & y_T(t) & z_T \end{pmatrix}^T$. The respective horizontal positions of the target of interest and of the center of the array at time t are denoted by $P_T(t) = \begin{pmatrix} x_T(t) & y_T(t) \end{pmatrix}^T$ and $P_O(t) = \begin{pmatrix} x_O(t) & y_O(t) \end{pmatrix}^T$. The sea bottom depth (assumed to be a constant) is denoted as D. The source is said to be endfire to the line array if its trajectory is in the same line as the array (which implies that the array and the source are at the same depth, and share the same route). It is broadside to the antenna if it navigates in the vertical plane orthogonal to the line array and passing by the acoustic center of the array. The sensor array detects the line of sight of the target; more precisely, ad hoc array processing (or spatial filtering) delivers at time t the cosine of the conical angle $c_a(t)$ given by $\cos(c_a(t)) = \cos(\theta(t) - h(t))\cos(\phi(t)) \triangleq m(t)$, where $\theta(t)$ and $\phi(t)$ are, respectively, the azimuth (or bearing) and the elevation of the path along which the sound emitted by the source propagates. The angle $h(t)$ is the heading of the sensor array. Denoting the relative position coordinates of the source with reference to the acoustic center of the array by $x_{OT}(t) = x_T(t) - x_O(t)$ and $y_{OT}(t) = y_T(t) - y_O(t)$, we have $\theta(t) = arctan(x_{OT}(t), y_{OT}(t))$. Figure 1 displays the different angles and the two actors (the observer reduced to the linear array, and the target).

The ray of the sound (or signal) emitted by the source can be reflected by the bottom and/or the surface or travels in the surface or deep channel. The sound–speed profile makes the paths curve. In this paper, we will consider that the target is in the near field (the distance between the source and the array is less than 20 km), and the bottom depth is in the range 2000–5000 m. Due to the large curvature of the ray (about 80 km), we will approximate the path of the sound as a set of zigzags defined by the reflections on the bottom or on the surface. So, we implicitly use the Snell law widely employed in geometrical optics. An image-source is created whose depth ζ_T will be called "image-depth". A path is then defined by the triplet (δ, n_B, n_S), where

- δ indicates the direction of the path of the sound emitted by the source: if the path is toward the surface, $\delta = -1$, otherwise $\delta = +1$,
- n_B is the number of bottom reflections, and
- n_S is the number of surface reflections.

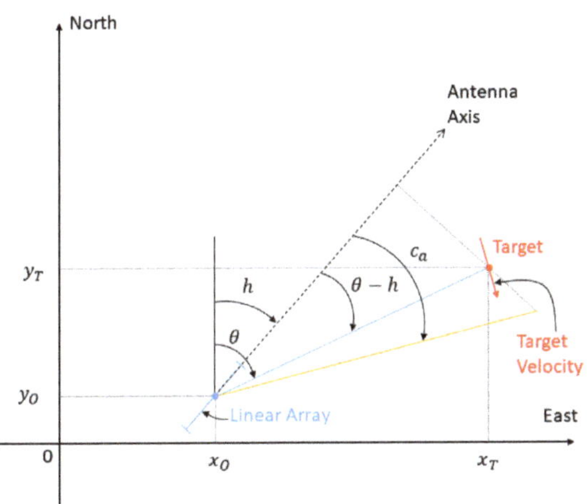

Figure 1. A typical scenario, viewed from the sky.

Figure 2 illustrates three different paths

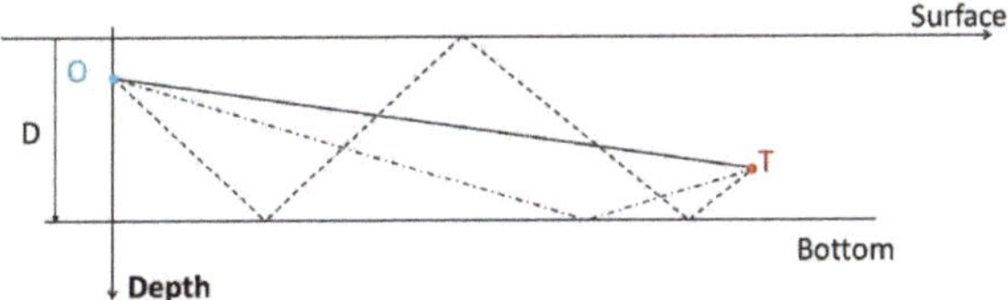

Figure 2. Three examples of ray paths: the solid line represents the direct path $(\delta, n_B, n_S) = (+1, 0, 0)$, the dashed-dotted line represents the bottom reflected path $(\delta, n_B, n_S) = (+1, 1, 0)$, and the dashed line represents the bottom-surface-bottom reflected path $(\delta, n_B, n_S) = (+1, 2, 1)$.

We have to consider the depth difference between the array and the image-source defined by $\zeta_{OT} \triangleq \zeta_T - z_O$ if the ray has been reflected (by the sea bottom or by the surface), or $\zeta_{OT} \triangleq z_T - z_O$ if the sound wave uses the direct path.

A general expression of ζ_{OT} based on the triplet (δ, n_B, n_S) is given by $\zeta_{OT}(\delta, n_B, n_S) = -2\delta n_B(-1)^{n_S + n_B} D - z_O + (-1)^{n_S + n_B} z_T$. Note that, given the path, the link between $\zeta_{OT}(\delta, n_B, n_S)$ and z_T is linear: $\zeta_{OT}(\delta, n_B, n_S) = a z_T + b$, the constants being a function of the triplet (δ, n_B, n_S), D, and z_O. Moreover, $\zeta_{OT}(\delta, n_B, n_S)$ is null if and only if the antenna and the target are navigating at the same depth $(z_T = z_O)$, and sound is traveling in the direct path. In this case, $\cos(\phi(t)) = 1$. For the sake of simplicity of the notations, we will simply subsequently denote ζ_{OT} instead of $\zeta_{OT}(\delta, n_B, n_S)$.

For the above examples, we have $\zeta_{OT}(1, 0, 0) = z_T - z_O$ (direct path), $\zeta_{OT}(+1, 1, 0) = 2D - (z_T + z_O)$ (bottom-reflected path), and $\zeta_{OT}(+1, 2, 1) = 4D - (z_T + z_O)$ (bottom-surface-bottom reflected path). Note that $\zeta_{OT}(\delta, n_B, n_S)$ can be negative (the image-source is above the surface). Consequently, the cosine of the elevation is $\cos(\phi(t)) = \dfrac{\sqrt{x_{OT}^2(t) + y_{OT}^2(t)}}{\sqrt{x_{OT}^2(t) + y_{OT}^2(t) + \zeta_{OT}^2(\delta, n_B, n_S)}}$.

Figure 3a displays the cone of ambiguity, defined by the set of sources sharing the same $\cos(\phi(t))$. In Figure 3b, we plot a direct ray and a bottom-bounced ray, which allows us to figure out the various angles with which we will work.

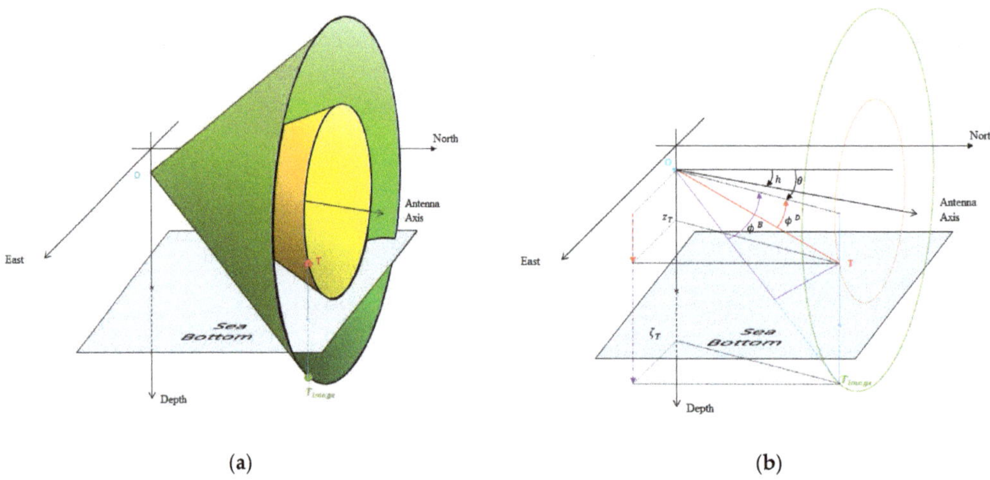

Figure 3. Cones of ambiguity. (**a**) The cones that the target belongs to, and the one that the image-target belongs to. (**b**) Example of conical angles of the target and of the image-target, for a bottom-reflected ray: ϕ^D and ϕ^B are the elevations of the direct path and of the bottom-reflected path, respectively.

We assume that the source is moving in constant velocity (CV) motion during the scenario. Our challenge is to estimate its trajectory, i.e., the state vector defining it, $X \triangleq \left(\begin{array}{ccccc} x_T(t^*) & y_T(t^*) & z_T & \dot{x}_T & \dot{y}_T \end{array} \right)^T$, for a chosen t^*, from noisy measurements.

We consider two situations:

1. Only one ray is detected by the array during the scenario; in this case, we have at each time t a measurement $m(t)$, given the path along which the wave propagates.
2. Two rays (traveling on two different paths) arrive at the sensor's antenna. In this case, the available measurement at time t is a couple of measurements, say $(m_1(t), m_2(t))$, given the two paths along which the wave propagates.

After the spatial filtering, the antenna supplies a noisy measurement of $m(t)$ or a noisy measurement of $(m_1(t), m_2(t))$. The noisy measurements are regularly acquired at $t_k = (k-1)\Delta t, k \in \{1, \ldots, N\}$, for a fixed sampling time Δt.

Before attempting to estimate X, we must answer several questions:

1. Is the vector X observable from the set of measurements $\{m(t), t \in [0, T]\}$? Note that, in TMA problems, observability is often analyzed in continuous time (see [15,17], for example), even though the noisy measurements are given in discrete time.
2. If not, what are the ghost targets (those which could be detected at the same set of measurements $\{m(t), t \in [0, T]\}$)?
3. How do we make X observable or with which new information?
4. Is the vector X observable from the set of couples $\{(m_1(t), m_2(t)), t \in [0, T]\}$?

For the cases where X is observable, we have then to compute the asymptotical performance of an unbiased estimator (given by the CRLB [24]), and the performance of our estimators in terms of bias and the covariance matrix. It is worth noting that using the FIM to prove observability can lead to a wrong conclusion [25]. This why we use an analytic approach.

3. TMA from One Ray

In this section, we consider the case where the array collects the cosine of a conical angle, the path of the ray being known by the operator. We start by analyzing the observability of the trajectory of the source of interest.

3.1. Observability Analysis

Theorem 1. *Let a linear antenna measure the cosine of a conical angle in the direction of a source, both in CV motion. The path of the sound emitted by the source is known, as is also the sea bottom depth.*

1. *If the target is broadside to the antenna, then the set of ghost targets is composed of virtual sources broadside to the antenna.*
2. *If the target is endfire to the antenna, the set of ghost targets is composed of virtual sources endfire to the antenna.*
3. *If the target has the same heading as the array (but is not endfire to it), then the set of ghost targets is composed of virtual targets with the same heading as the antenna. More precisely, the ghost image of each ghost target is moving on a cylinder whose axis is the antenna axis, and whose radius is a positive scalar β. The relative ghost target velocity is equal to β times the target's velocity. The initial distance between the ghost image and the center of the antenna is equal to β times the initial distance between the target-image and the center of the antenna.*
4. *In any other case, for a chosen image-depth ζ_G, the set of ghost targets is composed of virtual targets whose motion relative to the array is defined by $P_{OG}(t) = \beta P_{OT}(t)$ or $P_{OG}(t) = \beta S P_{OT}(t)$, where S is the 2D axial symmetry around the line of the array, and β is a positive scalar. The scalar β is equal to $\left|\frac{\zeta_{OG}}{\zeta_{OT}}\right|$ if $\zeta_{OT} \neq 0$. If $\zeta_{OT} = 0$ (which can happen with a direct path only), β can have any positive value.*

Preamble: In the following proof, we choose $t^* = 0$. Instead of working with the state vector $X = \begin{pmatrix} x_T(0) & y_T(0) & z_T & \dot{x}_T & \dot{y}_T \end{pmatrix}^T$, we will use the relative state vector of the image source, which is $Y \triangleq \begin{pmatrix} x_{OT}(0) & y_{OT}(0) & \zeta_{OT} & \dot{x}_{OT} & \dot{y}_{OT} \end{pmatrix}^T$. The reason for this is that we are able to recover X from Y without ambiguity.

We will prove this theorem in the special case where the heading of the antenna is equal to $0°$, and the value $y_{OT}(t)$ is positive. This can be easily obtained with an ad hoc rotation of the whole scenario. This will simplify the expression of the measurement, without loss of generality.

Proof of Theorem 1. We are seeking the ghost target whose horizontal position at time t is $\begin{pmatrix} x_G(t) & y_G(t) \end{pmatrix}^T$, detected in the same cosine of the conical angle, that is $\frac{y_{OT}(t)}{\sqrt{x_{OT}^2(t)+y_{OT}^2(t)+\zeta_{OT}^2}} = \frac{y_{OG}(t)}{\sqrt{x_{OG}^2(t)+y_{OG}^2(t)+\zeta_{OG}^2}}$, with $x_{OG}(t) = x_G(t) - x_O(t)$, $y_{OG}(t) = y_G(t) - y_O(t)$, and ζ_{OG} is the image-depth of the ghost target. This equality is equivalent to

$$\frac{y_{OT}^2(t)}{x_{OT}^2(t) + y_{OT}^2(t) + \zeta_{OT}^2} = \frac{y_{OG}^2(t)}{x_{OG}^2(t) + y_{OG}^2(t) + \zeta_{OG}^2} \tag{1}$$

Note that because the target is moving (as is the ghost target also), the denominators of the left term and of the right term of (1) are two polynomial functions of degree 2.

Case 1: $y_{OT}(t)$ is a zero function, i.e., $\forall t\, y_{OT}(t) = 0$.

This means the source is broadside to the antenna: $Y_T = \begin{pmatrix} x_{OT}(0) & 0 & \zeta_{OT} & \dot{x}_{OT} & 0 \end{pmatrix}^T$.

In this case, $y_{OT}(t) = 0, \forall t \in [0, T]$. Hence, the set of ghost targets is composed of the virtual targets broadside to the antenna: $Y_G = \begin{pmatrix} x_{OG}(0) & 0 & \zeta_{OG} & \dot{x}_{OG} & 0 \end{pmatrix}^T$.

Case 2: $y_{OT}(t)$ is not a zero function.

If $\dot{y}_{OT} = 0$, then $y_{OT}(t)$ is a constant. To respect the degrees of the terms of (1), $y_{OG}(t)$ is a constant too.

If $\dot{y}_{OT} \neq 0$, then there is a root, say $\tilde{t}$, such as $y_{OT}(\tilde{t}) = 0$, since $y_{OT}(t)$ is a polynomial function of degree 1. Consequently, $y_{OG}(\tilde{t}) = 0$, and $\forall t \neq \tilde{t}, y_{OG}(t) \neq 0$.

We deduce that, in both cases ($\dot{y}_{OT} = 0$, and $\dot{y}_{OT} \neq 0$), there exists a positive value β such that $y_{OG}(t) = \beta y_{OT}(t)$.

$$(1) \Leftrightarrow \left\{ \left[x_{OG}^2(t) + y_{OG}^2(t) + \zeta_{OG}^2 \right] - \beta^2 \left[x_{OT}^2(t) + y_{OT}^2(t) + \zeta_{OT}^2 \right] \right\} y_{OT}^2(t) = 0$$
$$\Leftrightarrow x_{OG}^2(t) + \zeta_{OG}^2 = \beta^2 \left[x_{OT}^2(t) + \zeta_{OT}^2 \right] \tag{2}$$

The quantity $x_{OT}^2(t) + \zeta_{OT}^2$ can be equal to zero at any time, or at one time or never.

Subcase 1: $\forall t$, $x_{OT}^2(t) + \zeta_{OT}^2 = 0$.

Then, $x_{OT}(t) = 0$, $\forall t$ and $\zeta_{OT} = 0$. Note that this case is the one when the target is traveling in the endfire to the array and at the same depth as the antenna and the path is the direct one. For the same reason, $x_{OG}(t) = 0$, $\forall t$ and $\zeta_{OG} = 0$. The set of ghost targets is hence composed of virtual targets traveling in the endfire to the array and at the same depth as the antenna.

Subcase 2: $\exists \check{t}$ such that $x_{OT}^2(\check{t}) + \zeta_{OT}^2 \neq 0$.

We deduce from (2) that

$$x_{OG}^2(0) = \beta^2 x_{OT}^2(0) + \beta^2 \zeta_{OT}^2 - \zeta_{OG}^2 \tag{3}$$

$$x_{OG}(0)\dot{x}_{OG} = \beta^2 x_{OT}(0)\dot{x}_{OT} \tag{4}$$

$$\dot{x}_{OG}^2 = \beta^2 \dot{x}_{OT}^2 \tag{5}$$

If $\dot{x}_{OT} = 0$, then

$$Y_G = \left(\ \pm\sqrt{\beta^2 x_{OT}^2(0) + \beta^2 \zeta_{OT}^2 - \zeta_{OG}^2} \quad \beta y_{OT}(0) \quad \zeta_{OG} \quad 0 \quad \beta \dot{y}_{OT} \ \right)^T, \text{ for any posi-}$$

tive constant β and any positive constant ζ_{OG} less than $\sqrt{\beta^2 x_{OT}^2(0) + \beta^2 \zeta_{OT}^2}$. Note that, when $\dot{y}_{OT} = 0$, the target is motionless relative to the center of the array (both have the same velocity); and when $\dot{y}_{OT} \neq 0$, the target has the same heading as the array.

If $\dot{x}_{OT} \neq 0$, then squaring the elements of (4), and using (5), we draw from (3) that $\beta^2 \zeta_{OT}^2 = \zeta_{OG}^2$. If $\zeta_{OT} = 0$, then $\zeta_{OG} = 0$, and the scalar β can take any positive value; else $\beta = \frac{|\zeta_{OG}|}{|\zeta_{OT}|}$. In both cases, the trajectory of a ghost target is defined by the state vector

$$Y_G = \left(\ \pm\beta x_{OT}(0) \quad \beta y_{OT}(0) \quad \beta \zeta_{OT} \quad \pm\beta \dot{x}_{OT} \quad \beta \dot{y}_{OT} \ \right)^T. \quad \square$$

Remark 1.

1. *When the source and the observer are at the same depth, and the path is direct, Theorem 1 recovers the conclusions given in [26].*

2. *The cases (1), (2) and (3) of Theorem 1 are "rare events", since the events of dealing with a source in endfire, broadside or having the same heading as the antenna during the scenario occur with a probability equal to 0. However, when the target has a trajectory close to one of these special cases, the estimates will have a poor behavior.*

3. *For case (4), when the detected ray is not a direct path, for example, when the ray is bottom-reflected, a hypothesis about the source is sufficient to obtain one solution, corresponding to a ghost target. Indeed, if we suppose that the depth of the target is z_{As} (whereas the true value is z_T), then we have $\beta = \frac{2D-(z_{As}+z_O)}{2D-(z_T+z_O)}$, whose biggest value $\beta_{Max} = \frac{2D-z_O}{2D-(z_T+z_O)}$, and the minimum value is $\beta_{Min} = \frac{2D-(z_{Max}+z_O)}{2D-(z_T+z_O)}$, where z_{Max} is the largest depth of a submarine vehicle. Typically, in deep water, $D \geq 4000$ m. A reasonable choice of z_{Max} could be 400 m. We can then have a range of β: $[\beta_{Min}, \beta_{Max}] = \left[\frac{7600-z_O}{8000-(z_T+z_O)}, \frac{8000-z_O}{8000-(z_T+z_O)} \right]$. For instance, when the depths of the antenna and the target are, respectively, 200 and 100 m, we have $[\beta_{Min}, \beta_{Max}] = [0.974, 1.013]$. In this way, we bound the set of ghost targets, and we can expect that the bias induced by a wrong choice of z_{As} is very low.*

4. *For case (4) again, with a direct path, if the target is not at the same depth as the antenna, $\beta = \frac{z_{As}-z_O}{z_T-z_O}$. Because β is a positive number, $z_{As} - z_O$ has the same sign as $z_T - z_O$: if $z_T > z_O$, then $z_O < z_{As} \leq z_{Max}$, and $[\beta_{Min}, \beta_{Max}] = \left[0, \frac{z_{Max}-z_O}{z_T-z_O} \right]$; if $z_T < z_O$, then*

$0 \leq z_{As} < z_O$, *and* $[\beta_{Min}, \beta_{Max}] = \left[0, \frac{z_O}{z_T - z_O}\right]$. *In both cases, the range* $[\beta_{Min}, \beta_{Max}]$ *is too wide to be useful. If the target and the antenna are at the same depth,* β *can take any positive value.*

3.2. Estimation of the Trajectory

We run 500 Monte Carlo simulations for a typical scenario described as follows:

The observer starts from $\begin{pmatrix} 0 & 0 \end{pmatrix}^T$ at the depth $z_O = 200$ m. Its speed and heading are, respectively, 5 m/s and $0°$. The initial position of the target is $\begin{pmatrix} 5000 & 7000 \end{pmatrix}^T$ and its depth is $z_T = 100$ m. Its route is $45°$ and its speed is 4 m/s.

- The measurements are collected every 4 s ($\Delta t = 4$ s). The scenario lasts 20 min.
- The sea bottom depth is 4000 m. The detected ray is a bottom-reflected ray.
- The assumed target depth is $z_{As} = 200$ m (whereas the true one is 100 m).
- First, the measurements have been corrupted with an additive Gaussian noise whose standard deviation is $\sigma = 1.7 \times 10^{-2}$.

Then, we choose the least squares estimator, which is identical to the maximum likelihood estimator with these assumptions. Note that, in open literature about TMA, the confidence regions are given by the confidence ellipsoid obtained with the covariance matrix of the estimate. Since the maximum likelihood estimate is asymptotically efficient under nonrestrictive conditions, we use here the Cramér–Rao lower bound to compute such confidence regions.

The result of the simulation is presented in Table 1 and illustrated in Figure 4. Obviously, even if the assumption made on the target's depth makes the state vector observable, it remains inestimable: the hugeness of the diagonal elements of the CRLB does not allow this kind of TMA to be employed. We note in Figure 4 that the cloud of horizontal estimates is hyperbola-shaped. This is because the state vector is "weakly" estimable. The parametric equation of this hyperbola is

$$\begin{cases} x(\omega) = \zeta_{As}\sinh(\omega) \\ y(\omega) = \frac{\zeta_{As}m}{\sqrt{1-m^2}} \cosh(\omega) \end{cases} \text{, with } \zeta_{As} = 2D - (z_{As} + z_O), \text{ and } m = \frac{y_{OT}(0)}{\sqrt{x_{OT}^2(0) + y_{OT}^2(0) + \zeta_{OT}^2}}$$

Table 1. Performance of the estimator of the reduced state vector when $\sigma = 1.7 \times 10^{-2}$, in terms of bias, sample standard deviation and the one given by the square root of the diagonal of the CRLB.

X_r	Bias	σ_{samp}	σ_{CRLB}
5000 m	-3525	6962	13,356
7000 m	-2367	4052	5599
2.83 m/s	-1.37	1.81	4.35
2.83 m/s	0.53	1.62	2.75

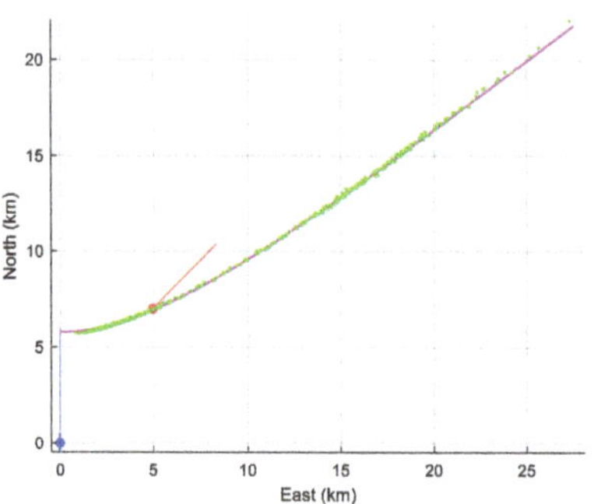

Figure 4. The cloud of estimated position (in green), a piece of the hyperbola (intersection of the cone of ambiguity and the plane $z = z_{As} (= 200 \text{ m})$, for $\sigma = 1.7 \times 10^{-2}$.

We further reduced the standard deviation to $\sigma = 1.7 \times 10^{-4}$ in order to appreciate the behavior of the MLE. With this (unrealistic) value, the MLE is efficient, as shown in Table 2 and in Figure 5 (which validates our observability analysis).

Table 2. Performance of the estimator of the reduced state vector with $\sigma = 1.7 \times 10^{-4}$.

X_r	Bias	σ_{samp}	σ_{CRLB}
5000 m	60.40	138.67	133.56
7000 m	88.20	58.42	55.99
2.83 m/s	0.043	0.044	0.044
2.83 m/s	0.037	0.028	0.028

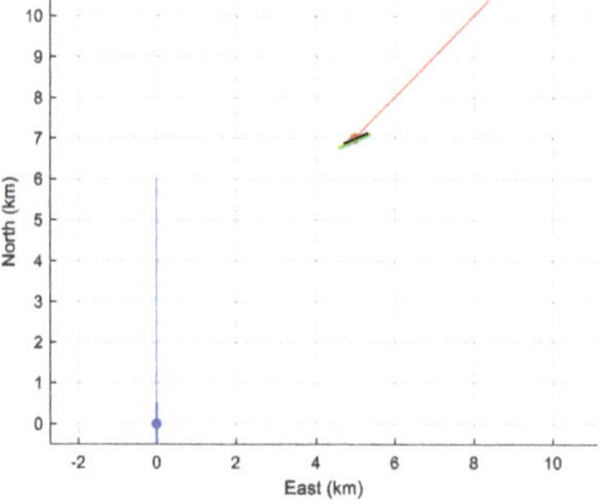

Figure 5. The cloud of estimated position (in green) for $\sigma = 1.7 \times 10^{-4}$. The cloud is no longer hyperbola-shaped. The small black segment is the 90%-confidence ellipsoid.

Our conclusion is that the state vector is not estimable, even though it is observable with an assumption on the target's depth.

This is why we propose to maneuver the antenna in order to render the state vector observable with no assumption on the target's depth, and to augment the information about it.

4. TMA with One Ray When the Array Maneuvers

In this section, the antenna maneuvers, i.e., it changes its own heading. We start by proving that the state vector is observable (without any assumption on the target's depth). Then, we have recourse to perform Monte Carlo simulations to evaluate the performance of the MLE.

4.1. Observability Analysis

Theorem 2. *Suppose the antenna's trajectory is composed of two successive legs at constant velocity (however with the same speed). Let the target be in CV motion. The linear array acquires the conical angles of the wave emitted from the target, the path of the ray being known as well as the sea bottom depth. If the target is broadside or endfire to the antenna during a leg, then there is at most a ghost target. Otherwise, there is no ghost target.*

Due to its length, the proof of this theorem is given in the Appendix A.

4.2. Estimation

In this subsection, we present the result of 500 Monte Carlo simulations that are run to illustrate the behavior of the proposed estimators. First, we give the scenario used here.

The center of the array and the initial position of the source are, respectively, at $(\ 0 \ \ 0 \ \ 200 \ )^T$ and $(\ 5000 \ \ 7000 \ \ 100 \ )^T$ at the very beginning of the scenario. The speed of the array is a constant along the scenario and is equal to 5 m/s. The trajectory of the array is composed of two legs linked by an arc of a circle. The first leg lasts 1 min 40 s, during which the array's heading is $135°$. Then, the array turns to the right with a turn rate equal to $20°$/min to adopt a new heading equal to $270°$. The duration of the maneuver is hence equal to 6 min 44 s. The second leg lasts 5 min, so the total duration of the scenario is 13 min and 20 s. Meanwhile, the target is navigating with a heading equal to $45°$ and a speed of 4 m/s. The bottom depth is $D = 4000$ m.

The state vector we have to estimate is hence $X = (\ 5000 \ \ 7000 \ \ 100 \ \ 2.83 \ \ 2.83 \ )^T$.

The array is assumed to measure the cosines of the conical angles of the bottom-reflected path given by

$$m(t_k) = \frac{y_{OT}(t_k)}{\sqrt{x_{OT}^2(t_k) + y_{OT}^2(t_k) + [2D - (z_T + z_O)]^2}} + \varepsilon_k$$

Measurements are acquired every $\Delta t = 4$ s, with $t_k = (k-1)\Delta t$.

The noise vector ε_k is assumed to be Gaussian, 0-mean and its standard deviation equal to $\sigma = 1.7 \times 10^{-2}$. The vectors ε_k are also assumed to be temporally independent.

Again, we choose the least squares estimator.

4.2.1. Estimation of X

The 500 obtained estimates of the initial horizontal position are plotted in Figure 6, together with the trajectory of the target, the 90%-confidence ellipse and the trajectory of the array. Again, the view is from the sky.

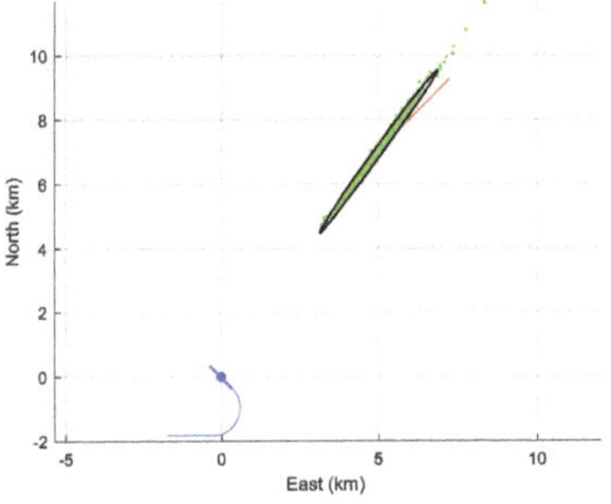

Figure 6. The cloud of the 500 initial positions estimates and the 90%-confidence ellipse.

The performance of the estimator (bias and standard deviation of each component) is presented in Table 3.

Table 3. Performance of the estimator of the plain state vector.

X	Bias	σ_{samp}	σ_{CRLB}
5000 m	−44.77	854.72	868.12
7000 m	−68.16	1162.1	1173.60
100 m	7.14	558.55	545.99
2.83 m/s	0.092	1.67	1.65
2.83 m/s	0.194	2.72	2.68

A convenient way to evaluate the behavior of an estimator is to compute the so-called normalized estimation error squared (NEES) [27], defined as $N_l = (\hat{X}_l - X)^T F (\hat{X}_l - X)$, where F is the FIM, and $\hat{X}_l$ is the estimate computed at the l-th simulation. If $\hat{X}_l$ is Gaussian-distributed with X as the mathematical expectation and the CRLB as the covariance matrix, then N_l is chi-square distributed with d degrees of freedom (χ_d^2), where d is the dimension of X (here 5). From the central limit theorem, the averaged NEES $N_S \triangleq \frac{1}{N_{Sim}} \sum_{l=1}^{N_{Sim}} N_l$ is approximately Gaussian; its mathematical expectation is d, and its standard deviation is equal to $\sqrt{\frac{2d}{N_{Sim}}}$.

From our simulations, we obtain $N_S = 5.34$.

In conclusion, the estimator can be declared efficient. However, the minimum standard deviation of the target's depth is not compatible with the physical constraints: with the standard deviation given in Table 3, the target could be up above the sea surface! Therefore, a palliative of this is to impose a depth on the target. Indeed, we saw in Section 3.1 that a supposed depth creates a small bias in estimation of the horizontal position of the target.

4.2.2. Estimation of X Reduced When the Depth of the Target Is Fixed

Now, the third component of X does not have to be estimated. The new state vector is the denoted as $X_r = \begin{pmatrix} x_T(0) & y_T(0) & \dot{x}_T & \dot{y}_T \end{pmatrix}^T$. We impose that $z_{As} = 200$ m (whereas the true depth is still 100 m). Hence, we introduce a bias.

Figure 7 displays the position's estimates in the same manner as Figure 6. The bias is not visible to the naked eye. However, Table 4 reveals this bias, which may be acceptable in a real situation. Even though the averaged NEES (=7.31) is out of its 90% confidence interval, its value remains acceptable.

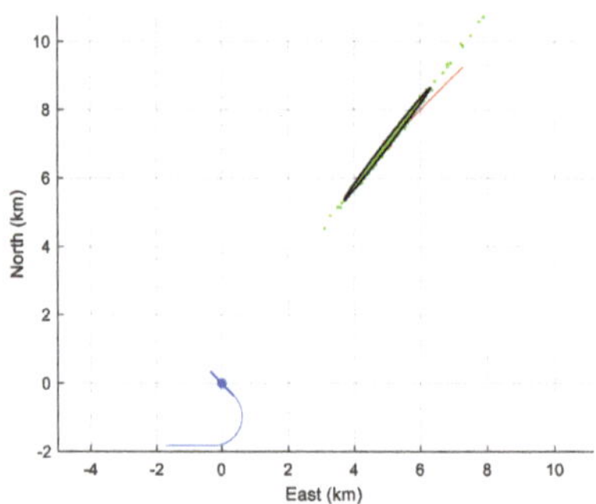

Figure 7. The cloud of the 500 initial positions estimates with the reduced state vector and the 90%-confidence ellipse.

Table 4. Performance of the estimator of the reduced state vector.

X_r	Bias	σ_{samp}	σ_{CRLB}
5000 m	65.40	655.61	606.41
7000 m	93.05	831.75	762.56
2.83 m/s	0.034	1.71	1.58
2.83 m/s	0.063	2.76	2.52

The main interest of assuming the depth to be known is to economize on the CPU time, and reduce the standard deviation of the remaining components to estimate. We are in the presence of the well-known bias–variance tradeoff.

4.2.3. Estimation of the Reduced State Vector by the Conventional BOTMA

In such a scenario, the conventional BOTMA can be run by neglecting the site effect, so by imposing that $\cos(\phi(t)) = 1$, $\forall t$. The (incorrect) noise-free measurement model is then

$$\cos(\alpha(t)) = \cos(\theta(t) - h(t)).$$

The results are plotted in Figure 8. Obviously, a huge bias appears, leading to an averaged NEES equal to 1960. More precisely, the bias on the components of the reduced state vector is $\begin{pmatrix} -3062.8 & -2319.9 & 14.4 & 15.8 \end{pmatrix}^T$, rendering the BOTMA inoperative. Clearly, the conventional BOTMA cannot be recommended for the near field. This justifies a posteriori the interest in taking the site effect and the nature of the wave ray into account, as previously pointed out in the introduction of [23].

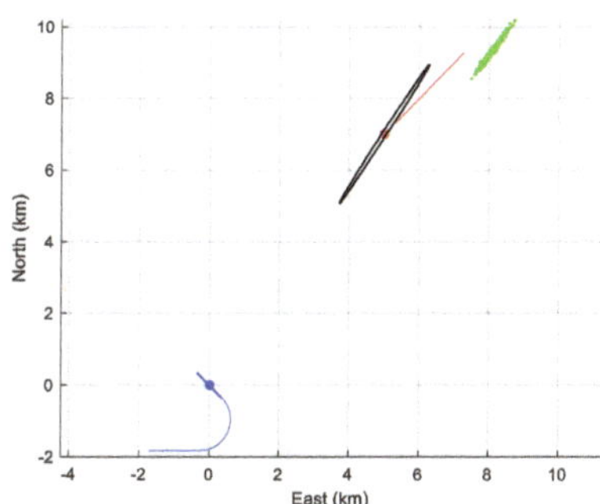

Figure 8. The cloud (in green) of the 500 initial positions estimates given by the classic BOTMA together with the 90%-confidence ellipse.

5. TMA from the Direct Path and the Bottom-Reflected Path

We assume in this section that the sound wave emitted by the target travels on the direct path and the bottom-reflected path.

5.1. Observability

Theorem 3. *Let a linear antenna and a source both be in CV motion.*

The antenna acquires the cosines of the conical angles of the direct path and of the bottom-reflected path.

1. *If the target is broadside to the array, then the set of ghost targets is uncountable: it is composed of all the (virtual) targets at broadside to the array.*
2. *If the target is endfire to the antenna, the set of ghost targets is composed of virtual sources at endfire to the antenna.*
3. *If the route of the antenna and the route of the target are parallel, then the set of ghost targets is uncountable: at each depth z_G, there are two ghost targets moving on a cylinder whose axis*

is the antenna axis, and the radius is a positive scalar $\beta = \sqrt{\frac{D-z_G}{D-z_T}}$. The relative ghost target velocity is equal to β times the target's velocity. The initial distance between the ghost image and the center of the antenna is equal to β times the initial distance between the ghost image and the center of the antenna.

4. *If the route of the antenna and the route of the target are not parallel, then there are three ghost targets whose motion relative to the antenna is $P_{OG}(t) = SP_{OT}(t)$, $P_{OG}(t) = \beta P_{OT}(t)$, and $P_{OG}(t) = \beta SP_{OT}(t)$, where S is the matrix of the axial symmetry around the line of the antenna, and $\beta \triangleq \frac{D-z_O}{D-z_T}$. If the depth of the antenna is equal to the depth of the source, then there is one single ghost target given by $P_{OG}(t) = SP_{OT}(t)$.*

Proof of Theorem 3. With no loss of generality, we will again assume that the axis of the sensor array is pointed toward north and that the target is in the half-space where the second component y of any vector is positive. A convenient rotation helps us in this case. So the noise-free measurements at time t are $m_1(t) = \dfrac{y_{OT}(t)}{\sqrt{x_{OT}^2(t)+y_{OT}^2(t)+z_{OT}^2}}$, and

$m_2(t) = \dfrac{y_{OT}(t)}{\sqrt{x_{OT}^2(t)+y_{OT}^2(t)+[2D-(z_T+z_O)]^2}}$.

We have to seek a five-dimensional state vector $X_G = \begin{pmatrix} x_G(0) & y_G(0) & z_G & \dot{x}_G & \dot{y}_G \end{pmatrix}^T$ defining the trajectory of a ghost target, i.e., producing the same noise-free measurement as X, that is $m_1(t) = \dfrac{y_{OG}(t)}{\sqrt{x_{OG}^2(t)+y_{OG}^2(t)+z_{OG}^2}}$, and $m_2(t) = \dfrac{y_{OG}(t)}{\sqrt{x_{OG}^2(t)+y_{OG}^2(t)+[2D-(z_G+z_O)]^2}}$.

hence satisfying the two following equalities (in time):

$$\frac{y_{OT}(t)}{\sqrt{x_{OT}^2(t) + y_{OT}^2(t) + z_{OT}^2}} = \frac{y_{OG}(t)}{\sqrt{x_{OG}^2(t) + y_{OG}^2(t) + z_{OG}^2}} \tag{6}$$

$$\frac{y_{OT}(t)}{\sqrt{x_{OT}^2(t) + y_{OT}^2(t) + [2D - (z_T + z_O)]^2}} = \frac{y_{OG}(t)}{\sqrt{x_{OG}^2(t) + y_{OG}^2(t) + [2D - (z_G + z_O)]^2}} \tag{7}$$

under the constraint that z_G is in $[0, D]$.

Case 1: $y_{OT}(t)$ is a zero function, i.e., $\forall t \; y_{OT}(t) = 0$.

The target is broadside to the antenna, so any ghost targets will be too (see Case 1 in the proof of theorem 1).

Case 2: $y_{OT}(t)$ is not a zero function.

From Case 2 of the proof of theorem 1, there is a positive scalar β such that $y_{OG}(t) = \beta y_{OT}(t)$.

$$(6) \Leftrightarrow \sqrt{x_{OG}^2(t) + y_{OG}^2(t) + z_{OG}^2} = \beta\sqrt{x_{OT}^2(t) + y_{OT}^2(t) + z_{OT}^2}$$
$$\Leftrightarrow [x_{OG}^2(t) + y_{OG}^2(t) + z_{OG}^2] = \beta^2[x_{OT}^2(t) + y_{OT}^2(t) + z_{OT}^2] \tag{8}$$

$$(7) \Leftrightarrow [x_{OG}^2(t) + y_{OG}^2(t) + [2D - (z_G + z_O)]^2] = \beta^2[x_{OT}^2(t) + y_{OT}^2(t) + [2D - (z_T + z_O)]^2] \tag{9}$$

Subtracting (9) from (8), we get $z_{OG}^2 - [2D - (z_G + z_O)]^2 = \beta^2\left[z_{OT}^2 - [2D - (z_T + z_O)]^2\right]$.

Now, we simplify the expressions of these two terms:

$$z_{OG}^2 - [2D - (z_G + z_O)]^2 = -4(D - z_G)(D - z_O)$$

$$z_{OT}^2 - [2D - (z_T + z_O)]^2 = -4(D - z_T)(D - z_O)$$

We deduce from this that

$$\beta = \sqrt{\frac{D - z_G}{D - z_T}} \tag{10}$$

Note that $\beta = 1$ iif $z_G = z_T$.

$$(8) \Leftrightarrow x_{OG}^2(t) + y_{OG}^2(t) + z_{OG}^2 = \beta^2[x_{OT}^2(t) + y_{OT}^2(t) + z_{OT}^2]$$
$$\Leftrightarrow x_{OG}^2(t) - \beta^2 x_{OT}^2(t) = \beta^2 z_{OT}^2 - z_{OG}^2 \tag{11}$$

Since $x_{OG}^2(t) - \beta^2 x_{OT}^2(t)$ is a polynomial function of degree 2, (11) is equivalent to

$$x_{OG}^2(0) = \beta^2 x_{OT}^2(0) + \beta^2 z_{OT}^2 - z_{OG}^2 \tag{12}$$

$$x_{OG}(0)\dot{x}_{OG} = \beta^2 x_{OT}(0)\dot{x}_{OT} \tag{13}$$

$$\dot{x}_{OG}^2 = \beta^2 \dot{x}_{OT}^2 \tag{14}$$

<u>First case $\dot{x}_{OT} = 0$</u>

Equation (14) implies that $\dot{x}_{OG} = 0$.

Consequently, for any z_G in $[0, D]$, the vector $X_{OG} = \left(\pm\sqrt{\beta^2 x_{OT}^2(0) + \beta^2 z_{OT}^2 - z_{OG}^2} \quad \beta y_{OT}(0) \quad z_{OG} \quad 0 \quad \beta \dot{y}_{OT} \right)^T$ (with $\beta = \sqrt{\frac{D - z_G}{D - z_T}}$) defines the trajectory of a ghost target.

<u>Second case $\dot{x}_{OT} \neq 0$</u>

Using (14), and squaring the terms of (13), we get $x_{OG}^2(0) = \beta^2 x_{OT}^2(0)$.

Reporting this in (12), we obtain finally $\beta^2 z_{OT}^2 = z_{OG}^2$, i.e.,

$$\beta^2 = \left(\frac{z_{OG}}{z_{OT}} \right)^2 \tag{15}$$

If $z_T = z_O$, then $z_G = z_O$. In this case, $\beta = 1$, and consequently $y_{OG}(t) = y_{OT}(t)$ and $x_{OG}^2(t) = x_{OT}^2(t)$ from (11). The source's trajectory is observable up to the axial symmetry around the (Oy)-axis.

Equations (10) and (15) give us $\frac{D - z_G}{D - z_T} = \left(\frac{z_{OG}}{z_{OT}} \right)^2$.

The unknown z_G is hence a root of the following equation of degree 2:

$(z_G - z_O)^2 - \frac{(z_T - z_O)^2}{D - z_T}(D - z_G) = 0$ which can be expanded as follows: $z_G^2 + z_G\left[-2z_O + \frac{(z_T - z_O)^2}{D - z_T} \right] - \frac{D(z_T - z_O)^2}{D - z_T} + z_O^2 = 0.$

Of course, z_T is a root of this equation. For this value, $z_G = z_T$, hence $\beta = 1$.

The second root (z_G itself) is hence $2z_O - z_T - \frac{(z_T - z_O)^2}{D - z_T} \triangleq z_G$. We can check readily that $z_G - z_O = z_O - z_T - \frac{(z_T - z_O)^2}{D - z_T} = \frac{(z_O - z_T)(D - z_T) - (z_T - z_O)^2}{D - z_T}.$

Hence, $\frac{z_G - z_O}{z_T - z_O} = \frac{z_O - D}{D - z_T}$ (which is negative).

We deduce from this that:

1. when the target's depth is larger than the array's depth, there is a ghost whose depth is smaller than the array's depth, and vice versa.
2. β, which is a positive coefficient, is equal to $\frac{D - z_O}{D - z_T}$, or 1.

   Therefore, we have identified three ghost targets:

   the first one is defined by $X_{OG} = \left(-x_{OT}(0) \quad y_{OT}(0) \quad z_{OT} \quad -\dot{x}_{OT} \quad \dot{y}_{OT} \right)^T$,

   the second is defined by $X_{OG} = \left(\beta x_{OT}(0) \quad \beta y_{OT}(0) \quad -\beta z_{OT} \quad \beta \dot{x}_{OT} \quad \beta \dot{y}_{OT} \right)^T$,

   and the third by $X_{OG} = \left(-\beta x_{OT}(0) \quad \beta y_{OT}(0) \quad -\beta z_{OT} \quad -\beta \dot{x}_{OT} \quad \beta \dot{y}_{OT} \right)^T$. $\square$

Remark 2. *Most of the time, the depth of a submarine vehicle is under the operational constraint: values of z_T are in $[0, z_{Max}]$ and $z_{Max} \ll D$. For example, $z_{Max} = 400\ m$, while $D = 4000\ m$.*

The proof of the previous theorem must be adapted to this new constraint.

First, we use the fact that the function $u \mapsto f(u) \triangleq 2z_O - u - \frac{(u - z_O)^2}{D - u}$ is an involution, i.e., $f(f(u)) = u$.

Since $f(0) = 2z_O - \frac{z_O^2}{D}$, $f\left(2z_O - \frac{z_O^2}{D}\right) = 0$.

Now the question is: what are the values of z_O for which the following inequality holds: $2z_O - \frac{z_O^2}{D} \leq z_{Max}$, the greatest value of z_O guaranteeing that $2z_O - \frac{z_O^2}{D-z_T} \leq z_{Max}$ is $D - D\sqrt{1 - \frac{z_{Max}}{D}}$ (which is less than z_{Max}).

If $z_O > D - D\sqrt{1 - \frac{z_{Max}}{D}}$, then $z_G > z_{Max}$. In this case, there is a unique ghost target given by $X_G = \begin{pmatrix} -x_T(0) & y_T(0) & z_T & -\dot{x}_T & \dot{y}_T \end{pmatrix}^T$.

If $z_O \leq D - D\sqrt{1 - \frac{z_{Max}}{D}}$, then $z_G \leq z_{Max}$. In this case, there are three ghost targets:

one is defined by $X_G = \begin{pmatrix} -x_T(0) & y_T(0) & z_T & -\dot{x}_T & \dot{y}_T \end{pmatrix}^T$,

the second is defined by $X_G = \begin{pmatrix} \beta x_T(0) & \beta y_T(0) & f(z_T) & \beta \dot{x}_T & \beta \dot{y}_T \end{pmatrix}^T$,

and the third by $X_G = \begin{pmatrix} -\beta x_T(0) & \beta y_T(0) & f(z_T) & -\beta \dot{x}_T & \beta \dot{y}_T \end{pmatrix}^T$.

Note that the operational constraint allows us to benefit from the following range: $\frac{D-z_{Max}}{D} \leq \beta \leq \frac{D}{D-z_{Max}}$. For example, when $z_{Max} = \frac{D}{10}$, $0.9 \leq \beta \leq 1.11$. Consequently, the ghost target is very close to the target of interest.

5.2. Estimation of the Trajectory

This section is devoted to the estimation of the target's trajectory, or in other words, the estimation of X with $t^* = 0$ (the first time). Before going into detail, we compute the so-called Cramér–Rao lower bound to evaluate the asymptotical performance of any unbiased estimator.

We have considered two typical scenarios. In both, the array is assumed motionless (or, more realistically, all the mobiles are referenced to it) at the depth $z_O = 200$ m, and the state vector defining the target's trajectory is given by the state vector $X = \begin{pmatrix} 5000 & 7000 & 100 & 2.83 & 2.83 \end{pmatrix}^T$. The standard deviation of the measurement is $\sigma = 1.7 \times 10^{-2}$. The total duration of the scenario is 5 min, and the sampling time is $\Delta t = 4$ s; consequently, the number of measurement couples is $N = 75$.

In the first scenario, the bottom depth is $D = 2000$ m, while in the second, $D = 4000$ m.

Note that in the first scenario, $\beta = 0.89$, and in the second one, $\beta = 0.97$. The ghost target is hence very close to the target of interest.

5.2.1. Estimability

As pointed out in Section 1, the state vector X is "estimable" if its asymptotical performance given by the CRLB is compatible with the physical constraints. Typically, if the minimum standard deviation defined by the square root of the third diagonal element of the CRLB (hence of the depth) is much larger than the depth, then X is declared non-estimable.

1. First scenario

For this scenario, the square root of the diagonal of the CRLB $\sigma_{CRLB} = \begin{pmatrix} 1.16 \times 10^6 & 1.59 \times 10^6 & 8.22 \times 10^5 & 637.9 & 646.1 \end{pmatrix}^T$.

2. Second scenario

With the bottom depth, things are not much better, since $\sigma_{CRLB} = \begin{pmatrix} 6.59 \times 10^5 & 8.96 \times 10^5 & 9.73 \times 10^5 & 352.6 & 362.2 \end{pmatrix}^T$.

In both cases, the minimum standard deviations are huge. We can conclude that the state vector is not estimable. Computations of minimum standard deviations were made for various scenarios; in all, the state vector is not estimable.

A palliative of this is to fix the depth of the source at an arbitrary and realistic value, say z_{As}, and compute the CRLB of the reduced state vector $X_r \triangleq \begin{pmatrix} x_T(0) & y_T(0) & \dot{x}_T & \dot{y}_T \end{pmatrix}^T$

when we assume that $z_T = z_{As}$. For example, for $z_{As} = 300$ m, the minimum standard deviations are

$$\sigma_{CRLB} = \left(\begin{array}{cccc} 281.17 & 319.37 & 1.78 & 2.02 \end{array}\right)^T \text{ for the first scenario, and}$$

$$\sigma_{CRLB} = \left(\begin{array}{cccc} 130.1 & 115.3 & 0.80 & 0.71 \end{array}\right)^T \text{ for the sec ond one.}$$

Therefore, we propose to estimate the state vector with this hypothesis ($z_{As} = 300$ m). In so doing, we introduce a bias. The next subsection gives us the result of the 500 Monte Carlo simulations.

5.2.2. Monte Carlo simulations

The computation of the maximum likelihood estimator (MLE) is made with the Gauss-Newton routine. No numerical issue was encountered.

1. First scenario

The performance of the MLE is summarized in Table 5. We have numerically computed the bias and the empirical standard deviation (given, respectively in the second and third column of the table). We can see that the empirical standard deviation is very close to that given by the CRLB. However, as expected, the MLE is biased (of course, there is no bias if we choose $z_{As} = z_T$). In Figure 9, the 90% confidence ellipse is drawn, together with the cloud of the 500 estimates (in pink).

Table 5. Performance of the estimator of the reduced state vector.

X_r	Bias	σ_{samp}	σ_{CRLB}
5000 m	401.12	281.85	281.17
7000 m	557.24	330.87	319.37
2.83 m/s	0.13	1.58	1.78
2.83 m/s	0.12	1.81	2.02

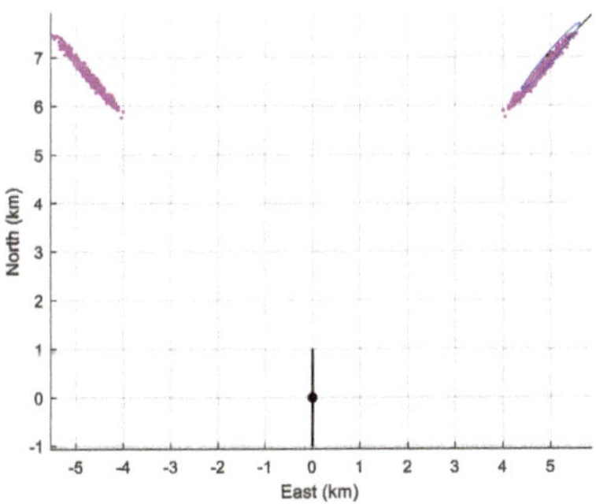

Figure 9. The location of the sensor array (in black), the cloud of the 500 estimates and the 90%-confidence ellipse when $D = 2000$ m, $z_{As} = 300$ m, and $z_T = 100$ m. The symmetrical cloud is plotted too.

2. Second scenario: Bottom depth $D = 4000$ m.

Again, the performance is presented in Table 6. The bias of the estimator is similar to the one obtained for the first scenario. Only the empirical standard deviations of $\left(\begin{array}{cc} x_T(0) & y_T(0) \end{array}\right)^T$ are larger than that computed from the CRLB. However, Figure 10 shows us that the cloud of estimates is close to the true value and not spread.

Table 6. Performance of the estimator.

X_r	Bias	σ_{samp}	σ_{CRLB}
5000 m	306.81	219.28	130.08
7000 m	432.46	276.61	115.26
2.83 m/s	0.18	0.74	0.80
2.83 m/s	0.18	0.66	0.71

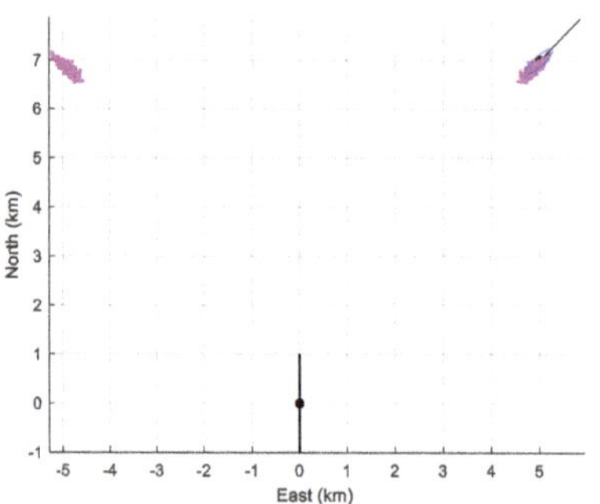

Figure 10. The location of the sensor array (in black), the cloud of the 500 estimates of the initial positions, and the 90%-confidence ellipse when $D = 4000$ m, $z_{As} = 300$ m, and $z_{As} = 100$ m, together with the symmetrical cloud.

What is remarkable is the short duration and still the very good performance (in terms of accuracy) of the result. Numerous simulations (not reported here) were performed; all confirm the correct performance of the MLE. The shortness of the scenario is crucial, because everything that we propose here works properly under the condition that the sea bottom is a plane. During a short scenario, this assumption is likely.

6. Conclusions

In this paper, conical-angle TMA has been addressed, and various multipaths of sound have been taken into account. The sensor is a line array. Observability was analyzed deeply, allowing all the existing ghost targets to be identified. The main results are that, if the array detects one ray (corresponding to one path), the trajectory is not observable: the set of ghost targets is composed of trajectories that are homothetic to the trajectory of the target of interest, and their symmetrical images by the axial symmetry around the line array. If the array detects two rays (corresponding to two different paths), the number of ghost targets is reduced to three (except when the target is endfire or broadside to the antenna). When the antenna maneuvers, the target's trajectory is observable (apart from the special scenario where there is one single ghost target). Even for "observable" scenarios, the depth of the target is not estimable (its asymptotical standard deviation is huge). In these cases, we give a non-restrictive palliative that allows us to provide estimates close to the truth.

In the future, in this context, many problems remain to be faced: identification of the paths, maneuvering targets, and fusion of data collected by other sensors, as in [28]. The problem of seeking a "good" maneuver of the observer, as it was solved in a 2D environment [29–31], will be addressed in the future. Some of these problems are already under investigation.

Author Contributions: Conceptualization, A.L. and A.-C.P.; methodology, C.J.; software, A.L.; validation, A.L., A.-C.P., C.J. and D.L.; formal analysis, A.L.; investigation, A.L.; resources, D.L.; data curation, A.L.; writing—original draft preparation, C.J.; writing—review and editing, A.L. and C.J.; visualization, A.-C.P.; supervision, A.-C.P.; project administration, not applicable; funding acquisition, not applicable. All authors have read and agreed to the published version of the manuscript.

Funding: This research received no external funding.

Institutional Review Board Statement: Not applicable.

Informed Consent Statement: Not applicable.

Appendix A

Proof of Theorem 2. The proof is made when the first leg is towards North (as previously).
Hence, $\quad V_1 = \begin{bmatrix} 0 \\ v \end{bmatrix}, \quad V_2 = \begin{bmatrix} v\sin(\alpha) \\ v\cos(\alpha) \end{bmatrix}, \quad S_1 = \begin{bmatrix} -1 & 0 \\ 0 & 1 \end{bmatrix},$
and $S_2 = \begin{bmatrix} -\cos(2\alpha) & \sin(2\alpha) \\ \sin(2\alpha) & \cos(2\alpha) \end{bmatrix}$.

Moreover, we assume $\alpha \neq k\pi$.

From Theorem 1, we have to consider the four following cases for each leg:

- the target is broadside to the antenna,
- the target is endfire to the antenna,
- the target has the same heading as the array (but is not endfire to it),
- the other cases.

Note that if the target is in case (1) during the first leg, then in case (2) during the second one (provided that this situation is possible), the conclusion about observability will be the same as if the target is in case (2) during the first leg, then in case (1) during the second leg. To be convinced of this, we just have to reverse the time in the equation. This remark allows us to shorten the proof.

Case 1: the target is broadside to the antenna during the first leg.

Hence, $P_T(0) = \begin{bmatrix} x_T(0) \\ 0 \end{bmatrix}$, and $V_{OT} = \begin{bmatrix} c_T \\ 0 \end{bmatrix}$ during the first leg, which implies
$P_{OT}(t) = \begin{bmatrix} x_T(0) + tc_T \\ 0 \end{bmatrix}$ for $t \leq \tau$. The ghost targets are also in the broadside, hence
$P_{OG}(t) = \begin{bmatrix} x_G(0) + tc_G \\ 0 \end{bmatrix}$ and $V_{OG} = \begin{bmatrix} c_G \\ 0 \end{bmatrix}$ for $t \leq \tau$.

Can the target be endfire to the antenna? If so, the target has the same heading as the antenna during the second leg or, in other words, $V_T - V_2 = \lambda V_2$, and $P_{OT}(t)$, which is equal to $P_{OT}(t) = P_{OT}(\tau) + (t - \tau)(V_T - V_2)$, is collinear with V_2, whenever $t \geq \tau$. The first condition cannot be satisfied since $V_T = \begin{bmatrix} c_T \\ v \end{bmatrix}$, and $V_2 = \begin{bmatrix} \pm v \\ 0 \end{bmatrix}$. There is no ghost.

We skip the case where the target is in case (3) during the second leg. This will be treated later. Therefore, we now have to consider the other cases during the second leg. There are two possibilities for the ghost targets: those whose trajectories are defined by (i) $P_{OG}(t) = \beta P_{OT}(t)$, and those whose trajectories are given by (ii) $P_{OG}(t) = \beta S_2 P_{OT}(t)$, both for $t \geq \tau$.

The derivative of (i) is $V_G - V_2 = \beta V_T - \beta V_2$, hence $V_G = \beta V_T + (1 - \beta)V_2$.
$$\Leftrightarrow \begin{bmatrix} c_G \\ v \end{bmatrix} = \beta \begin{bmatrix} c_T \\ v \end{bmatrix} + (1 - \beta)v \begin{bmatrix} \sin(\alpha) \\ \cos(\alpha) \end{bmatrix},$$
which implies that $(1 - \beta)\cos(\alpha) = 1 - \beta$. Since $\cos(\alpha) \neq 1$, $\beta = 1$. There is no ghost given by (i).

The derivative of (ii) is $V_G - V_2 = \beta S_2 V_T - \beta V_2$, hence $V_G = \beta S_2 V_T + (1 - \beta)V_2$.
$$\Leftrightarrow \begin{bmatrix} c_G \\ v \end{bmatrix} = \beta \begin{bmatrix} -c_T \cos(2\alpha) + v\sin(2\alpha) \\ c_T \sin(2\alpha) + v\cos(2\alpha) \end{bmatrix} + (1 - \beta)v \begin{bmatrix} \sin(\alpha) \\ \cos(\alpha) \end{bmatrix}.$$
We deduce that $\beta = v\dfrac{1 - \cos(\alpha)}{c_T \sin(2\alpha) + v\cos(2\alpha) - v\cos(\alpha)}$.

One ghost exists if $c_T \sin(2\alpha) + v\cos(2\alpha) - v\cos(\alpha)$ is a positive quantity. If so, we then compute c_G. There is one ghost at most.

Case 2: the target is endfire to the antenna during the first leg.

Hence, $P_T(0) = \begin{bmatrix} 0 \\ y_T(0) \end{bmatrix}$, and $V_{OT} = \begin{bmatrix} 0 \\ c_T \end{bmatrix}$, which implies that $P_{OT}(t) = \begin{bmatrix} 0 \\ y_T(0) + tc_T \end{bmatrix}$ for $t \leq \tau$. During this first leg, the ghost targets are also endfire to the antenna, so $P_{OG}(t) = \begin{bmatrix} 0 \\ y_G(0) + tc_G \end{bmatrix}$ for $t \leq \tau$, and $V_G - V_1 = \begin{bmatrix} 0 \\ c_G \end{bmatrix}$.

Again, we skip the case where the target is in case (3) during the second leg. This will be treated later. So, we now have to consider the other cases during the second leg. There are two possibilities for the ghost targets: those whose trajectories are defined by (i) $P_{OG}(t) = \beta P_{OT}(t)$ and those whose trajectories are given by (ii) $P_{OG}(t) = \beta S_2 P_{OT}(t)$, both for $t \geq \tau$.

The derivative of (i) is $V_G - V_2 = \beta V_T - \beta V_2$, hence $V_G = \beta V_T + (1 - \beta)V_2$.

$$\Leftrightarrow \begin{bmatrix} 0 \\ c_G + v \end{bmatrix} = \beta \begin{bmatrix} 0 \\ c_T + v \end{bmatrix} + (1 - \beta)v \begin{bmatrix} \sin(\alpha) \\ \cos(\alpha) \end{bmatrix}.$$

We deduce that $\beta = 1$. There is no ghost.

Now, differentiating (ii) gives us $V_G - V_2 = \beta S_2 V_T - \beta V_2$, hence $V_G = \beta S_2 V_T + (1 - \beta)V_2$.

$$\Leftrightarrow \begin{bmatrix} 0 \\ c_G + v \end{bmatrix} = \beta(c_T + v) \begin{bmatrix} \sin(2\alpha) \\ \cos(2\alpha) \end{bmatrix} + (1 - \beta)v \begin{bmatrix} \sin(\alpha) \\ \cos(\alpha) \end{bmatrix}.$$
$$\Rightarrow \beta(c_T + v)\sin(2\alpha) + (1 - \beta)v\sin(\alpha) = 0.$$

We deduce that $\beta = -v\dfrac{\sin(\alpha)}{c_T \sin(2\alpha) + v \sin(2\alpha) - v \sin(\alpha)}$. One ghost exists if $c_T \sin(2\alpha) + v \sin(2\alpha) - v \sin(\alpha)$ is a negative quantity. If so, we then compute c_G. There is one ghost at most.

Case 3: the target has the same heading as the array (but is not endfire to it)

As in case (2), $V_{OT} = \begin{bmatrix} 0 \\ c_T \end{bmatrix}$, but here, the first component of $P_{OT}(t)$ is not zero.

Hence, $V_T = \begin{bmatrix} 0 \\ c_T + v \end{bmatrix}$, and the target cannot be endfire to the antenna during the second leg. In this case, $V_{OG} = \begin{bmatrix} 0 \\ \beta c_T \end{bmatrix}$, hence $V_G = \begin{bmatrix} 0 \\ \beta c_T + v \end{bmatrix}$.

Can the target be broadside to the antenna? The answer is positive if $V_2 \perp V_T$ and V_{OT} is collinear to $P_{OT}(t)$, when $t \geq \tau$. The first condition implies that $V_2 = \begin{bmatrix} \pm v \\ 0 \end{bmatrix}$. Since $P_{OT}(t) = P_{OT}(\tau) + (t - \tau)(V_T - V_2)$, the second condition is satisfied if $P_{OT}(\tau)$ is collinear to $V_T - V_2$. This is not the case when the first component of $P_{OT}(\tau)$ is zero, while the first component of $V_T - V_2$ is $\pm v$.

So, we now have to consider the other cases during the second leg. There are two possibilities for the ghost targets: those whose trajectories are defined by (i) $P_{OG}(t) = \beta P_{OT}(t)$ and those whose trajectories are given by (ii) $P_{OG}(t) = \beta S_2 P_{OT}(t)$, both for $t \geq \tau$.

The derivative of (i) is $V_G - V_2 = \beta V_T - \beta V_2$, hence $V_G = \beta V_T + (1 - \beta)V_2$ or, in other words,

$$\begin{bmatrix} 0 \\ \beta c_T + v \end{bmatrix} = \begin{bmatrix} 0 \\ \beta(c_T + v) \end{bmatrix} + (1 - \beta)v \begin{bmatrix} \sin(\alpha) \\ \cos(\alpha) \end{bmatrix}.$$

We conclude that $\beta = 1$, i.e., there is no ghost.

If $P_{OG}(t) = \beta S_2 P_{OT}(t)$, then $V_G = \beta S_2 V_T + (1 - \beta)V_2$

$$\begin{bmatrix} 0 \\ \beta c_T + v \end{bmatrix} = \beta(c_T + v) \begin{bmatrix} \sin(2\alpha) \\ \cos(2\alpha) \end{bmatrix} + (1 - \beta)v \begin{bmatrix} \sin(\alpha) \\ \cos(\alpha) \end{bmatrix}.$$

This implies that $\alpha = 0$, which must be rejected by assumption. There is no ghost.

The other cases:

In the other cases, the motion of ghost targets is defined during the first leg by when $t \leq \tau$,

$$P_{OG}(t) = \beta_1 P_{OT}(t) \tag{A1}$$

$$\text{or } P_{OG}(t) = \gamma_1 S_1 P_{OT}(t) \tag{A2}$$

and during the second leg by
when $t \geq \tau$,

$$P_{OG}(t) = \beta_2 P_{OT}(t) \tag{A3}$$

$$\text{or } P_{OG}(t) = \gamma_2 S_2 P_{OT}(t) \tag{A4}$$

Hence, at time τ, the position of a ghost target is

$$P_{OG}(\tau) = \beta_1 P_{OT}(\tau) \tag{A5}$$

$$\text{or } P_{OG}(\tau) = \gamma_1 S_1 P_{OT}(\tau) \tag{A6}$$

$$\text{and } P_{OG}(\tau) = \beta_2 P_{OT}(\tau) \tag{A7}$$

$$\text{or } P_{OG}(\tau) = \gamma_2 S_2 P_{OT}(\tau) \tag{A8}$$

Of course, (A5) and (A6) are not compatible, and neither are (A7) and (A8).

Now, let us show that (A5) is not compatible with (A8):

Indeed, if $P_{OG}(\tau) = \beta_1 P_{OT}(\tau) = \gamma_2 S_2 P_{OT}(\tau)$, then

$$\frac{\beta_1}{\gamma_2} P_{OT}(\tau) = S_2 P_{OT}(\tau) \tag{A9}$$

Equation (A9) implies that $P_{OT}(\tau)$ is an eigenvector of S_2, with the eigenvalue $\frac{\beta_1}{\gamma_2}$. Since $\frac{\beta_1}{\gamma_2}$ is positive, this eigenvalue is equal to 1, i.e., $\gamma_2 = \beta_1$. Hence, $P_{OT}(\tau)$ is in the second leg.

Hence, the set of ghost targets is reduced to those whose positions at time τ are given by (A5) or (A6), and (A7). Now suppose that a ghost target satisfies (A6) and (A7). By the same computation, we conclude that $P_{OT}(\tau)$ is in the first leg, which is impossible since $P_{OT}(\tau)$ is in the second leg.

We have proven that (A5) and (A7) only are compatible. It follows that a ghost target verifies these two equalities (given by (A1) and (A3)):

$P_{OG}(\tau) = \beta_1 P_{OT}(\tau) = \beta_2 P_{OT}(\tau)$.

Hence, $\beta_1 = \beta_2$.

Now taking the derivative of the two members of (A1) and of (A3), we obtain

$V_G = \beta_1(V_T - V_1) + V_1 = \beta_1(V_T - V_2) + V_2$, which is equivalent to

$(\beta_1 - 1)(V_2 - V_1) = 0$.

Since $V_2 \neq V_1$, $\beta_1 = 1$.

Putting this value into (A1) or (A3), we finally get $P_G(t) = P_T(t)$. The "ghost" is the target of interest. In conclusion, there is no ghost target. $\square$

References

1. Nardone, S.; Lindgren, A.; Gong, K. Fundamental properties and performance of conventional bearings-only target motion analysis. *IEEE Trans. Autom. Control* **1984**, *29*, 775–787. [CrossRef]
2. Song, T.L.; Um, T.Y. Practical guidance for homing missiles with bearings-only measurements. *IEEE Trans. Aerosp. Electron. Syst.* **1996**, *32*, 434–443. [CrossRef]
3. Clavard, J.; Pillon, D.; Pignol, A.-C.; Jauffret, C. Target Motion Analysis of a Source in a Constant Turn from a Nonmaneuvering Observer. *IEEE Trans. Aerosp. Electron. Syst.* **2013**, *49*, 1760–1780. [CrossRef]
4. Farina, A. Target tracking with bearingsOnly measurements. *Signal. Process* **1999**, *78*, 61–78. [CrossRef]
5. Jauffret, C.; Pillon, D.; Pignol, A.-C. Bearings-Only Maneuvering Target Motion Analysis from a Nonmaneuvering Platform. *IEEE Trans. Aerosp. Electron. Syst.* **2010**, *46*, 1934–1949. [CrossRef]
6. Zhang, Y.; Lan, J.; Mallick, M.; Li, X.R. Bearings-Only Filtering Using Uncorrelated Conversion Based Filters. *IEEE Trans. Aerosp. Electron. Syst.* **2021**, *57*, 882–896. [CrossRef]
7. Aidala, V.; Hammel, S. Utilization of modified polar coordinates for bearings-only tracking. *IEEE Trans. Autom. Control* **1983**, *28*, 283–294. [CrossRef]
8. Arulampalam, S.; Clark, M.; Vinter, R. Performance of the shifted Rayleigh filter in single-sensor bearings-only tracking. In Proceedings of the 2007 10th International Conference on Information Fusion, Quebec, QC, Canada, 9–12 July 2007; pp. 1–6.
9. Laneuville, D.; Jauffret, C. Recursive Bearings-Only TMA via Unscented Kalman Filter: Cartesian vs. Modified Polar Coordinates. In Proceedings of the 2008 IEEE Aerospace Conference, Big Sky, MT, USA, 1–8 March 2008; pp. 1–11.

10. Arulampalam, M.S.; Ristic, B.; Gordon, N.; Mansell, T. Bearings-Only Tracking of Maneuvring Targets Using Particle Filters. *Eurasip J. App. Sig. Process.* **2004**, *15*, 2351–2365.
11. Zhang, Y.J.; Xu, G.Z. Bearings-Only Target Motion Analysis via Instrumental Variable Estimation. *IEEE Trans. Signal. Process.* **2010**, *58*, 5523–5533. [CrossRef]
12. Doğançay, K. On the efficiency of a bearings-only instrumental variable estimator for target motion analysis. *Signal. Process.* **2005**, *85*, 481–490. [CrossRef]
13. Chan, Y.T.; Rea, T.A. Bearings-only tracking using data fusion and instrumental variables. In Proceedings of the Proceedings of the Third International Conference on Information Fusion, Paris, France, 10–13 July 2000.
14. Kirubarajan, T.; Bar-Shalom, Y.; Lerro, D. Bearings-only tracking of maneuvering targets using a batch-recursive estimator. *IEEE Trans. Aerosp. Electron. Syst.* **2001**, *37*, 770–780. [CrossRef]
15. Jauffret, C.D.; Pillon, D. Observability in Passive Target Motion Analysis. *IEEE Trans. Aerosp. Electro. Syst.* **1996**, *AES-32*, 1290–1300. [CrossRef]
16. Le Cadre, J.; Jauffret, C. Discrete-time observability and estimability analysis for bearings-only target motion analysis. *IEEE Trans. Aerosp. Electron. Syst.* **1997**, *33*, 178–201. [CrossRef]
17. Nardone, S.C.; Aidala, V.J. Observability Criteria for Bearings-Only Target Motion Analysis. *IEEE Trans. Aerosp. Electron. Syst.* **1981**, *AES-17*, 162–166. [CrossRef]
18. Urick, R.J. *Principles of Underwater Sound*; McGraw-Hill: New York, NY, USA, 1983.
19. Van Trees, H.L. *Optimum Array Processing: Part IV of Detection, Estimation, and Modulation Theory*; Wiley Interscience: New York, NY, USA, 2002.
20. Abraham, D.A. *Underwater Acoustic Signal Processing: Modeling, Detection, and Estimation*; Modern Acoustic and Signal Processing; Springer: Berlin/Heidelberg, Germany, 2019.
21. Gong, K.F. *Multipath Target Motion Analysis: Properties and Implication of the Multipath Process*; in Technical Report 6687; NUSC: Newport, RI, USA, 1982.
22. Blanc-Benon, P.; Jauffret, C. Target Motion Analysis from Bearing and Time-delay Measurements: The Use of Multipath. *IEEE Trans. Aerosp. Electron. Syst.* **1997**, *AES-33*, 813–824. [CrossRef]
23. Oh, R.; Song, T.L.; Choi, J.W. Batch Processing through Particle Swarm Optimization for Target Motion Analysis with Bottom Bounce Underwater Acoustic Signals. *Sensors* **2020**, *20*, 1234. [CrossRef] [PubMed]
24. Steven, S.M. *Fundamentals if Statistical Signal Processing—Estimation Theory*; Prentice Hall: Englehood Cliffs, NJ, USA, 1993.
25. Jauffret, C. Observability and fisher information matrix in nonlinear regression. *IEEE Trans. Aerosp. Electron. Syst.* **2007**, *43*, 756–759. [CrossRef]
26. Pignol, A.-C.; Jauffret, C.; Pillon, D. A statistical fusion for a leg-by-leg bearings-only TMA without observer maneuver. In Proceedings of the 13th International Conference on Information Fusion, Edinburgh, UK, 26–29 July 2010; Volume 6, pp. 1–8. [CrossRef]
27. Bar-Shalom, Y.; Rong Li, X.; Kirubarajan, T. *Estimation and Tracking: Principles, Techniques and Software*; Artech House: Boston, MA, USA, 1993.
28. Payan, J.; Lebon, A.; Perez, A.-C.; Jauffret, C.; Laneuville, D. Passive Target Motion Analysis by Fusion of Linear Arrays and Sonobuoys in a Cluttered Environment. *IEEE Trans. Aerosp. Electron. Syst.* **2021**, 1. [CrossRef]
29. Le Cadre, J.-P.; Gauvrit, H. Optimization of the observer motion for bearings-only target motion analysis. In Proceedings of the 1st Australian Data Fusion Symposium, Adelaide, SA, Australia, 21–22 November 1996; pp. 190–195.
30. Fawcett, J. Effect of course maneuvers on bearings-only range estimation. *IEEE Trans. Acoust. Speech, Signal. Process* **1988**, *36*, 1193–1199. [CrossRef]
31. Passerieux, J.M.; Van Cappel, D. Optimal observer maneuver for bearings-only tracking. *IEEE Trans. Aerosp. Electron. Syst.* **1998**, *34*, 777–788. [CrossRef]

Article

A New Pseudolinear Filter for Bearings-Only Tracking without Requirement of Bias Compensation

Shizhe Bu [1,2], Aiqiang Meng [1,2] and Gongjian Zhou [1,2,*]

1   Department of Electronic Engineering, Harbin Institute of Technology, Harbin 150001, China; bushizhe@hit.edu.cn (S.B.); maq.0417@foxmail.com (A.M.)
2   Key Laboratory of Marine Environmental Monitoring and Information Processing, Ministry of Industry and Information Technology, Harbin 150001, China
*   Correspondence: zhougj@hit.edu.cn

Abstract: In bearings-only tracking systems, the pseudolinear Kalman filter (PLKF) has advantages in stability and computational complexity, but suffers from correlation problems. Existing solutions require bias compensation to reduce the correlation between the pseudomeasurement matrix and pseudolinear noise, but incomplete compensation may cause a loss of estimation accuracy. In this paper, a new pseudolinear filter is proposed under the minimum mean square error (MMSE) framework without requirement of bias compensation. The pseudolinear state-space model of bearings-only tracking is first developed. The correlation between the pseudomeasurement matrix and pseudolinear noise is thoroughly analyzed. By splitting the bearing noise term from the pseudomeasurement matrix and performing some algebraic manipulations, their cross-covariance can be calculated and incorporated into the filtering process to account for their effects on estimation. The target state estimation and its associated covariance can then be updated according to the MMSE update equation. The new pseudolinear filter has a stable performance and low computational complexity and handles the correlation problem implicitly under a unified MMSE framework, thus avoiding the severe bias problem of the PLKF. The posterior Cramer–Rao Lower Bound (PCRLB) for target state estimation is presented. Simulations are conducted to demonstrate the effectiveness of the proposed method.

Keywords: bearings-only tracking; pseudolinear estimation; correlation analysis; MMSE framework

Citation: Bu, S.; Meng, A.; Zhou, G. A New Pseudolinear Filter for Bearings-Only Tracking without Requirement of Bias Compensation. *Sensors* **2021**, *21*, 5444. https://doi.org/10.3390/s21165444

Academic Editors: Mahendra Mallick and Ratnasingham Tharmarasa

Received: 24 June 2021
Accepted: 11 August 2021
Published: 12 August 2021

Publisher's Note: MDPI stays neutral with regard to jurisdictional claims in published maps and institutional affiliations.

1. Introduction

Target tracking has been researched for decades with a wide range of applications in civilian and military areas. It refers to estimate a moving target's state using the noise-corrupted measurements collected by one or more sensors at fixed locations or on moving platforms [1–5]. The typical measurements include target range, Doppler velocity and bearing angles, while in passive bearings-only tracking (BOT) systems [6–11], the sensors listen for signals emitted by a target and only acquire the bearing data.

Bearings-only tracking has been under intensive investigation in recent decades, and the main challenge is the intrinsic nonlinearities in measurement equations. Early research used the extended Kalman filter (EKF) to estimate target state in Cartesian coordinates, but this filter shows poor performance due to premature collapse of the error covariance matrix [12]. Later on, a modified polar coordinate EKF (MPEKF) was developed in [13] to improve the stability. However, both the EKF and MPEKF require good initialization to avoid divergence. The unscented Kalman filter [14] (UKF) and particle filter [15–17] (PF) are also applied for bearings-only tracking. The UKF has better estimation performance than the EKF, but still faces the divergence problems. The PF can exhibit a good performance but at the price of heavy computation load.

Another basic and famous recursive Bayesian estimator is the pseudolinear Kalman filter [18] (PLKF). The PLKF solves the bearings-only tracking problem by converting the nonlinear measurement equation to the pseudolinear equation and then applying the

Kalman filter (KF) to produce target state estimates. The PLKF is superior in computational complexity and robust to initialization errors compared to the above nonlinear filtering methods [19,20]. However, the pseudomeasurement matrix is a function of the noisy bearing measurements and correlated with the pseudolinear noise. This makes the PLKF exhibit a bias which can be severe in unfavorable geometries and degrades the tracking performance [21].

Several methods have been presented to improve the performance of the PLKF by compensating or reducing the pseudolinear estimation bias. In [22], a modified pseudo-linear estimator (MPLE) is developed to reduce the bias by defining the target motion parameters in a new coordinate system related to modified polar coordinates. In [23], a bias-compensated PLKF (BC-PLKF) method is developed to compensate for the bias of the PLKF. The estimate of the cross-term that contains the pseudomeasurement matrix and pseudolinear noise is calculated and then subtracted from the PLKF estimate to generate the final state estimate. The unbiasedness of methods in [22,23] can only be guaranteed under the assumption of small measurement noise, and their performances will be adversely affected at large measurement noise.

The well-known instrumental variable [24] (IV) estimation is also applied for reducing the bias of the PLKF. The essential step of the IV approach is the formulation of the so-called IV matrix, which is statistically independent of the pseudolinear noise and is strongly correlated with pseudomeasurement matrix. Several IV-based estimators are developed in [25–27], but they do not have closed-form solutions and require good initialization to guarantee convergence. The methods in [28,29] utilize the bias compensated estimator to construct an IV matrix, and then implement the IV estimation procedure to obtain asymptotically unbiased estimates. However, the correlation between the IV matrix and the pseudomeasurement matrix can be weaken in the presence of large measurement noise and in unfavorable geometries, which can lead to the estimation performance degradation. To maintain a strong correlation between the IV matrix and pseudomeasurement matrix, an IV Kalman filter (IVKF) based on selective-angle-measurement [30] (SAM) strategy is presented in [23], resulting in the SAM-IVKF method. According to the SAM threshold, the IVKF is implemented in the case of small measurement noise, and the BC-PLKF is selected in the large measurement noise. Benefit from the SAM strategy, the SAM-IVKF has a better tracking performance than the BC-PLKF and the IVKF methods, and is robust to the measurement noise and initialisation errors. However, the SAM-IVKF method is a hybrid method, and its theoretical framework is not unified. In addition, the SAM-IVKF method utilizes a empirical scheme to select the SAM threshold, and the threshold values will have a great influence on the tracking performance [31].

The above methods rely on bias compensation to reduce the correlation between the pseudomeasurement matrix and pseudolinear noise, which can improve the performance of the PLKF. As variants of the PLKF, they all show a stable performance and low computational complexity. However, the bias compensation is not always perfect, especially in case of large measurement noise and unfavorable geometries, which will lead to loss of estimation accuracy and consistency.

In this paper, we propose a new pseudolinear filter under the MMSE framework without requirement of bias compensation. Inspired by the methods in [32,33], we make a thorough analysis about the correlations between the pseudomeasurement matrix and pseudolinear noise, and evaluate their impacts on the estimation results. First, the pseudolinear state-space model of bearings-only tracking problems is formulated. Under the MMSE framework, we provide the expression of each step in the filtering process according to its definition. It is found out that in the step of measurement prediction covariance calculation, the correlation between pseudomeasurement matrix and pseudolinear noise will cause the cross-items that contain the two components to be nonzero matrices, which is different from the traditional linear state-space model. Similar situations can be found in the step of calculating the covariance between the state and measurement. Accordingly, we thoroughly analyze the correlation between the pseudomeasurement matrix and pseudolinear noise, and incorporate their cross-covariance into the corresponding processing steps of the filtering process to account for their effects on the estimation results, resulting in the new pseudolinear filter. The proposed method inherits the merits of the PLKF and implicitly handles the correlation under the MMSE framework, which guarantees its low computational complexity and the stable estimation accuracy. The superiority of the proposed method is illustrated by numerical simulations.

The rest of this paper is organized as follows. In Section 2, the bearings-only tracking problem is formulated. In Section 3, the pseudolinear state-space model is constructed and the new pseudolinear filter is presented in detail. The posterior Cramer–Rao lower bound [34,35] (PCRLB) of the state estimation is derived in Section 4. Section 5 presents the simulation results, followed by conclusions in Section 6.

2. Problem Formulation

The problem of bearings-only target tracking by a single moving sensor in the two-dimensional (2D) plane is shown in Figure 1.

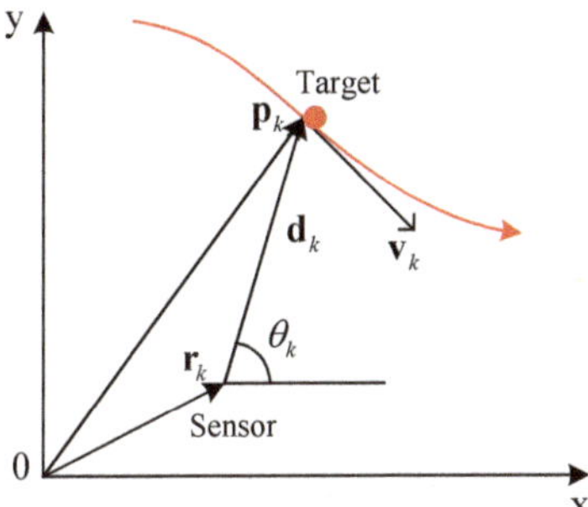

Figure 1. 2D bearings-only target tracking geometry.

As shown in Figure 1, $\mathbf{p}_k = [p_{x,k},\ p_{y,k}]^T$ and $\mathbf{v}_k = [v_{x,k},\ v_{y,k}]^T$ are the position and velocity of the target at time instant k, respectively, which constitute the unknown target state vector $\mathbf{x}_k = [p_{x,k}, p_{y,k}, v_{x,k}, v_{y,k}]^T$. $\mathbf{r}_k = [r_{x,k}, r_{y,k}]^T$ is the position of the sensor and assumed to be precisely known at each time instant, and $\mathbf{d}_k$ is the distance vector pointing from the sensor to the target. We assume that the target follows the nearly constant velocity (NCV) motion [1] in the whole paper. The target state equation is given by

$$\mathbf{x}_k = \mathbf{F}\mathbf{x}_{k-1} + \mathbf{w}_{k-1} \tag{1}$$

where $\mathbf{F}$ is the target state transition matrix, and $\mathbf{w}_{k-1}$ is the zero-mean Gaussian white process noise with known covariance $\mathbf{Q}_{k-1}$. The matrices $\mathbf{F}$ and $\mathbf{Q}_{k-1}$ are given by

$$\mathbf{F} = \begin{bmatrix} 1 & 0 & T & 0 \\ 0 & 1 & 0 & T \\ 0 & 0 & 1 & 0 \\ 0 & 0 & 0 & 1 \end{bmatrix} \tag{2}$$

$$\mathbf{Q}_{k-1} = \begin{bmatrix} \frac{T^3}{3}q_x & 0 & \frac{T^2}{2}q_x & 0 \\ 0 & \frac{T^3}{3}q_y & 0 & \frac{T^2}{2}q_y \\ \frac{T^2}{2}q_x & 0 & Tq_x & 0 \\ 0 & \frac{T^2}{2}q_y & 0 & Tq_y \end{bmatrix} \tag{3}$$

where T is the sampling interval, and q_x and q_y are power spectral densities of the process noise in the x-coordinate and y-coordinate, respectively.

According to the geometric relationship in Figure 1, the true bearing angle at time instant k is $\theta_k = \tan^{-1}(p_{y,k} - r_{y,k}, \ p_{x,k} - r_{x,k})$, which is corrupted by the independent Gaussian noise n_k with zero mean and variance σ_θ^2. The bearing measurement equation is given by

$$\theta_k^z = \theta_k + n_k = \tan^{-1}\left(p_{y,k} - r_{y,k}, \ p_{x,k} - r_{x,k}\right) + n_k \tag{4}$$

where θ_k^z is the bearing measurement at time instant k. To ensure the target is observable, the sensor needs to maneuver while collecting the bearing measurements [28].

Equations (1) and (4) formulate the state-space model for bearings-only target tracking, and the objective is to estimate the target state using the noise-corrupted bearing measurement at each time instant. Due to the nonlinearity of the bearing measurement equation, the KF cannot be used to obtain the target state estimation. The nonlinear filtering methods such as the EKF and UKF are intuitive solutions for bearings-only target tracking but can lead to instability problems. The PLKF is an attractive alternative due to its stable performance and low computational complexity. However, this method suffers from the correlation problem and results in severe bias problem. Accordingly, it is necessary to find an effective method to solve the PLKF correlation problem, which will be investigated in the next section.

3. The New Pseudolinear Filter

In this section, we propose a new pseudolinear filter under the MMSE framework for bearings-only target tracking, which is referred as the pseudolinear-MMSE (PL-MMSE). The PL-MMSE does not require analysis of the PLKF bias caused by the correlation, and performs the bias compensation procedure. Instead, the proposed method evaluates the correlation between the pseudomeasurement matrix and pseudolinear noise, and their cross-covariance is involved in the filtering process to account for their effects on the state and covariance update. The pseudolinear state-space model is presented in the next section, followed by the filtering process of the PL-MMSE.

3.1. The Pseudolinear State-Space Model

To be able to apply linear filtering method to the bearings-only tracking, the bearing measurement equation must be linearized. According to the geometry in Figure 1, the bearing measurement equation in (4) is rewritten as

$$\frac{\sin(\theta_k^z - n_k)}{\cos(\theta_k^z - n_k)} \overset{\Delta}{=} \frac{\sin \theta_k}{\cos \theta_k} \overset{\Delta}{=} \frac{p_{y,k} - r_{y,k}}{p_{x,k} - r_{x,k}} \tag{5}$$

where $\sin \theta_k = \dfrac{p_{y,k} - r_y}{\|\mathbf{d}_k\|}$ and $\cos \theta_k = \dfrac{p_{x,k} - r_x}{\|\mathbf{d}_k\|}$. The symbol $\|\cdot\|$ denotes the Euclidean norm, and the distance vector $\mathbf{d}_k$ is a function of the target position and sensor position. That is

$$\mathbf{d}_k = \mathbf{C}\mathbf{x}_k - \mathbf{r}_k, \quad \mathbf{C} = \begin{bmatrix} 1 & 0 & 0 & 0 \\ 0 & 1 & 0 & 0 \end{bmatrix}. \tag{6}$$

We expand (5) according to the triangle formula and have

$$\begin{aligned} r_{x,k} \sin \theta_k^z \cos n_k - r_{y,k} \cos \theta_k^z \cos n_k = p_{x,k} \sin \theta_k^z \cos n_k - p_{y,k} \cos \theta_k^z \cos n_k \\ - (p_{x,k} - r_{x,k}) \cos \theta_k^z \sin n_k - (p_{y,k} - r_{y,k}) \sin \theta_k^z \sin n_k \end{aligned} . \tag{7}$$

Dividing both sides of (7) by $\cos n_k$, we have

$$\begin{aligned} r_{x,k} \sin \theta_k^z - r_{y,k} \cos \theta_k^z = p_{x,k} \sin \theta_k^z - p_{y,k} \cos \theta_k^z \\ - \left[(p_{x,k} - r_{x,k}) \cos \theta_k^z + (p_{y,k} - r_{y,k}) \sin \theta_k^z \right] \frac{\sin n_k}{\cos n_k} \end{aligned} . \tag{8}$$

Substituting $\theta_k^z = \theta_k + n_k$ into the third item in the right side of (8), we have

$$
\begin{aligned}
\frac{1}{\cos n_k} &\left[(p_{x,k} - r_{x,k}) \cos \theta_k^z + (p_{y,k} - r_{y,k}) \sin \theta_k^z \right] \\
&= (p_{x,k} - r_{x,k})(\cos \theta_k - \sin \theta_k \tan n_k) + (p_{y,k} - r_{y,k})(\sin \theta_k + \cos \theta_k \tan n_k) \\
&= (p_{x,k} - r_{x,k})\Big(\frac{p_{x,k} - r_{x,k}}{\|\mathbf{d}_k\|} - \frac{p_{y,k} - r_{y,k}}{\|\mathbf{d}_k\|} \tan n_k \Big) \\
&\quad + (p_{y,k} - r_{y,k})\Big(\frac{p_{y,k} - r_{y,k}}{\|\mathbf{d}_k\|} + \frac{p_{x,k} - r_{x,k}}{\|\mathbf{d}_k\|} \tan n_k \Big) \\
&= \frac{(p_{x,k} - r_{x,k})^2 + (p_{y,k} - r_{y,k})^2}{\|\mathbf{d}_k\|} = \|\mathbf{d}_k\|
\end{aligned}
\tag{9}
$$

After the manipulations in (5)–(9), the nonlinear bearing measurement equation in (4) is converted into a pseudolinear function of the state vector $\mathbf{x}_k$

$$
z_k = \mathbf{H}_k \mathbf{x}_k + \eta_k
\tag{10}
$$

where the measurement z_k, the pseudomeasurement matrix $\mathbf{H}_k$, and the pseudolinear noise η_k are, respectively, given by

$$
z_k = r_{x,k} \sin \theta_k^z - r_{y,k} \cos \theta_k^z.
\tag{11}
$$

$$
\mathbf{H}_k = [\sin \theta_k^z, \ -\cos \theta_k^z, \ 0, \ 0].
\tag{12}
$$

$$
\eta_k = -\|\mathbf{d}_k\| \sin n_k.
\tag{13}
$$

The bearing noise n_k is assumed to be zero-mean Gaussian white variable. In this case, one has [36]

$$
E[\cos n_k] = e^{-\frac{\sigma_\theta^2}{2}}.
\tag{14a}
$$

$$
E[\sin n_k] = 0.
\tag{14b}
$$

$$
E[\cos^2 n_k] = \frac{1 + e^{-2\sigma_\theta^2}}{2}.
\tag{14c}
$$

$$
E[\sin^2 n_k] = \frac{1 - e^{-2\sigma_\theta^2}}{2}.
\tag{14d}
$$

$$
E[\cos n_k \sin n_k] = 0.
\tag{14e}
$$

Accordingly, the mean μ_k and the variance R_k of the pseudolinear noise η_k are

$$
\mu_k = E[\eta_k] = -\|\mathbf{d}_k\| E[\sin n_k] = 0.
\tag{15a}
$$

$$
R_k = E[\eta_k^2] = \|\mathbf{d}_k\|^2 E[\sin^2 n_k] = \frac{1 - e^{-2\sigma_\theta^2}}{2} \|\mathbf{d}_k\|^2.
\tag{15b}
$$

Using the pseudolinear measurement, Equation (10), the pseudolinear state-space model for bearings-only target tracking can be described as

$$
\mathbf{x}_k = \mathbf{F}\mathbf{x}_{k-1} + \mathbf{w}_{k-1}.
\tag{16a}
$$

$$
z_k = \mathbf{H}_k \mathbf{x}_k + \eta_k.
\tag{16b}
$$

The pseudomeasurement matrix $\mathbf{H}_k$ and the pseudolinear noise η_k in (16b) both are functions of bearing noise n_k, which causes $\mathbf{H}_k$ and η_k to be correlated. This leads to the significant difference between the pseudolinear state-space model (16a)–(16b) and the traditional linear state-space model. The PLKF ignores the correlation between $\mathbf{H}_k$ and η_k, and directly applies the KF to estimate the target state according to the pseudolinear state-space model. To solve this problem, we analyze the correlation between $\mathbf{H}_k$ and

η_k, and incorporate their cross-covariance into the filtering process under the MMSE framework to account for the effects on the estimation results.

3.2. Filtering Process

In this subsection, the filtering process of the PL-MMSE is presented under the MMSE framework, including the prediction stage, covariance calculation stage and update stage.

3.2.1. Prediction Stage

According to Equation (16a), the one-step predicted state is

$$\hat{\mathbf{x}}_{k|k-1} = E[\mathbf{x}_k | Z^{k-1}] = E[\mathbf{F}\mathbf{x}_{k-1} + \mathbf{w}_{k-1} | Z^{k-1}] = \mathbf{F}\hat{\mathbf{x}}_{k-1|k-1} \tag{17}$$

where Z^{k-1} denotes the sequence of measurements available at time instant $k-1$. Subtracting the above from (16a) yields the state prediction error

$$\tilde{\mathbf{x}}_{k|k-1} = \mathbf{x}_k - \hat{\mathbf{x}}_{k|k-1} = \mathbf{F}\tilde{\mathbf{x}}_{k-1|k-1} + \mathbf{w}_{k-1}. \tag{18}$$

The state prediction covariance is

$$\begin{aligned}
\mathbf{P}_{k|k-1} &= E\left[\tilde{\mathbf{x}}_{k|k-1} \cdot \tilde{\mathbf{x}}_{k|k-1}^T \middle| Z^{k-1}\right] = E\left[(\mathbf{F}\tilde{\mathbf{x}}_{k-1|k-1} + \mathbf{w}_{k-1})(\mathbf{F}\tilde{\mathbf{x}}_{k-1|k-1} + \mathbf{w}_{k-1})^T \middle| Z^{k-1}\right] \\
&= \mathbf{F}\mathbf{P}_{k-1|k-1}\mathbf{F}^T + \mathbf{Q}_{k-1}
\end{aligned} \tag{19}$$

where the cross-items contains $\tilde{\mathbf{x}}_{k-1|k-1}$ and $\mathbf{w}_{k-1}$ vanish since the process noise $\mathbf{w}_{k-1}$ is independent zero-mean Gaussian white variable.

The predicted measurement $\hat{z}_{k|k-1}$ is obtained by taking the expected value of (16b) conditioned on Z^{k-1}. That is,

$$\begin{aligned}
\hat{z}_{k|k-1} &= E\left[z_k \middle| Z^{k-1}\right] = E\left[\mathbf{H}_k\mathbf{x}_k + \eta_k \middle| Z^{k-1}\right] \\
&= \mathbf{H}_k \cdot E\left[\mathbf{x}_k \middle| Z^{k-1}\right] + E\left[\eta_k \middle| Z^{k-1}\right] = \mathbf{H}_k\hat{\mathbf{x}}_{k|k-1}
\end{aligned} \tag{20}$$

where $\mathbf{H}_k$ is a deterministic vector according to (12), which can be pulled out from the expectation operator. Subtracting the above from (16b) yields the measurement prediction error

$$\tilde{z}_{k|k-1} = z_k - \hat{z}_{k|k-1} = \mathbf{H}_k\mathbf{x}_k + \eta_k - \mathbf{H}_k\hat{\mathbf{x}}_{k|k-1} = \mathbf{H}_k\tilde{\mathbf{x}}_{k|k-1} + \eta_k. \tag{21}$$

3.2.2. Covariance Calculation Stage

In this part, we will present the steps to calculate the measurement prediction covariance P_{zz}, and the covariance P_{xz} between the state and measurement. As discussed in Section 1, the correlation between $\mathbf{H}_k$ and η_k will cause the cross-terms in P_{zz} that contains the two components to be nonzero matrices. Similar situations can be found in P_{xz}. Accordingly, we will first present the expression of P_{xz} and P_{zz} according to their definitions. By splitting the bearing noise term from the pseudomeasurement matrix and performing some algebraic manipulations, the estimates of the cross-terms that contains $\mathbf{H}_k$ and η_k can be calculated.

The measurement prediction covariance is

$$\begin{aligned}
P_{zz} &= E[\tilde{z}_{k|k-1} \tilde{z}_{k|k-1}^T] = E\left[(\mathbf{H}_k\tilde{\mathbf{x}}_{k|k-1} + \eta_k)(\mathbf{H}_k\tilde{\mathbf{x}}_{k|k-1} + \eta_k)^T\right] \\
&= E[\mathbf{H}_k\tilde{\mathbf{x}}_{k|k-1}\tilde{\mathbf{x}}_{k|k-1}^T\mathbf{H}_k^T + \mathbf{H}_k\tilde{\mathbf{x}}_{k|k-1}\eta_k^T + \eta_k\tilde{\mathbf{x}}_{k|k-1}^T\mathbf{H}_k^T + \eta_k^2] \\
&= \mathbf{H}_k\mathbf{P}_{k|k-1}\mathbf{H}_k^T + E[\mathbf{H}_k\tilde{\mathbf{x}}_{k|k-1}\eta_k^T] + E[\eta_k\tilde{\mathbf{x}}_{k|k-1}^T\mathbf{H}_k^T] + R_k
\end{aligned} \tag{22}$$

where $\mathbf{H}_k$ and η_k both are functions of bearing noise n_k and correlated with each other, causing the second and third items in (22) to be nonzero matrices, which should be calculated as follows.

First, split n_k out of $\mathbf{H}_k$. That is,

$$
\begin{aligned}
\mathbf{H}_k &= [\sin(\theta_k + n_k),\ -\cos(\theta_k + n_k),\ 0,\ 0] \\
&= [(\sin\theta_k \cos n_k + \cos\theta_k \sin n_k),\ (-\cos\theta_k \cos n_k + \sin\theta_k \sin n_k),\ 0,\ 0]\ . \\
&= \cos n_k [\sin\theta_k,\ -\cos\theta_k,\ 0,\ 0] + \sin n_k [\cos\theta_k,\ \sin\theta_k,\ 0,\ 0]
\end{aligned}
\tag{23}
$$

For simplicity, we denote

$$
\mathbf{H}_{1,k} = [\ \sin\theta_k \quad -\cos\theta_k \quad 0 \quad 0\ ].
\tag{24a}
$$

$$
\mathbf{H}_{2,k} = [\ \cos\theta_k \quad \sin\theta_k \quad 0 \quad 0\ ].
\tag{24b}
$$

and (23) can be rewritten as

$$
\mathbf{H}_k = \cos n_k \mathbf{H}_{1,k} + \sin n_k \mathbf{H}_{2,k}.
\tag{25}
$$

Second, we substitute $\sin\theta_k = \dfrac{p_{y,k} - r_y}{\|\mathbf{d}_k\|}$ and $\cos\theta_k = \dfrac{p_{x,k} - r_x}{\|\mathbf{d}_k\|}$ into (25). Since the target position $\mathbf{p}_k = [p_{x,k},\ p_{y,k}]^T$ is unavailable in practice, the approximate forms $p_{x,k} = \hat{\mathbf{x}}_{k|k-1}(1) + \tilde{\mathbf{x}}_{k|k-1}(1)$ and $p_{y,k} = \hat{\mathbf{x}}_{k|k-1}(2) + \tilde{\mathbf{x}}_{k|k-1}(2)$ are used as substitutes.

Therefore, the second cross-term of (22), i.e., $E[\mathbf{H}_k \tilde{\mathbf{x}}_{k|k-1} \eta_k^T]$, can be rewritten as

$$
\begin{aligned}
E\left[\mathbf{H}_k \tilde{\mathbf{x}}_{k|k-1} \eta_k^T\right] &= E\left[(\cos n_k \mathbf{H}_{1,k} + \sin n_k \mathbf{H}_{2,k})\tilde{\mathbf{x}}_{k|k-1}(-\|\mathbf{d}_k\|\sin n_k)\right] \\
&= E\left[-\sin^2 n_k \|\mathbf{d}_k\| \mathbf{H}_{2,k}\tilde{\mathbf{x}}_{k|k-1}\right] \\
&= E\left[-\sin^2 n_k\right] E\left\{\|\mathbf{d}_k\|[\cos\theta_k,\ \sin\theta_k,\ 0,\ 0]\tilde{\mathbf{x}}_{k|k-1}\right\} \\
&= \frac{e^{-2\sigma_\theta^2} - 1}{2} \\
&\quad \times E\left\{\|\mathbf{d}_k\|\left[\frac{\hat{\mathbf{x}}_{k|k-1}(1) + \tilde{\mathbf{x}}_{k|k-1}(1) - r_{x,k}}{\|\mathbf{d}_k\|},\ \frac{\hat{\mathbf{x}}_{k|k-1}(2) + \tilde{\mathbf{x}}_{k|k-1}(2) - r_{y,k}}{\|\mathbf{d}_k\|},\ 0,\ 0\right]\tilde{\mathbf{x}}_{k|k-1}\right\} \\
&= \frac{e^{-2\sigma_\theta^2} - 1}{2} E\left\{\left[\hat{\mathbf{x}}_{k|k-1}(1) + \tilde{\mathbf{x}}_{k|k-1}(1) - r_{x,k},\ \hat{\mathbf{x}}_{k|k-1}(2) + \tilde{\mathbf{x}}_{k|k-1}(2) - r_{y,k},\ 0,\ 0\right]\tilde{\mathbf{x}}_{k|k-1}\right\} \\
&= \frac{e^{-2\sigma_\theta^2} - 1}{2}\left[P_{k|k-1}(1,1) + P_{k|k-1}(2,2)\right]
\end{aligned}
\tag{26}
$$

where $P_{k|k-1}(j,j)$ represents the element located at the jth row and jth column of $P_{k|k-1}$.

The third item $E[\eta_k \tilde{\mathbf{x}}_{k|k-1}^T \mathbf{H}_k^T]$ of (22) is the transpose of $E[\mathbf{H}_k \tilde{\mathbf{x}}_{k|k-1} \eta_k^T]$ and given by

$$
E\left[\eta_k \tilde{\mathbf{x}}_{k|k-1}^T \mathbf{H}_k^T\right] = \frac{e^{-2\sigma_\theta^2} - 1}{2}\left[P_{k|k-1}(1,1) + P_{k|k-1}(2,2)\right].
\tag{27}
$$

The measurement noise variance R_k in (22) is $\dfrac{1 - e^{-2\sigma_\theta^2}}{2}\|\mathbf{d}_k\|^2$ according to (15b), where the distance vector $\mathbf{d}_k = \mathbf{C}\mathbf{x}_k - \mathbf{r}_k$ is unavailable due to the unknown state $\mathbf{x}_k$. To overcome this, the predicted distance vector $\hat{\mathbf{d}}_{k|k-1} = \mathbf{C}\hat{\mathbf{x}}_{k|k-1} - \mathbf{r}_k$, approximated from $\mathbf{d}_k$, is used as a substitute.

Substituting all the required matrices into (22), the measurement prediction covariance can be rewritten as

$$
P_{zz} = \mathbf{H}_k P_{k|k-1} \mathbf{H}_k^T + (e^{-2\sigma_\theta^2} - 1)\left[\mathbf{P}_{k|k-1}(1,1) + \mathbf{P}_{k|k-1}(2,2)\right] + \frac{1 - e^{-2\sigma_\theta^2}}{2}\left\|\hat{\mathbf{d}}_{k|k-1}\right\|^2.
\tag{28}
$$

Similar to the derivation of P_{zz}, the covariance P_{xz} between the state and measurement is given by

$$
\begin{aligned}
\mathbf{P}_{xz} &= E\left[\tilde{\mathbf{x}}_{k|k-1}\,\tilde{z}_{k|k-1}^T\right] = E\left[\tilde{\mathbf{x}}_{k|k-1}\left(\mathbf{H}_k\tilde{\mathbf{x}}_{k|k-1}+\eta_k\right)^T\right] \\
&= E\left[\tilde{\mathbf{x}}_{k|k-1}\,\tilde{\mathbf{x}}_{k|k-1}^T\left(\cos n_k\mathbf{H}_{1,k}+\sin n_k\mathbf{H}_{2,k}\right)^T-\sin n_k\|\mathbf{d}_k\|\tilde{\mathbf{x}}_{k|k-1}\right] \\
&= E[\cos n_k]E\left[\tilde{\mathbf{x}}_{k|k-1}\,\tilde{\mathbf{x}}_{k|k-1}^T\,\mathbf{H}_{1,k}^T\right] = e^{-\frac{\sigma_\theta^2}{2}}\mathbf{P}_{k|k-1}\mathbf{H}_{1,k}^T
\end{aligned}
\tag{29}
$$

where $\mathbf{H}_{1,k}$ relies on the true bearing angle θ_k as shown in (24a), which is unavailable in practice. To make the result useful, we replace the unknown true bearing angle θ_k with the predicted angle $\hat{\theta}_{k|k-1}$ computed from $\hat{\mathbf{x}}_{k|k-1}$. That is,

$$
\hat{\theta}_{k|k-1} = \tan^{-1}\left(\hat{\mathbf{x}}_{k|k-1}(2)-r_{y,k},\ \hat{\mathbf{x}}_{k|k-1}(1)-r_{x,k}\right).
\tag{30}
$$

$$
\hat{\mathbf{H}}_{1,k} = \left[\sin\hat{\theta}_{k|k-1},\ -\cos\hat{\theta}_{k|k-1},\ 0,\ 0\right].
\tag{31}
$$

We can rewrite the covariance between the state and measurement as

$$
\mathbf{P}_{xz} = e^{-\frac{\sigma_\theta^2}{2}}\mathbf{P}_{k|k-1}\hat{\mathbf{H}}_{1,k}^T.
\tag{32}
$$

3.2.3. Update Stage

Based on the above results, the state and covariance update equations at time instant k under the MMSE framework can be given by

$$
\hat{\mathbf{x}}_{k|k} = \hat{\mathbf{x}}_{k|k-1}+\mathbf{P}_{xz}P_{zz}^{-1}(z_k-\hat{z}_{k|k-1}).
\tag{33}
$$

$$
\mathbf{P}_{k|k} = \mathbf{P}_{k|k-1}-\mathbf{P}_{xz}P_{zz}^{-1}\mathbf{P}_{xz}^T.
\tag{34}
$$

The filtering process of the PL-MMSE is carried out in the pseudolinear state-space model, thereby ensuring the low complexity and stable performance like the PLKF method. Meanwhile, the PL-MMSE incorporates the cross-covariance between the pseudomeasurement matrix and pseudolinear noise into its filtering process, so the correlation problem can be handled implicitly under a unified MMSE estimation framework. This can avoid the severe bias in estimation results of the PLKF. Additionally, the PL-MMSE does not require bias compensation like the previous reduced-bias methods, thus avoiding the loss of estimation accuracy caused by the possible incomplete compensation.

4. Lower Bound of Performance

Since the bearing measurement, Equation (4), is nonlinear, the optimal solution to the bearings-only tracking problem cannot be derived analytically. A theoretical lower bound of performance would be helpful to assess the level of approximation introduced by the proposed method. The PCRLB on the variance of estimation error provides the performance limit for any unbiased estimator of a fixed parameter, which is derived briefly as follows.

The lower bound on the estimation error is determined by the Fisher information matrix $\mathbf{J}_k$, and the covariance of $\hat{\mathbf{x}}_{k|k}$ is bounded by

$$
E\{(\hat{\mathbf{x}}_{k|k}-\mathbf{x}_k)(\hat{\mathbf{x}}_{k|k}-\mathbf{x}_k)^T\} \ge \mathbf{J}_k^{-1} = \mathbf{PCRLB}_{\mathbf{x}_k}.
\tag{35}
$$

The general frame work for derivation of PCRLB of an unbiased estimator for nonlinear discrete-time system is described in [34], and the information matrix can be calculated by recursion [34,37]

$$
\mathbf{J}_k = \mathbf{D}_{k-1}^{22}-\mathbf{D}_{k-1}^{21}\left[\mathbf{J}_{k-1}+\mathbf{D}_{k-1}^{11}\right]^{-1}\mathbf{D}_{k-1}^{12}
\tag{36}
$$

where

$$\mathbf{D}_{k-1}^{11} = \mathbf{F}^T \mathbf{Q}_{k-1}^{-1} \mathbf{F} \tag{37a}$$

$$\mathbf{D}_{k-1}^{12} = -\mathbf{F}^T \mathbf{Q}_{k-1}^{-1} = \left[\mathbf{D}_{k-1}^{21} \right]^T \tag{37b}$$

$$\mathbf{D}_{k-1}^{22} = \mathbf{Q}_{k-1}^{-1} + E_{\mathbf{x}_k} \left\{ \left[\nabla_{\mathbf{x}_k} \theta_k{}^T \right] \left[\sigma_\theta^2 \right]^{-1} \left[\nabla_{\mathbf{x}_k} \theta_k{}^T \right]^T \right\} \tag{37c}$$

where the expectation $E\{\cdot\}$ in (37c) is taken over $\mathbf{x}_k$, $\nabla_{\mathbf{x}_k}$ is the gradient operator, and $\left[\nabla_{\mathbf{x}_k} \theta_k{}^T \right]^T$ is the Jacobian matrix of θ_k evaluated at the true target state $\mathbf{x}_k$. For simplicity, we denote $\mathbf{V}_k = \left[\nabla_{\mathbf{x}_k} \theta_k{}^T \right]^T$, which is given by

$$\mathbf{V}_k = \left[\frac{\partial \theta_k}{\partial p_{x,k}}, \ \frac{\partial \theta_k}{\partial p_{y,k}}, \ \frac{\partial \theta_k}{\partial v_{x,k}}, \ \frac{\partial \theta_k}{\partial v_{x,k}} \right] \tag{38}$$

and the entries of $\mathbf{V}_k$ are

$$\frac{\partial \theta_k}{\partial p_{x,k}} = - \frac{p_{y,k} - r_{y,k}}{\|\mathbf{p}_k - \mathbf{r}_k\|^2}. \tag{39a}$$

$$\frac{\partial \theta_k}{\partial p_{y,k}} = \frac{p_{x,k} - r_{x,k}}{\|\mathbf{p}_k - \mathbf{r}_k\|^2}. \tag{39b}$$

$$\frac{\partial \theta_k}{\partial v_{x,k}} = \frac{\partial \theta_k}{\partial v_{y,k}} = 0. \tag{39c}$$

Using the matrix inversion lemma, we can show that (36) and (37a)−(37c) are equivalent to the following recursion:

$$\mathbf{J}_k = \left[\mathbf{Q}_{k-1} + \mathbf{F} \mathbf{J}_{k-1}^{-1} \mathbf{F}^T \right]^{-1} + \frac{1}{\sigma_\theta^2} E_{\mathbf{x}_k} \left\{ \mathbf{V}_k^T \mathbf{V}_k \right\}. \tag{40}$$

The PCRLBs of the target state components are calculated as the corresponding diagonal elements of the inverse information matrix

$$\mathbf{PCRLB}_{\mathbf{x}_k}(j, j) = \left[\mathbf{J}_k^{-1} \right]_{jj} \tag{41}$$

where $\mathbf{PCRLB}_{\mathbf{x}_k}(j, j)$ denotes the PCRLB of the jth component of the state $\mathbf{x}_k$, and $[\cdot]_{jj}$ represents the element located at the jth row and jth column of a matrix.

The recursion in (40) can be implemented based on Monte Carlo simulation averaging over multiple realizations of the target trajectory. Given the initial information matrix, we can calculate the PCRLB through the recursion in (40). In practice, the recursion can be initialized with the inverse of the initial covariance matrix of the filtering method as $\mathbf{J}_k = \mathbf{P}_{1|1}^{-1}$, which will be presented in Section 5.2.

5. Simulation Results

Simulations and performance comparisons are presented in this section to evaluate the effectiveness of the proposed method. The proposed PL-MMSE method is compared with the existing PLKF [18], BC-PLKF [23] and SAM-IVKF [23] methods. In the simulations, M = 10,000 Monte Carlo runs are carried out on the given experiment, and the number of the sampling time instants is set as $N = 150$ in each run.

5.1. Performance Metrics

Several performance metrics are introduced in this subsection to evaluate the performance of these methods, including the root mean square errors (RMSEs), the bias norms (BNorms) and the normalized estimation error squared [1] (NEES). The PCRLB is also used as the performance benchmark to quantify the best achievable accuracy.

The position and velocity RMSEs are, respectively, defined by

$$\text{RMSE}_k^p = \sqrt{\frac{1}{M} \sum_{i=1}^{M} \left\| \hat{\mathbf{p}}_{k|k}^i - \mathbf{p}_k^i \right\|^2} \tag{42a}$$

$$\text{RMSE}_k^v = \sqrt{\frac{1}{M} \sum_{i=1}^{M} \left\| \hat{\mathbf{v}}_{k|k}^i - \mathbf{v}_k^i \right\|^2} \tag{42b}$$

where $\mathbf{p}_k^i = \mathbf{C}\mathbf{x}_k^i$ and $\hat{\mathbf{p}}_{k|k}^i = \mathbf{C}\hat{\mathbf{x}}_{k|k}^i$ are the true and estimated target positions at the ith Monte Carlo run at time instant k, $\mathbf{v}_k^i = \mathbf{D}\mathbf{x}_k^i$ and $\hat{\mathbf{v}}_{k|k}^i = \mathbf{D}\hat{\mathbf{x}}_{k|k}^i$ are the true and estimated target velocities, where

$$\mathbf{D} = \begin{bmatrix} 0 & 0 & 1 & 0 \\ 0 & 0 & 0 & 1 \end{bmatrix}. \tag{43}$$

The position and velocity BNorms are, respectively, defined by

$$\text{BNorm}_k^p = \left\| \frac{1}{M} \sum_{i=1}^{M} (\hat{\mathbf{p}}_{k|k}^i - \mathbf{p}_k^i) \right\|. \tag{44a}$$

$$\text{BNorm}_k^v = \left\| \frac{1}{M} \sum_{i=1}^{M} (\hat{\mathbf{v}}_{k|k}^i - \mathbf{v}_k^i) \right\|. \tag{44b}$$

The position and velocity PCRLBs at time instant k are the square root of the sum of the corresponding elements on the diagonal of $\mathbf{PCRLB}_{\mathbf{x}_k}$, which are, respectively, given by

$$\text{PCRLB}_k^p = \sqrt{\mathbf{PCRLB}_{\mathbf{x}_k}(1,1) + \mathbf{PCRLB}_{\mathbf{x}_k}(2,2)} \tag{45a}$$

$$\text{PCRLB}_k^v = \sqrt{\mathbf{PCRLB}_{\mathbf{x}_k}(3,3) + \mathbf{PCRLB}_{\mathbf{x}_k}(4,4)} \tag{45b}$$

where $\mathbf{PCRLB}_{\mathbf{x}_k}(j,j)$ has been defined in (41).

To compare the average performance of these methods at different noise levels, the time-averaged RMSEs, BNorms and PCRLBs are also utilized. The time-averaged RMSEs, BNorms and PCRLBs are, respectively, defined by

$$\text{RMSE}_{avg}^p = \sqrt{\frac{1}{MU} \sum_{i=1}^{M} \sum_{k=L}^{N} \left\| \hat{\mathbf{p}}_{k|k}^i - \mathbf{p}_k^i \right\|^2} \tag{46a}$$

$$\text{BNorm}_{avg}^p = \frac{1}{U} \sum_{k=L}^{N} \left(\left\| \frac{1}{M} \sum_{i=1}^{M} (\hat{\mathbf{p}}_{k|k}^i - \mathbf{p}_k^i) \right\| \right) \tag{46b}$$

$$\text{RMSE}_{avg}^v = \sqrt{\frac{1}{MU} \sum_{i=1}^{M} \sum_{k=L}^{N} \left\| \hat{\mathbf{v}}_{k|k}^i - \mathbf{v}_k^i \right\|^2} \tag{46c}$$

$$\text{BNorm}_{avg}^v = \frac{1}{U} \sum_{k=L}^{N} \left(\left\| \frac{1}{M} \sum_{i=1}^{M} (\hat{\mathbf{v}}_{k|k}^i - \mathbf{v}_k^i) \right\| \right) \tag{46d}$$

$$\text{PCRLB}_{avg}^p = \sqrt{\mathbf{PCRLB}_{avg}(1,1) + \mathbf{PCRLB}_{avg}(2,2)} \tag{46e}$$

$$\mathrm{PCRLB}_{avg}^{v} = \sqrt{\mathbf{PCRLB}_{avg}(3,3) + \mathbf{PCRLB}_{avg}(4,4)} \tag{46f}$$

where

$$\mathbf{PCRLB}_{avg} = \frac{1}{U} \sum_{k=L}^{N} \mathbf{PCRLB}_{x_k}. \tag{47}$$

Here, $U = N - L + 1$, where L is an offset parameter to make the time-averaged performance metrics unaffected by the initial estimation errors. We set $L = 60$ in the simulations.

5.2. Simulation Parameters

We consider a single target tracking problem with a single bearings-only sensor in the 2D plane. To perform an objective and fair performance comparison with the SAM-IVKF, we use the same target-observer geometry as in [23]. Specifically, the trajectory of the sensor is five constant-velocity legs with the end position of each leg set to $[60,0]^T$ m, $[0,7.5]^T$ m, $[60,15]^T$ m, $[0,22.5]^T$ m, $[60,30]^T$ m and $[0,77.5]^T$ m. The sensor trajectory is depicted in Figure 2 and its initial position is marked with Pentagram. The sensor collects bearing measurements at regular time instants $t_k = kT$, $k \in \{1, 2, \ldots, 150\}$ with the sampling interval being $T = 0.1$ s. The measurement noise n_k is assumed to be i.i.d. with known variance σ_θ^2, whose value will be given next. The moving target takes a constant velocity $[0,12]^T$ m/s starting from the position $[30,42]^T$ m. The power spectral densities of the process noise are set to $q_x = q_y = 0.2$ m^2/s^3. In the simulations, track initialization is obtained as in [38] by generating a target state estimate $\hat{\mathbf{x}}_{1|1}$ from a Gaussian distribution around the true target state $\mathbf{x}_1$ with the covariance given by $\mathbf{P}_{1|1} = \rho^2 \mathrm{diag}(2.6^2, 2.6^2, 0.26^2, 0.26^2)$. The variable ρ controls the level of track initialization error and is used to evaluate the performance of the tracking algorithms with respect to initialization error levels.

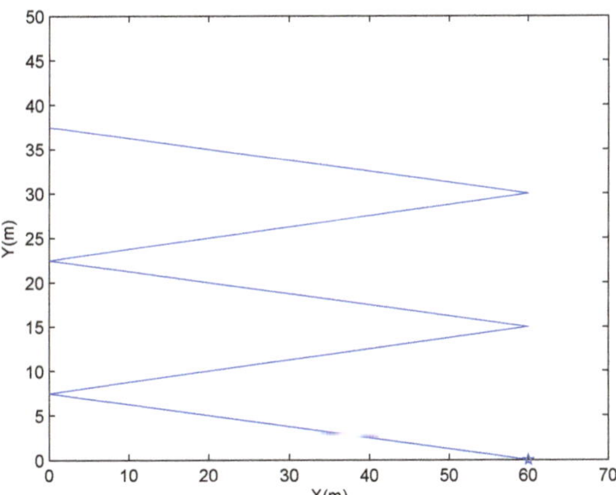

Figure 2. Sensor trajectory.

5.3. Tracking Performance

The tracking performance of the proposed PL-MMSE, the PLKF, the BC-PLKF and the SAM-PLKF for different measurement noise levels is compared in this subsection. In practice, the initialization error ρ is often proportional to the measurement noise σ_θ, and their values in each level are given in Table 1. According to [23], the performance of the SAM-IVKF depends on the selection of the threshold κ. Therefore, the SAM-IVKF under the thresholds $\kappa = 4, 3$ and 2 is compared with the proposed PL-MMSE method.

Table 1. The bearing noise standard deviation and initialization error in each level.

Level	1	2	3	4	5	6	7	8	9	10
σ_θ (degree)	1	2	3	4	5	6	7	8	9	10
ρ		1	2	3	4	5	6	7	8	9	10

The time-averaged RMSEs and BNorms of the target position and velocity estimates versus the bearing noise standard deviations are shown in Figures 3–5. The SAM-IVKF under the thresholds $\kappa = 4, 3$ and 2 are plotted in this figures, respectively. Additionally, the evolution of RMSEs and BNorms of the target position and velocity estimates for $\sigma_\theta = 7°$ and $\kappa = 4, 3$ and 2 are presented in Figures 6–8, respectively. The PCRLB is used to quantify the best achievable accuracy.

As shown in Figures 3–5, the PLKF suffers from severe bias problem and provides unsatisfactory RMSEs and BNorms performances at both small and large bearing noise levels. As the noise standard deviation σ_θ increases, the time-averaged RMSEs and BNorms performances of the PLKF deteriorate rapidly. The previously developed BC-PLKF and SAM-IVKF methods compensate for the bias and have a performance improvement compared to the PLKF. The BC-PLKF shows a good performance at small noise levels, but the bias cannot be completely compensated under large bearing noise ($\sigma_\theta \geq 7°$), which leads to a decrease in performance. Benefit from the SAM strategy, the time-averaged RMSEs and BNorms of the SAM-IVKF are lower than those of the BC-PLKF under the best choice of the threshold, i.e., $\kappa = 4$, as shown in Figure 3. However, the performance of the SAM-IVKF may vary greatly as the SAM thresholds change, and its performance is even worse than the BC-PLKF under the threshold $\kappa = 2$, as shown in Figure 5.

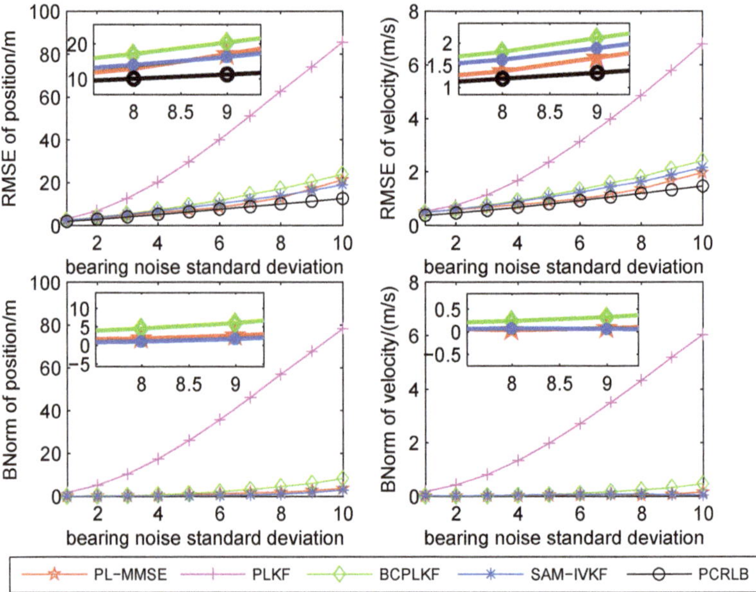

Figure 3. Time-averaged RMSEs and BNorms versus bearing noises for the PLKF, the BC-PLKF and the SAM-IVKF with $\kappa = 4$, as well as, the proposed PL-MMSE.

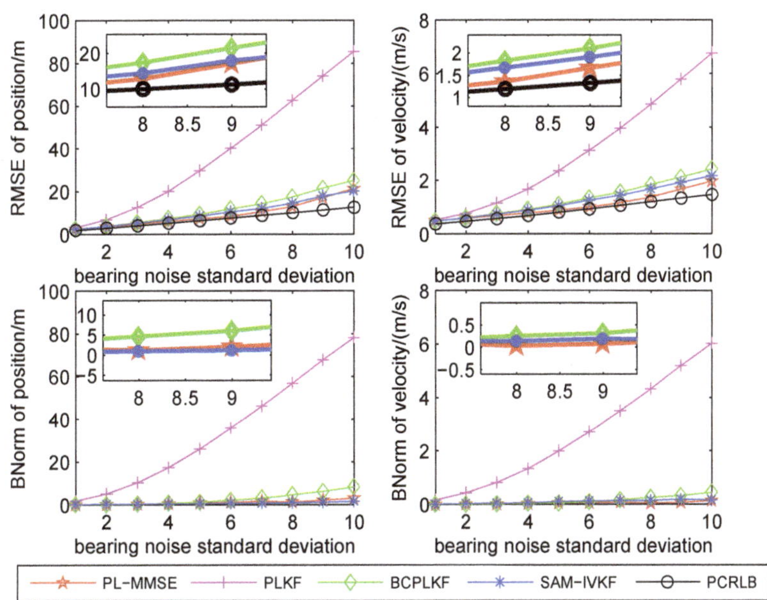

Figure 4. Time-averaged RMSEs and BNorms versus bearing noises for the PLKF, the BC-PLKF and the SAM-IVKF with $\kappa = 3$, as well as, the proposed PL-MMSE.

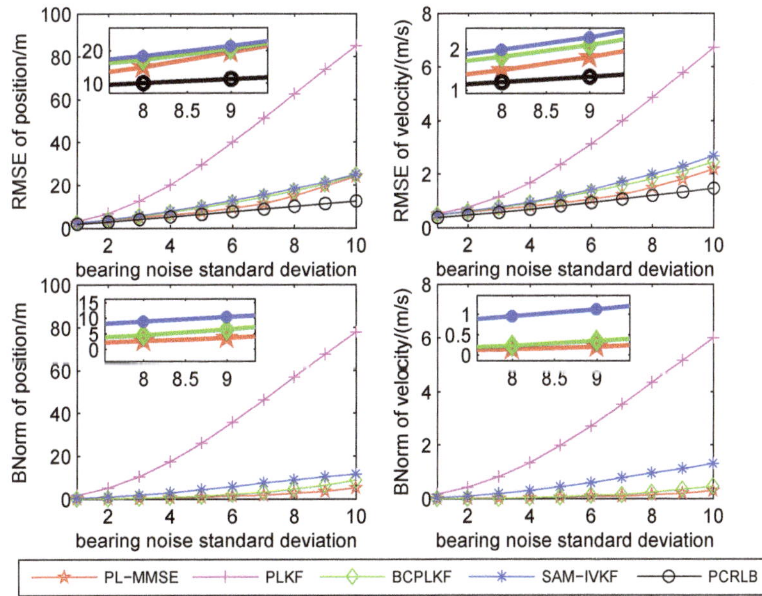

Figure 5. Time-averaged RMSEs and BNorms versus bearing noises for the PLKF, the BC-PLKF and the SAM-IVKF with $\kappa = 2$, as well as, the proposed PL-MMSE.

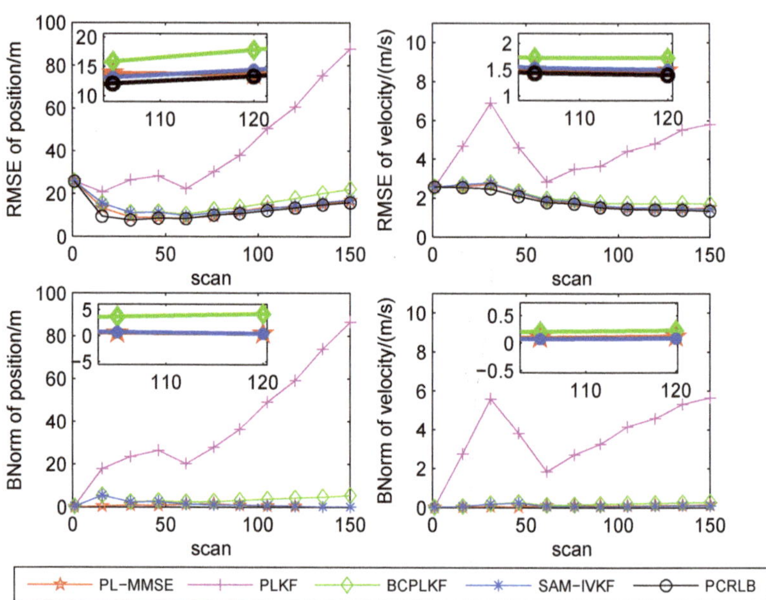

Figure 6. RMSEs and BNorms versus time k for $\sigma_\theta = 7°$ for the PLKF, the BC-PLKF and the SAM-IVKF with $\kappa = 4$, as well as, the proposed PL-MMSE.

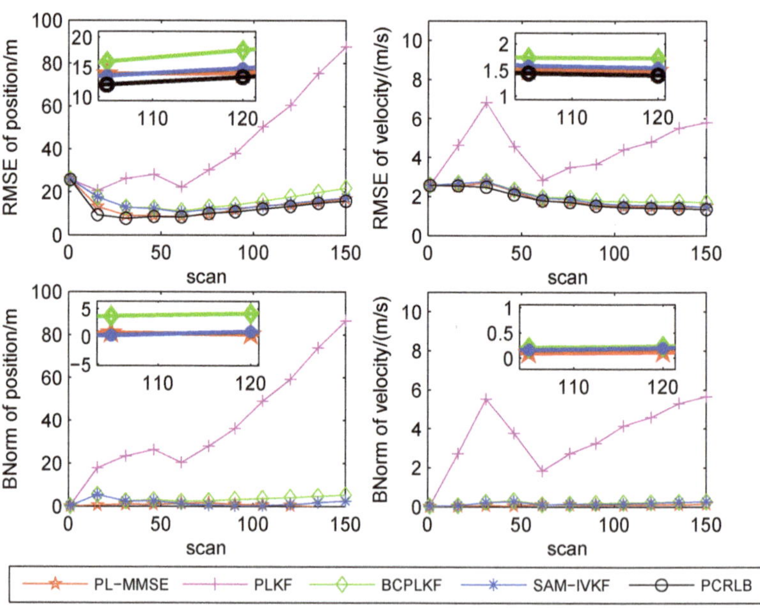

Figure 7. RMSEs and BNorms versus time k for $\sigma_\theta = 7°$ for the PLKF, the BC-PLKF and the SAM-IVKF with $\kappa = 3$, as well as, the proposed PL-MMSE.

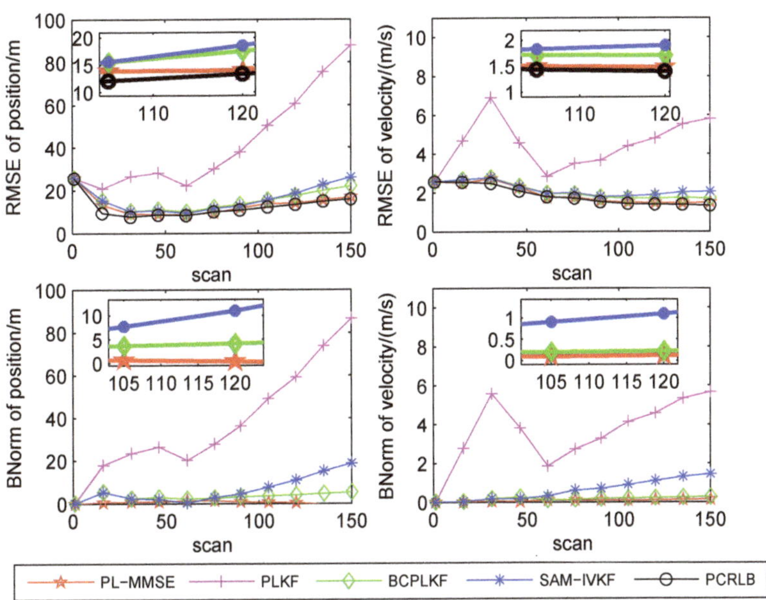

Figure 8. RMSEs and BNorms versus time k for $\sigma_\theta = 7°$ for the PLKF, the BC-PLKF and the SAM-IVKF with $\kappa = 2$, as well as, the proposed PL-MMSE.

The proposed PL-MMSE method also has a significant performance improvement compared to the PLKF, and shows good stability at both small and large bearing noise levels. The target state RMSEs of the PL-MMSE approach the PCRLB. Compared with the BC-PLKF, the PL-MMSE has lower time-averaged RMSE and BNorm values, and still maintains a good and stable performance at large bearing noise. Additionally, the PL-MMSE provides the comparable RMSE and BNorm performance to the best choice SAM-IVKF (i.e., $\kappa = 4$) as shown in Figure 3. As mentioned above, the SAM-IVKF relies on the selection of the thresholds. When the SAM thresholds change (i.e., $\kappa = 3, 2$), there are visible degradations in the performance of the SAM-IVKF, as shown in Figures 4 and 5, which is inferior to the proposed PL-MMSE. The selection of the SAM threshold depends on experience and there is no guarantee that the best threshold can be selected in practical applications. As a contrast, the PL-MMSE is derived under the unified MMSE estimation framework and does not depend on experience to choose any parameter, so it has stable performance. The PL-MMSE handle the correlation problem implicitly under the MMSE framework and does not require bias compensation as performed in BC-PLKF, so its estimation accuracy still can be guaranteed at large bearing noise level. Figures 6–8 show the RMSEs and BNorms of the target position and velocity estimates versus time instant k for $\sigma_\theta = 7°$ with threshold $\kappa = 4, 3$ and 2, respectively. Similar results can be found in Figures 6–8 as in Figures 3–5.

For the purposes of computational complexity comparison, the methods are executed on the same platform and their averaged runtimes are presented in Table 2. For convenience, the averaged runtimes of the BC-PLKF, the SAM-IVKF and the PL-MMSE are normalized by that of the PLKF.

Table 2. Averaged runtimes.

Algorithm	PLKF	BC-PLKF	SAM-IVKF	PL-MMSE
Runtime	1	1.18	1.71	1.46

In Table 2, it can be seen that the PLKF has the lowest computational complexity, but it provides an unsatisfactory tracking performance in both small and large bearing noise levels. The BC-PLKF and SAM-IVKF have 18% and 71% longer runtimes compared to the PLKF due to the extra time to perform bias compensation steps. Similarly, the proposed PL-MMSE requires 46% longer runtimes than the PLKF to handle the correlation problems. Compared with the PLKF and the BC-PLKF, although sacrificing a bit of computational complexity, the PL-MMSE significantly outperforms the two methods in terms of the RMSE and BNorm performance. In addition, the SAM-IVKF requires larger computational complexity than the proposed PL-MMSE, and can only provide performance comparable to the PL-MMSE when selecting the best SAM threshold. Accordingly, the PL-MMSE is superior to the SAM-IVKF in both computational complexity and estimation accuracy.

In the following, the consistency of the methods is examined based on the evaluation of the NEES. The bearing noise standard deviation is $\sigma_\theta = 7^\circ$, and the SAM threshold of the SAM-IVKF is $\kappa = 4$. Here, we use the two-sided 95% probability region, where the upper and lower bounds are 3.46 and 4.57, respectively. The NEES results of the four methods are presented in Figure 9.

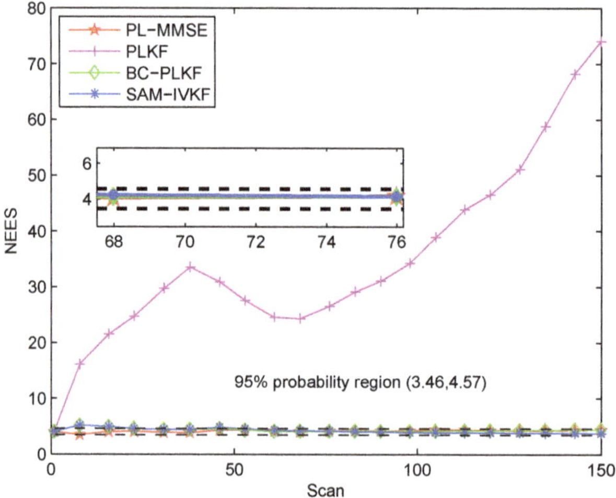

Figure 9. Consistency test of the PLKF, BC-PLKF, SAM-IVKF and the proposed PL-MMSE.

The results in Figure 9 show the inconsistency of the PLKF since its NEES values are outside the region of 95%. The NEES values of the BC-PLKF, SAM-IVKF and PL-MMSE all fall within the 95% probability region, which indicates that the three methods are consistent. Among these methods, the proposed PL-MMSE can provide the most stable and accurate tracking performance.

6. Conclusions

In this paper, a new pseudolinear filtering method is proposed under the MMSE framework to solve the bearings-only target tracking problem. The proposed PL-MMSE does not require performing bias compensation to solve the correlation problem of PLKF. Instead, the PL-MMSE analyzes the correlation between the pseudomeasurement matrix and the pseudolinear noise, and their cross-covariance is incorporated into the filtering process under the MMSE framework to account for their effects on estimation. Accordingly, the correlation problem can be handled implicitly in the filtering process of the PL-MMSE. Simulations show that the PL-MMSE meets the consistency requirement and can provide stable and accurate tracking performance, which is superior to the PLKF and BC-PLKF at both small and large bearing noise levels. Additionally, the PL-MMSE is comparable to the best choice SAM-IVKF, and is better than the SAM-IVKF in terms of computational complexity. These results verify the effectiveness of the proposed PL-MMSE to solve the bearings-only target tracking problem.

Author Contributions: Conceptualization, G.Z.; methodology, A.M.; software, S.B. and A.M.; validation, S.B.; formal analysis, A.M.; investigation, S.B.; resources, G.Z.; data curation, S.B.; writing—original draft preparation, A.M. and S.B.; writing—review and editing, G.Z.; visualization, S.B. and A.M.; supervision, G.Z.; project administration, G.Z.; funding acquisition, G.Z. All authors have read and agreed to the published version of the manuscript.

Funding: This research was funded by the National Natural Science Foundation of China grant number 61671181.

Institutional Review Board Statement: Not applicable.

Informed Consent Statement: Not applicable.

Data Availability Statement: Not applicable.

Conflicts of Interest: The authors declare no conflict of interest. The funders had no role in the design of the study; in the collection, analyses, or interpretation of data; in the writing of the manuscript, or in the decision to publish the results.

Abbreviations

The following abbreviations are used in this manuscript:

PLKF	Pseudolinear Kalman filter
MMSE	Minimum mean square error
PCRLB	Posterior Cramer–Rao Lower Bound
BOT	Bearings-only tracking
EKF	Extended Kalman filter
MPEKF	Modified polar coordinate extended Kalman filter
UKF	Unscented Kalman filter
PF	Particle filter
KF	Kalman filter
MPKF	Modified pseudolinear estimator
BC-PLKF	Bias-compensated pseudolinear Kalman filter
IV	Instrumental variable
IVKF	Instrumental variable Kalman filter
SAM	Selective-angle-measurement
SAM-IVKF	Selective-angle-measurement instrumental variable Kalman filter
2D	Two-dimensional
NCV	Nearly constant velocity
PL-MMSE	Pseudolinear-MMSE
RMSEs	Root mean square errors
BNorms	Bias Norms
NEES	Normalized estimation error squared

References

1. Bar-Shalom, Y.; Li, X.R.; Kirubarajan, T. *Estimation with Applications to Tracking and Navigation*; Wiley: New York, NY, USA, 2001.
2. He, Y.; Xiu, J.J.; Guan, X. *Radar Data Processing with Applications*; John Wiley & Sons: Singapore, 2016.
3. Bai, J.; Li, S.; Zhang, H.; Huang, L.B.; Wang, P. Robust Target Detection and Tracking Algorithm Based on Roadside Radar and Camera. *Sensors* **2021**, *21*, 1116. [CrossRef] [PubMed]
4. Luo, J.H.; Wang, Z.Y.; Chen, Y.P.; Wu, M.; Yang, Y. An Improved Unscented Particle Filter Approach for Multi-Sensor Fusion Target Tracking. *Sensors* **2020**, *20*, 6842. [CrossRef]
5. Shi, Y.F.; Qayyum, S.; Memon, S.A.; Khan, U.; Imtiaz, J.; Ullah, I.; Dancey, D.; Nawaz, R. A Modified Bayesian Framework for Multi-Sensor Target Tracking with Out-of-Sequence-Measurements. *Sensors* **2020**, *20*, 3821. [CrossRef]
6. Nardone, S.C.; Graham, M.L. A Closed-Form Solution to Bearings-Only Target Motion Analysis. *IEEE Trans. Ocean. Enfin.* **1997**, *22*, 168–178. [CrossRef]
7. Taghavi, E.; Tharmarasa, R.; Kirubarajan, T. Multisensor-Multitarget Bearing-Only Sensor Registration. *IEEE Trans. Aerosp. Electron. Syst.* **2016**, *52*, 1654–1666. [CrossRef]

8. Zhang, Y.J.; Xu, G.Z. Bearings-Only Target Motion Analysis via Instrumental Variable Estimation. *IEEE Trans. Signal Process.* **2010**, *58*, 5523–5533. [CrossRef]
9. Shi, Y.F.; Choi, J.W.; Xu, L.; Kim, J.; Ullah, I.; Khan, U. Distributed Target Tracking in Challenging Environments Using Multiple Asynchronous Bearing-Only Sensors. *Sensors* **2020**, *20*, 2671. [CrossRef]
10. Borisov, A.; Bosov, A.; Miller, B.; Miller, G. Passive Underwater Target Tracking: Conditionally Minimax Nonlinear Filtering with Bearing-Doppler Observations. *Sensors* **2020**, *20*, 2257. [CrossRef] [PubMed]
11. Karpenko, S.; Konovalenko, I.; Miller, A.; Miller, B.; Nikolaev, D. UAV Control on the Basis of 3D Landmark Bearing-Only Observations. *Sensors* **2015**, *15*, 29802–29820. [CrossRef]
12. Aidala, V.J. Kalman Filter Behavior in Bearings-Only Tracking Applications. *IEEE Trans. Aerosp. Electron. Syst.* **1979**, *AES-15*, 29–39. [CrossRef]
13. Aidala, V.; Hammel, S. Utilization of modified polar coordinates for bearings-only tracking. *IEEE Trans. Autom. Control* **1983**, *AC-28*, 283–294. [CrossRef]
14. Yang, R.; Ng, G.W.; Bar-Shalom, Y. Bearings-only tracking with fusion from heterogenous passive sensors: ESM/EO and acoustic. In Proceedings of the 18th International Conference on Information Fusion, Washington, DC, USA, 6–9 July 2015; pp. 1810–1816.
15. Karlsson, R.; Gustafsson, F. Recursive Bayesian estimation: bearing-only applications. *IEE Proc. Radar Sonar Navig.* **2005**, *152*, 305–313. [CrossRef]
16. Chang, D.C.; Fang, M.W. Bearing-only maneuvering mobile tracking with nonlinear filtering algorithms in wireless sensor networks. *IEEE Syst. J.* **2014**, *8*, 160–170. [CrossRef]
17. Hong, S.H.; Shi, Z.G.; Chen, K.S. Novel roughening algorithm and hardware architecture for bearings-only tracking using particle filter. *J. Electromagn. Waves Appl.* **2008**, *22*, 411–422. [CrossRef]
18. Lingren, A.; Gong, K. Position and Velocity Estimation Via Bearing Observations. *IEEE Trans. Aerosp. Electron. Syst.* **1978**, *AES-14*, 564–577. [CrossRef]
19. Miller, B.M.; Stepanyan, K.V.; Miller, A.B.; Andreev, K.V.; Khoroshenkikh, S.N. Optimal filter selection for UAV trajectory control problems. In Proceedings of the Conference Information Technology and System, Kaliningrad, Russia, 1–6 September 2013; pp. 327–333.
20. Lin, X.; Kirubarajan, T.; Bar-Shalom, Y.; Maskell, S. Comparison of EKF, pseudomeasurement, and particle filters for a bearing-only target tracking problem. *Proc. SPIE Int. Soc. Opt. Eng.* **2007**, *4728*, 240–250. [CrossRef]
21. Aidala, V.J.; Nardone, S.C. Biased Estimation Properties of the Pseudolinear Tracking Filter. *IEEE Trans. Aerosp. Electron. Syst.* **1982**, *AES-18*, 432–441. [CrossRef]
22. Holtsberg, A.; Holst, J.H. A nearly unbiased inherently stable bearings-only tracker. *IEEE J. Ocean. Eng.* **1993**, *18*, 138–141. [CrossRef]
23. Nguyen, N.H.; Dogancay, K. Improved Pseudolinear Kalman Filter Algorithms for Bearings-Only Target Tracking. *IEEE Trans. Signal Process.* **2017**, *65*, 6119–6134. [CrossRef]
24. Lindgren, A.G. Properties of a nonlinear estimator for determining position and velocity from angle-of-arrival measurements. In Processddings of the 14th Asilomar Conference on Circuits, Systems and Computers, Pacific Grove, CA, USA, 7–9 November 1980; pp. 394–401.
25. Nardone, S.; Lindgren, A.; Gong, K. Fundamental properties and performance of conventional bearings-only target motion analysis. *IEEE Trans. Autom. Control* **1984**, *29*, 775–787. [CrossRef]
26. Chen, Y.T.; Rudnicki, S.W. Bearings-only and Doppler-bearing tracking using instrumental variables. *IEEE Trans. Aerosp. Electron. Syst.* **1992**, *28*, 1076–1083. [CrossRef]
27. Cadre, J.P.L.; Jauffret, C. On the convergence of iterative methods for bearings-only tracking. *IEEE Trans. Aerosp. Electron. Syst.* **1999**, *35*, 801–818. [CrossRef]
28. Dogancay, K. Bias compensation for the bearings-only pseudolinear target track estimator. *IEEE Trans. Signal Process.* **2006**, *54*, 59–68. [CrossRef]
29. Doğançay, K. On the efficiency of a bearings-only instrumental variable estimator for target motion analysis. *Signal Proc.* **2005**, *85*, 481–490. [CrossRef]
30. Dogancay, K.; Arablouei, R. Selective angle measurements for a 3D-AOA instrumental variable TMA algorithm. In Proceedings of the 23rd European Signal Processing Conference, Nice, France, 31 August–4 September 2015; pp. 195–199.
31. Dogancay, K. 3D Pseudolinear Target Motion Analysis From Angle Measurements. *IEEE Trans. Signal Proc.* **2015**, *63*, 1570–1580. [CrossRef]
32. Zhou, G.; Yu, C.; Quan, T. A sequential tracking filter without requirement of measurement decorrelation. In Proceedings of the 2012 International Conference on Information Fusion, Singapore, 9–12 July 2012; pp. 2243–2248.
33. Zhou, G.; Xie, J.; Xu, R.; Quan, T. Sequential nonlinear tracking filter without requirement of measurement decorrelation. *J. Syst. Eng. Electron.* **2015**, *26*, 1135–1141. [CrossRef]
34. Tichavsky, P.; Muravchik, C.H.; Nehorai, A. Posterior Cramer—Rao Bounds for Discrete-Time Nonlinear Filtering. *IEEE Trans. Signal Proc.* **1998**, *46*, 1386–1396. [CrossRef]
35. Zhong, Z.W.; Meng, H.; Zhang, H.; Wang, X.Q. Performance Bound for Extended Target Tracking Using High Resolution Sensors. *Sensors* **2010**, *10*, 11618–11632. [CrossRef] [PubMed]

36. Lerro, D.; Bar-Shalom, Y. Tracking with debiased consistent converted measurements versus EKF. *IEEE Trans. Aerosp. Electron. Syst.* **1993**, *29*, 1015–1022. [CrossRef]
37. Ristic, B.; Zollo, S.; Arulampalam, S. Performance Bounds for Manoeuvring Target Tracking Using Asynchronous Multi-Platform Angle-Only Measurements. In Proceedings of the 4th International Conference on Information Fusion, Montréal, QC, Canada, 7–10 August 2001; pp. 1–8.
38. Bar-Shalom, Y.; Fortmann, T.E. *Tracking and Data Association*; Academic: San Diego, CA, USA, 1998.

Article

Closed-Form Pseudolinear Estimators for DRSS-AOA Localization

Jun Li [1,†], Kutluyil Dogancay [1,*,†] and Hatem Hmam [2]

[1] UniSA STEM, University of South Australia, Mawson Lakes Campus, Mawson Lakes, SA 5095, Australia; jun.li@mymail.unisa.edu.au
[2] Defence Science & Technology Group, Cyber and Electronic Warfare Division, Edinburgh, SA 5111, Australia; Hatem.Hmam@dst.defence.gov.au
* Correspondence: Kutluyil.Dogancay@unisa.edu.au
† These authors contributed equally to this work.

Abstract: This paper investigates the hybrid source localization problem using differential received signal strength (DRSS) and angle of arrival (AOA) measurements. The main advantage of hybrid measurements is to improve the localization accuracy with respect to a single sensor modality. For sufficiently short wavelengths, AOA sensors can be constructed with size, weight, power and cost (SWAP-C) requirements in mind, making the proposed hybrid DRSS-AOA sensing feasible at a low cost. Firstly the maximum likelihood estimation solution is derived, which is computationally expensive and likely to become unstable for large noise levels. Then a novel closed-form pseudolinear estimation method is developed by incorporating the AOA measurements into a linearized form of DRSS equations. This method eliminates the nuisance parameter associated with linearized DRSS equations, hence improving the estimation performance. The estimation bias arising from the injection of measurement noise into the pseudolinear data matrix is examined. The method of instrumental variables is employed to reduce this bias. As the performance of the resulting weighted instrumental variable (WIV) estimator depends on the correlation between the IV matrix and data matrix, a selected-hybrid-measurement WIV (SHM-WIV) estimator is proposed to maintain a strong correlation. The superior bias and mean-squared error performance of the new SHM-WIV estimator is illustrated with simulation examples.

Keywords: hybrid localization; differential received signal strength localization; bearings-only localization; maximum likelihood; pseudolinear estimator; least squares; instrumental variables

Citation: Li, J.; Dogancay, K.; Hmam, H. Closed-Form Pseudolinear Estimators for DRSS-AOA Localization. *Sensors* **2021**, *21*, 7159. https://doi.org/10.3390/s21217159

Academic Editors: Mahendra Mallick and Ratnasingham Tharmarasa

Received: 25 August 2021
Accepted: 26 October 2021
Published: 28 October 2021

Publisher's Note: MDPI stays neutral with regard to jurisdictional claims in published maps and institutional affiliations.

1. Introduction

Source localization plays an important role in wireless sensor networks, providing location information about sensor nodes and emitters from sensor measurements. Several sensor modalities have been considered for source localization such as angle of arrival (AOA), differential received signal strength (DRSS), time of arrival, time difference of arrival, and frequency difference of arrival. This paper develops new closed-form source localization methods using hybrid DRSS-AOA measurements, built on pseudolinear DRSS and AOA equations combined in a unique way to eliminate the undesirable nuisance parameter associated with pseudolinear DRSS equations.

Source localization and tracking using AOA measurements has been an active research area for several decades. The nonlinear relationship between source location and sensor measurements is the key challenge with AOA localization. This challenge is also shared to varying degrees by other sensor modalities. The pioneering work of Stansfield [1] established the basis for most AOA localization algorithms proposed to this day. The Stansfield estimator is a weighted least-squares estimator which requires prior knowledge of the source range from each AOA sensor. The maximum likelihood estimator (MLE) for AOA localization [2,3] solves a nonlinear optimization problem representing the log-likelihood function by using iterative algorithms such as the Gauss–Newton and Levenberg–Marquardt algorithms [4]. While the MLE enjoys asymptotic efficiency and unbiasedness, it is computationally expensive as a result of iterative computations and can suffer from divergence issues caused by poor initialization and threshold effect [5]. This makes the MLE unsuitable for most practical implementations.

The pseudolinear estimator (PLE) was developed as a closed-from alternative to the MLE, where the nonlinear estimation problem is converted into a linear problem, allowing for a computationally simple least squares solution [6]. An estimator identical to the PLE was also presented in [7]. Despite its simplicity, the PLE was discovered to produce biased estimates [3,8]. This led to an intensive research effort to reduce or eliminate the PLE bias (see, e.g., [9–18]). Among those, two ideas that have gained popularity are bias compensation and weighted instrumental variables (WIV) [13]. The bias compensation method is based on estimation and subtraction of bias, whereas the method of weighted instrumental variables reduces bias by introducing an instrumental variable matrix, which is statistically independent of measurement noise, into the WLS solution. In [19], a closed-form AOA localization algorithm is presented with no prior knowledge of AOA measurement variances. AOA-based self localization algorithms built on the PLE were developed in [20,21].

Received signal strength (RSS) localization offers a low-cost alternative to other localization systems as RSS measurements are readily available in most wireless systems. As different from RSS localization, DRSS localization methods use the differences between RSS measurements taken at pairs of sensor nodes, which eliminates the requirement for prior knowledge of transmit power at the source. This makes DRSS better suited for practical applications [22–24]. DRSS values, measured in dB, correspond to the ratio of source-sensor ranges from two sensors. Therefore, the DRSS source localization problem is reduced into a circular intersection problem where each circle represents a locus of possible source locations with the same range ratio from a pair of sensors as given by the corresponding DRSS measurement (the Apollonian circles theorem [25]). The research on DRSS localization has also focused on solving nonlinear and nonconvex optimization problems. Some of the existing solutions for DRSS localization include the MLE [22], weighted least-squares (WLS) [24,26,27], the generalized trust region subproblem (GTRS) estimator [24], semi-definite programming [24,28] and the PLE with bias reduction [29,30]. The derivation of DRSS equations and a summary of basic methods for DRSS localization are provided in [31].

Hybrid localization algorithms combining AOA and RSS measurements have been reported in the open literature. The work in [32–37] uses different linearization methods to convert both RSS and AOA equations into a linear form with a common unknown vector. The source location is easily obtained by using the WLS. However, the WLS estimates obtained from linearized RSS measurements have a bias problem, which has not been widely discussed in the current research. In contrast, for hybrid localization methods using TDOA-AOA measurements, besides the MLE and the WLS solution [38,39], the PLE with a bias reduction method has also been developed. For example, the work in [40] proposes bias compensation and weighted instrumental variable methods to reduce the bias.

Hybrid DRSS-AOA localization has not attracted much research despite the great potential it offers as a feasible and low-cost localization method compared with RSS-AOA and TDOA-AOA methods. The work in [41] proposes a hybrid RSS-AOA localization algorithm that treats the transmit power as unknown parameter, based on second-order cone programming relaxation techniques. In this paper, we present new hybrid localization algorithms using DRSS and AOA measurements based on the PLE and its instrumental variable variants. In DRSS localization, the knowledge of source transmit power is not required and therefore its estimation is not necessary. The conventional MLE is also derived, which is capable of achieving the Cramer–Rao lower bound (CRLB) with low bias, but has convergence problems and suffers from high computational complexity. The PLE is developed by converting the nonlinear measurement equations into linear form. The PLE is a closed-form estimator and has the advantage of low computational difficulty. In addition, the proposed PLE is free of nuisance parameter introduced into the linearized DRSS equations, thereby avoiding complications with constrained parameters in the solution vector. The PLE can be solved by using least squares (LS) and WLS. However, both these solutions have a bias problem due to the injection of measurement noise into the data matrix during the linearization process. The bias problem can be mitigated by introducing an instrumental variable matrix which correlates strongly with the data matrix and is independent of noise. However, when the measurement noise is high, the correlation between the instrumental variable matrix and data matrix is weakened. This can be remedied by adopting a selective measurement method when constructing the instrumental variable matrix, resulting in the selective-hybrid-measurement WIV (SHM-WIV) estimator. The SHM-WIV estimator is shown to outperform the MLE, LS, WLS and WIV estimators by way of simulation examples. The multipath effects on AOA and DRSS measurements are ignored even though shadowing effects are taken into account by lognormal noise on DRSS measurements.

The paper is organized as follows. Section 2 defines the hybrid localization problem addressed in this paper. The MLE and CRLB for the hybrid DRSS-AOA localization problem are presented in Section 3. In Section 4, linearized AOA and DRSS measurement equations are derived, and it is shown how the nuisance parameter present in the linearized DRSS measurement equation can be eliminated by incorporating the AOA measurements. The hybrid DRSS-AOA equation free of nuisance parameter is then solved using LS, WLS, WIV and SHM-WIV in Section 5. Comparative simulation results are demonstrated and discussed in Section 6. Concluding remarks are made in Section 7.

2. Problem Definition

We consider a 2D DRSS-AOA localization problem depicted in Figure 1, where the objective is to estimate the unknown source location $p = [x, y]^T$ from DRSS and AOA measurements collected by N sensors at fixed and known locations $r_i = [x_i, y_i]^T$, $i = 1, \ldots, N$. The distance between the source and a sensor is given by $d_i = \|d_i\|$ where $d_i = p - r_i$ and $\| \cdot \|$ denotes the Euclidean norm. Letting r_1 be the reference sensor location for DRSS measurements, we set $r_1 = 0$ after appropriate geometric translation with no loss of generality.

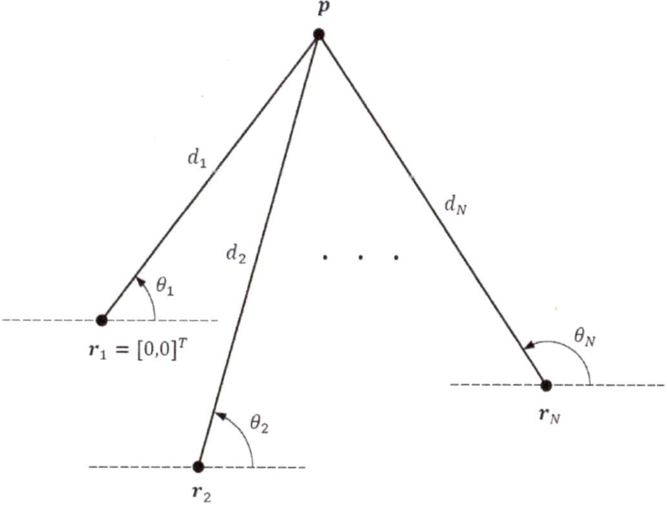

Figure 1. DRSS-AOA localization geometry.

The noisy AOA measurements at sensor i are given by

$$\tilde{\theta}_i = \theta_i + n_i, \quad i = 1, \ldots, N, \tag{1}$$

where $n_i \sim \mathcal{N}(0, \sigma_{\theta_i}^2)$ is an independent additive noise with zero mean and variance $\sigma_{\theta_i}^2$. The true angle θ_i is

$$\theta_i = \tan^{-1}(y - y_i, x - x_i), \quad \theta_i \in (-\pi, \pi] \tag{2}$$

where $\tan^{-1}$ is the 4-quadrant inverse tangent. The covariance matrix of the AOA measurements $[\tilde{\theta}_1, \ldots, \tilde{\theta}_N]$ is a diagonal matrix:

$$\mathbf{W}_{\text{AOA}} = \text{diag}(\sigma_{\theta_1}^2, \ldots, \sigma_{\theta_N}^2). \tag{3}$$

The power difference (DRSS) measurements with respect to the reference sensor at r_1 follow the propagation path loss model [24,26,29,30,42]

$$\tilde{p}_{i,1} = p_{i,1} + \epsilon_{i,1}, \quad i = 2, \ldots, N, \tag{4}$$

where $p_{i,1} = 10\gamma \log_{10} \frac{d_1}{d_i}$ is the true power difference between sensor i and the reference sensor (in dBm or dBW), γ is the path loss exponent which is assumed known a priori, and $\epsilon_{i,1} \sim \mathcal{N}(0, \sigma_{p_1}^2 + \sigma_{p_i}^2)$ is the log-normal noise representing shadowing effects with variance $\sigma_{p_1}^2 + \sigma_{p_i}^2$, which is the sum of RSS log-normal noise variances at r_1 and r_i. The covariance matrix of the DRSS measurements $[\tilde{p}_{2,1}, \ldots, \tilde{p}_{N,1}]$ is

$$\mathbf{W}_{\text{DRSS}} = \sigma_{p_1}^2 \mathbf{1}_{N-1} + \text{diag}(\sigma_{p_2}^2, \ldots, \sigma_{p_N}^2) \tag{5}$$

where $\mathbf{1}_{N-1}$ is an $(N-1) \times (N-1)$ matrix of ones.

The $(2N-1) \times 1$ hybrid measurement vector combining the AOA and DRRS measurements is

$$\tilde{\psi} = \psi + \beta, \tag{6}$$

where

$$\tilde{\psi} = [\tilde{\theta}_1, \ldots, \tilde{\theta}_N, \tilde{p}_{2,1}, \ldots, \tilde{p}_{N,1}]^T, \tag{7a}$$

$$\psi = [\theta_1, \ldots, \theta_N, p_{2,1}, \ldots, p_{N,1}]^T, \tag{7b}$$

$$\beta = [n_1, \ldots, n_N, \epsilon_{2,1}, \ldots, \epsilon_{N,1}]^T. \tag{7c}$$

The covariance matrix of β is a $(2N-1) \times (2N-1)$ block-diagonal matrix

$$\mathbf{W} = \mathbb{E}\{\beta\beta^T\} = \begin{bmatrix} \mathbf{W}_{\text{AOA}} & \mathbf{0} \\ \mathbf{0} & \mathbf{W}_{\text{DRSS}} \end{bmatrix}. \tag{8}$$

Observe that the AOA and DRSS measurement errors are not correlated. This is because the AOA measurement errors arise from thermal noise and possibly some interference at sensors while the log-normal noise in DRSS measurements is caused by shadowing. The two noise sources are physically independent phenomena.

3. Maximum Likelihood Estimator

The likelihood function for the hybrid measurements is a multivariate Gaussian pdf [43], which is given by

$$p(\tilde{\psi}|\hat{p}) = \frac{1}{(2\pi)^{(2N-1)/2}|\mathbf{W}|^{1/2}}$$
$$\times \exp\left\{ -\frac{1}{2}(\tilde{\psi} - \psi(\hat{p}))^T \mathbf{W}^{-1}(\tilde{\psi} - \psi(\hat{p}))\right\}, \tag{9}$$

where $|\cdot|$ denotes matrix determinant or scalar absolute value, and

$$\psi(\hat{p}) = [\theta_1(\hat{p}), \ldots, ,\theta_N(\hat{p}), p_{2,1}(\hat{p}), \ldots, p_{N,1}(\hat{p})]^T \tag{10}$$

is the $(2N-1) \times 1$ vector of DRSS-AOA estimates constructed by substituting the estimated source location $\hat{p} = [\hat{x}, \hat{y}]^T$ for the true source location p:

$$\theta_i(\hat{p}) = \tan^{-1}(\hat{y} - y_i, \hat{x} - x_i), \quad \theta_i(\hat{p}) \in (-\pi, \pi], \quad i = 1, \ldots, N, \tag{11a}$$

$$p_{i,1}(\hat{p}) = -10\gamma \log_{10} \frac{\|\hat{p} - r_i\|}{\|\hat{p}\|}, \quad i = 2, \ldots, N. \tag{11b}$$

The maximum likelihood estimate (MLE) of the source location is obtained by maximizing the log-likelihood function $\ln p(\tilde{\psi}|\hat{p})$ over $\hat{p}$, which is equivalent to

$$\hat{p}_{\text{ML}} = \underset{p \in \mathbb{R}^2}{\arg\min} \; h^T(p)W^{-1}h(p), \tag{12}$$

where

$$h(p) = \tilde{\psi} - \psi(p). \tag{13}$$

The nonlinear minimization problem in (12) can be solved numerically by an iterative search algorithm such as the steepest-descent, Levenberg–Marquardt, trust region and Gauss–Newton method [44]. In this paper, the Gauss–Newton method is adopted, which calculates the MLE using the following iterations:

$$\hat{p}(j+1) = \hat{p}(j) + (J^T(j)W^{-1}J(j))^{-1}J^T(j)W^{-1}h(\hat{p}(j)), \; j = 0, 1, \ldots \tag{14}$$

Here $J(j)$ is the $(2N-1) \times 2$ Jacobian matrix of $\psi(\hat{p})$ evaluated at $p = \hat{p}(j)$:

$$J(j) = [J_{\theta_1}^T(j), \ldots, J_{\theta_N}^T(j), J_{p_{2,1}}^T(j), \ldots, J_{p_{N,1}}^T(j)]^T, \tag{15}$$

where

$$J_{\theta_k}(j) = \frac{[-\sin\theta_i(\hat{p}(j)), \cos\theta_i(\hat{p}(j))]}{\|\hat{p}(j) - r_i\|}, \quad i = 1, \ldots, N \tag{16a}$$

$$\begin{aligned}
J_{p_{i,1}}(j) &= \frac{10\gamma}{\ln(10)} \left(\frac{(r_i - \hat{p}(j))^T}{\|r_i - \hat{p}(j)\|_2^2} + \frac{\hat{p}^T(j)}{\|\hat{p}(j)\|_2^2} \right) \\
&= \frac{10\gamma}{\ln(10)} \left[\begin{array}{c} -\frac{\cos\theta_i(\hat{p}(j))}{\|\hat{p}(j)-r_i\|} + \frac{\cos\theta_1(\hat{p}(j))}{\|\hat{p}(j)\|} \\ -\frac{\sin\theta_i(\hat{p}(j))}{\|\hat{p}(j)-r_i\|} + \frac{\sin\theta_1\hat{p}(j)}{\|\hat{p}(j)\|} \end{array} \right]^T, \quad i = 2, 3 \ldots, N.
\end{aligned} \tag{16b}$$

The GN is initialized by $\hat{p}(0)$, which needs to be selected carefully.

Being asymptotically efficient and unbiased, the MLE is often considered to be a benchmark in performance comparisons. However, the iterative methods used in MLE calculation can diverge if they are poorly initialized or the noise is too large, causing threshold effects (sharp degradation in estimation performance as the measurement noise increases above a threshold value). Furthermore, the MLE algorithms have a large computational complexity.

The Cramer–Rao lower bound for the hybrid DRSS-AOA localization problem is given by [43]

$$\text{CRLB} = \left(J_0^T W^{-1} J_0 \right)^{-1}, \tag{17}$$

where J_0 is the Jacobian matrix in (15) evaluated at the true source location p.

4. Pseudolinear Equations for Hybrid Measurements

4.1. Linearized AOA Equations

According to [6,40], the pseudolinear form for AOA measurements is

$$\mathbf{A}_{\theta_i} \mathbf{p} = b_{\theta_i} + e_{\theta_i}, \quad i = 1,\ldots,N \tag{18}$$

where

$$\mathbf{A}_{\theta_i} = [\sin \tilde{\theta}_i, -\cos \tilde{\theta}_i], \tag{19a}$$

$$b_{\theta_i} = [\sin \tilde{\theta}_i, -\cos \tilde{\theta}_i]\mathbf{r}_i, \tag{19b}$$

$$e_{\theta_i} = d_i \sin n_i \approx d_i n_i. \tag{19c}$$

The approximation in (19c) is valid for sufficiently small AOA measurement noise.

4.2. Linearized DRSS Equations

The DRSS measurement Equation (4) can be rewritten as

$$10^{\frac{\tilde{p}_{i,1}}{10\gamma}} d_i = 10^{\frac{\epsilon_{i,1}}{10\gamma}} d_1, \quad i = 2,\ldots,N. \tag{20}$$

Squaring both sides of the above equation yields

$$\mathbf{A}_{i,1} \mathbf{y} = b_i + e_i, \quad i = 2,\ldots,N \tag{21}$$

where

$$\mathbf{A}_{i,1} = [-2 \times 10^{\frac{\tilde{p}_{i,1}}{5\gamma}} \mathbf{r}_i^T, \ 10^{\frac{\tilde{p}_{i,1}}{5\gamma}} - 1] \tag{22a}$$

$$\mathbf{y} = \begin{bmatrix} \mathbf{p} \\ \|\mathbf{p}\|^2 \end{bmatrix} \tag{22b}$$

$$b_i = -10^{\frac{\tilde{p}_{i,1}}{5\gamma}} \|\mathbf{r}_i\|^2 \tag{22c}$$

$$e_i = \left(10^{\frac{\epsilon_{i,1}}{5\gamma}} - 1\right) d_1^2 \tag{22d}$$

A key challenge with the linearized DRSS Equation (21) is the presence of a nuisance parameter, viz., $\|\mathbf{p}\|^2$, in $\mathbf{y}$, that depends on the source location, thereby creating an undesirable nonlinear constraint in the solution. This constraint must be imposed on the estimate of $\mathbf{y}$ to assure good estimation performance.

4.3. Linearized DRSS-AOA Equations

Here we show how the nuisance parameter in the linearized DRSS equation can be eliminated by using hybrid DRSS-AOA measurements, leading to a linear matrix equation free of nuisance parameter and nonlinear constraints. To do this, first consider the noiseless DRSS equation

$$p_{i,1} = -10\gamma \log_{10} \frac{d_i}{d_1}, \quad i = 2,\ldots,N \tag{23}$$

which can be rewritten as

$$10^{-\frac{p_{i,1}}{10\gamma}} \|\mathbf{p}\| = d_i. \tag{24}$$

Next, consider the triangle formed by the corner points $\mathbf{p}$, $\mathbf{r}_1$ and $\mathbf{r}_i$, which shown in Figure 2. From the dot products $\mathbf{r}_i \cdot \mathbf{p}$ and $\mathbf{r}_i \cdot d_i$, we obtain

$$\cos \alpha_{1i} = \frac{\mathbf{r}_i^T \mathbf{p}}{\|\mathbf{r}_i\| \|\mathbf{p}\|} \Rightarrow \|\mathbf{p}\| = \frac{\mathbf{r}_i^T \mathbf{p}}{\|\mathbf{r}_i\| \cos \alpha_{1i}}, \tag{25a}$$

$$\cos \alpha_{2i} = -\frac{r_i^T d_i}{\|r_i\|\|d_i\|} \Rightarrow d_i = -\frac{r_i^T d_i}{\|r_i\| \cos \alpha_{2i}}. \tag{25b}$$

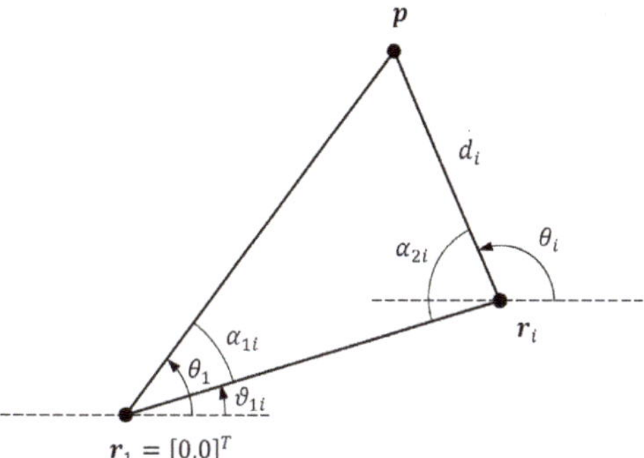

Figure 2. Triangle formed by corner points p, r_1 and r_i.

The angles of the triangle α_{1i} and α_{2i} are easily obtained from the AOA angles θ_1 and θ_i as follows:

$$\vartheta_{1i} = \angle r_i \tag{26a}$$

$$\alpha_{1i} = \theta_1 - \vartheta_{1i}, \quad -\pi < \alpha_{1i} \leq \pi \tag{26b}$$

$$\alpha_{2i} = \pi - \theta_i + \vartheta_{1i}, \quad -\pi < \alpha_{2i} \leq \pi \tag{26c}$$

where $\angle$ denotes the vector angle. Note that both α_{1i} and α_{2i} are wrapped to the interval $(-\pi, \pi]$.

Substituting (25a) and (25b) into (24) yields

$$10^{-\frac{P_{i,1}}{10\gamma}} \frac{r_i^T p}{\|r_i\| \cos \alpha_{1i}} = -\frac{r_i^T d_i}{\|r_i\| \cos \alpha_{2i}} \tag{27a}$$

$$10^{-\frac{P_{i,1}}{10\gamma}} r_i^T p \cos \alpha_{2i} = -r_i^T d_i \cos \alpha_{1i}. \tag{27b}$$

Finally, plugging $d_i = p - r_i$ into (27b), we obtain the following linearized DRSS equation incorporating AOA, which is free of nuisance parameter:

$$\bar{A}_{p_{i,1}} p = \bar{b}_{p_{i,1}} \tag{28}$$

where

$$\bar{A}_{p_{i,1}} = \left(10^{-\frac{P_{i,1}}{10\gamma}} \cos \alpha_{2i} + \cos \alpha_{1i} \right) r_i^T$$

and

$$\bar{b}_{p_{i,1}} = \|r_i\|^2 \cos \alpha_{1i}.$$

Replacing the true AOA and DRSS values with noisy measurements, (28) becomes

$$A_{p_{i,1}} p = b_{p_{i,1}} + e_{p_{i,1}}, \tag{29}$$

where $e_{p_{i,1}}$ is given by (A5) in Appendix A, and

$$\mathbf{A}_{p_{i,1}} = \left(10^{-\frac{\bar{P}_{i,1}}{10\gamma}} \cos \tilde{\alpha}_{2i} + \cos \tilde{\alpha}_{1i}\right) \mathbf{r}_i^T \tag{30a}$$

$$b_{p_{i,1}} = \|\mathbf{r}_i\|^2 \cos \tilde{\alpha}_{1i} \tag{30b}$$

$$\tilde{\alpha}_{1i} = \alpha_{1i} + n_1 \tag{30c}$$

$$\tilde{\alpha}_{2i} = \alpha_{2i} - n_i. \tag{30d}$$

As AOA and DRSS measurement errors are zero mean, we have

$$\mathbb{E}\{e_{p_{i,k}}\} \approx 0. \tag{31}$$

Stacking N AOA measurements and $N-1$ DRSS measurements, we obtain the linearized DRSS-AOA matrix equation:

$$\mathbf{A}p = b + e, \tag{32}$$

where

$$\mathbf{A} = [\mathbf{A}_{\theta_1}^T, \dots, \mathbf{A}_{\theta_N}^T, \mathbf{A}_{p_{2,1}}^T, \dots, \mathbf{A}_{p_{N,1}}^T]^T, \tag{33a}$$

$$b = [b_{\theta_1}, \dots, b_{\theta_N}, b_{p_{2,1}}, \dots, b_{p_{N,1}}]^T, \tag{33b}$$

$$e = [e_{\theta_1}, \dots, e_{\theta_N}, e_{p_{2,1}}, \dots, e_{p_{N,1}}]^T \tag{33c}$$

and

$$\mathbb{E}\{e\} \approx 0. \tag{34}$$

Note that (32) does not have a nuisance parameter. Therefore it can be solved without any constraint on the unknown vector as described in the next section.

5. Hybrid Pseudolinear Estimators

5.1. LS Solution and Bias Analysis

The least-squares solution for the linear matrix equation $\mathbf{A}p \approx b$ (see (32)) is [43]

$$\hat{p}_{\mathrm{LS}} = \arg\min_{p \in \mathbb{R}^2} \|\mathbf{A}p - b\|^2 \tag{35a}$$

$$= (\mathbf{A}^T\mathbf{A})^{-1}\mathbf{A}^T b. \tag{35b}$$

Substituting (32) into (35b), the least-squares estimate in terms of the pseudolinear noise vector e can be written as

$$\begin{aligned}
\hat{p}_{\mathrm{LS}} &= (\mathbf{A}^T\mathbf{A})^{-1}\mathbf{A}^T b \\
&= (\mathbf{A}^T\mathbf{A})^{-1}\mathbf{A}^T(\mathbf{A}p - e) \\
&= p - (\mathbf{A}^T\mathbf{A})^{-1}\mathbf{A}^T e.
\end{aligned} \tag{36}$$

The least-squares estimation bias is

$$\delta_{\mathrm{LS}} = \mathbb{E}\{\hat{p}_{\mathrm{LS}}\} - p = -\mathbb{E}\{(\mathbf{A}^T\mathbf{A})^{-1}\mathbf{A}^T e\}, \tag{37}$$

and the error covariance matrix of the estimate is

$$\begin{aligned}
\mathbf{C}_{\mathrm{LS}} &= \mathbb{E}\{(\hat{p}_{\mathrm{LS}} - p)(\hat{p}_{\mathrm{LS}} - p)^T\} \\
&= \mathbb{E}\{(\mathbf{A}^T\mathbf{A})^{-1}\mathbf{A}^T e e^T \mathbf{A}(\mathbf{A}^T\mathbf{A})^{-1}\}.
\end{aligned} \tag{38}$$

For sufficiently large N and under mild assumptions, Slutsky's theorem [45] allows (37) to be approximated by the product of expectations:

$$\delta_{\text{LS}} \approx -\mathbb{E}\left\{\frac{\mathbf{A}^T\mathbf{A}}{2N-1}\right\}^{-1}\mathbb{E}\left\{\frac{\mathbf{A}^T e}{2N-1}\right\}. \tag{39}$$

Using (33), the cross-correlation between $\mathbf{A}$ and e is

$$\mathbb{E}\{\mathbf{A}^T e\} = \sum_{i=1}^{N}\mathbb{E}\{\mathbf{A}_{\theta_i}^T e_{\theta_i}\} + \sum_{i=2}^{N}\mathbb{E}\{\mathbf{A}_{p_{i,1}}^T e_{p_{i,1}}\}. \tag{40}$$

According to (19a) and (19c), even for small AOA noise, we have [40]

$$\mathbb{E}\left\{\mathbf{A}_{\theta_i}^T e_{\theta_i}\right\} \approx \sigma_{\theta_i}^2 d_i \neq 0. \tag{41}$$

An approximate expression for $\mathbb{E}\{\mathbf{A}_{p_{i,1}}^T e_{p_{i,1}}\}$ can be derived from (30a) and (A4). Firstly, expanding the cosine terms of $\mathbf{A}_{p_{i,1}}$ and approximating $\mathbf{A}_{p_{i,1}}$ using (A3), we obtain

$$\begin{aligned}
\mathbf{A}_{p_{i,1}} \approx \Big(&C_{1i}\cos\alpha_{2i} + \cos\alpha_{1i} - C_{2i}\epsilon_{i,1}\cos\alpha_{2i} \\
&+ C_{1i}n_i\sin\alpha_{2i} - n_1\sin\alpha_{1i} \\
&- C_{1i}\frac{n_i^2}{2}\cos\alpha_{2i} + C_{3i}\epsilon_{i,1}^2\cos\alpha_{2i} \\
&- C_{2i}n_i\epsilon_{i,1}\sin\alpha_{2i} - \frac{n_1^2}{2}\cos\alpha_{1i} \\
&+ C_{2i}\frac{n_i^2}{2}\epsilon_{i,1}\cos\alpha_{2i} + C_{3i}n_i\epsilon_{i,1}^2\sin\alpha_{2i} \\
&- C_{3i}\frac{n_i^2}{2}\epsilon_{i,1}^2\cos\alpha_{2i}\Big)r_i^T.
\end{aligned} \tag{42}$$

Taking the expectation of $\mathbf{A}_{p_{i,1}}^T e_{p_{i,1}}$ yields

$$\begin{aligned}
\mathbb{E}\left\{\mathbf{A}_{p_{i,1}}^T e_{p_{i,1}}\right\} \approx \Big(&-C_{1i}^2\|r_i\|\frac{\sigma_{\theta_i}^2}{2}\cos^2\alpha_{2i} - C_{1i}\|r_i\|\frac{\sigma_{\theta_i}^2}{2}\cos\alpha_{2i}\cos\alpha_{1i} \\
&+ 3C_{1i}C_{3i}\|r_i\|\sigma_{p_{i,1}}^2\cos^2\alpha_{2i} + C_{3i}\|r_i\|\sigma_{p_{i,1}}^2\cos\alpha_{2i}\cos\alpha_{1i} \\
&+ C_{1i}^2\|d_i\|\frac{\sigma_{\theta_i}^2}{2}\cos^3\alpha_{2i} + C_{1i}\|d_i\|\frac{\sigma_{\theta_i}^2}{2}\cos^2\alpha_{2i}\cos\alpha_{1i} \\
&- 3C_{1i}C_{3i}\|d_i\|\sigma_{p_{i,1}}^2\cos^3\alpha_{2i} - C_{3i}\|d_i\|\sigma_{p_{i,1}}^2\cos^2\alpha_{2i}\cos\alpha_{1i} \\
&- C_{1i}^2\|d_i\|\sigma_{\theta_i}^2\sin^2\alpha_{2i}\cos\alpha_{2i} + C_{1i}\|r_i\|\frac{\sigma_{\theta_1}^2}{2}\cos\alpha_{2i}\cos\alpha_{1i} \\
&+ \|r_i\|\frac{\sigma_{\theta_1}^2}{2}\cos^2\alpha_{1i} - C_{1i}\|p\|\frac{\sigma_{\theta_1}^2}{2}\cos\alpha_{2i}\cos^2\alpha_{1i} \\
&- \|p\|\frac{\sigma_{\theta_1}^2}{2}\cos^3\alpha_{1i} + \|p\|\sigma_{\theta_1}^2\sin^2\alpha_{1i}\cos\alpha_{1i} \\
&+ C_{1i}^2\|r_i\|\sigma_{\theta_i}^2\sin^2\alpha_{2i} - \|r_i\|\sigma_{\theta_1}^2\sin^2\alpha_{1i}\Big)\|r_i\|r_i^T
\end{aligned} \tag{43}$$

It is clear that $\mathbb{E}\left\{\mathbf{A}_{p_{i,1}}^T e_{p_{i,1}}\right\}$ cannot be guaranteed to be zero for all $i = 2,\ldots,N$. Thus, $\mathbb{E}\left\{\mathbf{A}_{p_{i,1}}^T e_{p_{i,1}}\right\} \neq 0$.

Based on (41) and (43), we conclude that

$$\mathbb{E}\{\mathbf{A}^T e\} = \sum_{i=1}^{N} \mathbb{E}\left\{\mathbf{A}_{\theta_i}^T e_{\theta_i}\right\} + \sum_{i=2}^{N} \mathbb{E}\left\{\mathbf{A}_{p_{i,1}}^T e_{p_{i,1}}\right\} \neq \mathbf{0} \tag{44}$$

which means $\delta_{LS} \neq \mathbf{0}$ and the least-squares estimate (35b) is biased.

5.2. WLS Solution and Bias Analysis

The weighted least-squares estimate for p is obtained from [43]

$$\hat{p}_{\text{WLS}} = \arg\min_{p \in \mathbb{R}^2} \, (\mathbf{A}p - b)^T \mathbf{W}_{\text{PLE}}^{-1} (\mathbf{A}p - b) \tag{45a}$$

$$= (\mathbf{A}^T \mathbf{W}_{\text{PLE}}^{-1} \mathbf{A})^{-1} \mathbf{A}^T \mathbf{W}_{\text{PLE}}^{-1} b. \tag{45b}$$

where $\mathbf{W}_{\text{PLE}}$ is the weighting matrix that approximates the covariance of the noise vector e:

$$\mathbf{W}_{\text{PLE}} = \mathbb{E}\{ee^T\} = \begin{bmatrix} \mathbf{W}_{11} & \mathbf{W}_{12} & \mathbf{W}_{13} \\ \mathbf{W}_{12}^T & \mathbf{W}_{22} & \mathbf{W}_{23} \\ \mathbf{W}_{13}^T & \mathbf{W}_{23}^T & \mathbf{W}_{33} \end{bmatrix}. \tag{46}$$

The entries of $\mathbf{W}_{\text{PLE}}$ are given by

$$W_{11} = \mathbb{E}\{e_{\theta_{1,k}}^2\} \approx \|p\|^2 \sigma_{\theta_1}^2, \tag{47a}$$

$$\mathbf{W}_{12} = \mathbb{E}\{e_{\theta_{1,k}}[e_{\theta_{2,k}}, \dots, e_{\theta_{N,k}}]\} = \mathbf{0}_{1 \times (N-1)}, \tag{47b}$$

$$\mathbf{W}_{13} = \mathbb{E}\{e_{\theta_1}[e_{p_{2,1}}, \dots, e_{p_{N,1}}]\}$$

$$\approx \Big[\|r_2\|^2 \|p\| \sin \alpha_{12} - \|r_2\| \|p\|^2 \sin \alpha_{12} \cos \alpha_{12}, $$

$$\dots, \|r_N\|^2 \|p\| \sin \alpha_{1N} - \|r_N\| \|p\|^2 \sin \alpha_{1N} \cos \alpha_{1N} \Big] \sigma_{\theta_1}^2 \tag{47c}$$

$$\mathbf{W}_{22} = \mathbb{E}\{[e_{\theta_2}, \dots, e_{\theta_N}]^T [e_{\theta_2}, \dots, e_{\theta_N}]\}$$

$$\approx \text{diag}(\dots, \|d_i\|^2 \sigma_{\theta_i}^2, \dots)_{i=2,\dots,N}, \tag{47d}$$

$$\mathbf{W}_{23} = \mathbb{E}\{[e_{\theta_2}, \dots, e_{\theta_N}]^T [e_{p_{2,1}}, \dots, e_{p_{N,1}}]\}$$

$$\approx \text{diag}\left(\dots, \left(C_{1i} \|r_i\|^2 \|d_i\| \sin \alpha_{2i} \right.\right.$$

$$\left.\left. - C_{1i} \|r_i\| \|d_i\|^2 \sin \alpha_{2i} \cos \alpha_{2i} \right) \sigma_{\theta_i}^2, \dots \right)_{i=2,\dots,N} \tag{47e}$$

$$\mathbf{W}_{33} = \mathbb{E}\{[e_{p_{2,1}}, \dots, e_{p_{N,1}}]^T [e_{p_{2,1}}, \dots, e_{p_{N,1}}]\}$$

$$= [\beta_{2,1}, \dots, \beta_{N,1}]^T [\beta_{2,1}, \dots, \beta_{N,1}] \sigma_{\theta_1}^2$$

$$+ \text{diag}\left(\dots, \left(C_{1i} \|r_i\|^2 \sin \alpha_{2i} \right.\right.$$

$$\left.\left. - C_{1i} \|r_i\| \|d_i\| \sin \alpha_{2i} \cos \alpha_{2i} \right)^2 \sigma_{\theta_i}^2, \dots \right)_{i=2,\dots,N}$$

$$+ [\eta_{2,1}, \dots, \eta_{N,1}]^T [\eta_{2,1}, \dots, \eta_{N,1}] \mathbf{W}_{\text{DRSS}}. \tag{47f}$$

where

$$\beta_{2,i} = \|r_i\|^2 \sin \alpha_{1i} - \|r_i\| \|p\| \sin \alpha_{1i} \cos \alpha_{1i}, \tag{48a}$$

$$\eta_{2,i} = C_{2i} \|r_i\| \|d_i\| \cos^2 \alpha_{2i} - C_{2i} \|r_i\|^2 \cos \alpha_{2i}. \tag{48b}$$

Note that $\boldsymbol{d}_i$, $\boldsymbol{r}_i$, α_{1i}, α_{2i}, C_{1i} and C_{2i} require prior knowledge of the source location $\boldsymbol{p}$, which is not available. We replace $\boldsymbol{p}$ with $\hat{\boldsymbol{p}}_{\text{LS}}$ to calculate those terms.

Substituting (32) into (45b), we have

$$\hat{\boldsymbol{p}}_{\text{WLS}} = \boldsymbol{p} - (\mathbf{A}^T \mathbf{W}_{\text{PLE}}^{-1} \mathbf{A})^{-1} \mathbf{A}^T \mathbf{W}_{\text{PLE}}^{-1} e \tag{49}$$

whence the estimation bias is obtained as

$$\delta_{\text{WLS}} = \mathbb{E}\{\hat{\boldsymbol{p}}_{\text{WLS}}\} - \boldsymbol{p} = -\mathbb{E}\{(\mathbf{A}^T \mathbf{W}_{\text{PLE}}^{-1} \mathbf{A})^{-1} \mathbf{A}^T \mathbf{W}_{\text{PLE}}^{-1} e\}. \tag{50}$$

Similar to (39), for large N, (50) can be approximated as

$$\delta_{\text{WLS}} \approx -\mathbb{E}\left\{\frac{\mathbf{A}^T \mathbf{W}_{\text{PLE}}^{-1} \mathbf{A}}{2N-1}\right\}^{-1} \mathbb{E}\left\{\frac{\mathbf{A}^T \mathbf{W}_{\text{PLE}}^{-1} e}{2N-1}\right\}. \tag{51}$$

As $\mathbf{A}$ and e are correlated as shown in Section 5.1, $\mathbb{E}\{\mathbf{A}^T \mathbf{W}_{\text{PLE}}^{-1} e\} \neq \mathbf{0}$, which implies $\delta_{\text{WLS}} \neq \mathbf{0}$ and the WLS estimate is biased.

5.3. WIV Solution

The bias in the LS and WLS estimates can be significantly reduced by employing the method of instrumental variables. A weighted instrumental variable (WIV) estimator is obtained by introducing an IV matrix $\mathbf{G}$ which is strongly correlated with the matrix $\mathbf{A}$ while being statistically independent of e. The WIV solution is given by [45]

$$\hat{\boldsymbol{p}}_{\text{WIV}} = (\mathbf{G}^T \mathbf{W}_{\text{PLE}}^{-1} \mathbf{A})^{-1} \mathbf{G}^T \mathbf{W}_{\text{PLE}}^{-1} \boldsymbol{b}. \tag{52}$$

The IV matrix $\mathbf{G}$ is selected such that $\mathbb{E}\left\{\frac{\mathbf{G}^T \mathbf{W}_{\text{PLE}}^{-1} \mathbf{A}}{2N-1}\right\}$ is nonsingular and $\mathbb{E}\left\{\frac{\mathbf{G}^T \mathbf{W}_{\text{PLE}}^{-1} e}{2N-1}\right\} \to 0$ as $N \to \infty$ [46]. A practical IV matrix that meets these requirements can be constructed from an initial source location estimate, such as the LS or WLS estimate, as described below. This procedure is based on [47]. Consider the following row partitioning of the IV matrix $\mathbf{G}$:

$$\mathbf{G} = [\mathbf{G}_{\theta_1}^T, \ldots, \mathbf{G}_{\theta_N}^T, \mathbf{G}_{p_{2,1}}^T, \ldots, \mathbf{G}_{p_{N,1}}^T]^T, \tag{53}$$

where

$$\mathbf{G}_{\theta_i} = \left[\sin\hat{\theta}_i, \cos\hat{\theta}_i\right], \quad i = 1, \ldots, N \tag{54a}$$

$$\mathbf{G}_{p_{j,1}} = \left(10^{-\frac{p_{j,1}}{10\gamma}}\cos\hat{\alpha}_{2j} + \cos\hat{\alpha}_{1j}\right)\boldsymbol{r}_j^T, \quad j = 2, \ldots, N. \tag{54b}$$

Here the AOA, triangle angle and DRSS estimates are obtained from the initial source location estimate $\hat{\boldsymbol{p}} = [\hat{x}, \hat{y}]^T$ as

$$\hat{\theta}_i = \tan^{-1}(\hat{y} - y_i, \hat{x} - x_i), \quad \hat{\theta}_i \in (-\pi, \pi] \tag{55a}$$

$$\hat{\alpha}_{1j} = \hat{\theta}_1 - \vartheta_{1j}, \quad -\pi < \hat{\alpha}_{1i} \leq \pi \tag{55b}$$

$$\hat{\alpha}_{2j} = \pi - \hat{\theta}_j + \vartheta_{1j}, \quad -\pi < \hat{\alpha}_{2j} \leq \pi \tag{55c}$$

$$\hat{p}_{j,1} = 10\gamma \log_{10} \frac{\|\hat{\boldsymbol{p}}\|}{\|\hat{\boldsymbol{p}} - \boldsymbol{r}_j\|}. \tag{55d}$$

The bias of the WIV estimate is given by

$$\begin{aligned}
\delta_{\text{WIV}} &= \mathbb{E}\{\hat{\boldsymbol{p}}_{\text{WIV}}\} - \boldsymbol{p} \\
&= -\mathbb{E}\{(\mathbf{G}^T \mathbf{W}_{\text{PLE}}^{-1} \mathbf{A})^{-1} \mathbf{G}^T \mathbf{W}_{\text{PLE}}^{-1} e\}
\end{aligned} \tag{56}$$

which, for sufficiently large N, can be approximated as

$$\delta_{\text{WIV}} \approx -\mathbb{E}\left\{\frac{\mathbf{G}^T\mathbf{W}_{\text{PLE}}^{-1}\mathbf{A}}{2N-1}\right\}^{-1}\mathbb{E}\left\{\frac{\mathbf{G}^T\mathbf{W}_{\text{PLE}}^{-1}e}{2N-1}\right\} \tag{57}$$

where

$$\mathbb{E}\left\{\frac{\mathbf{G}^T\mathbf{W}_{\text{PLE}}^{-1}e}{2N-1}\right\} \approx 0.$$

As a result, $\delta_{\text{WIV}} \approx 0$ and, therefore, the WIV estimate is approximately unbiased.

5.4. SHM-WIV Solution

A selective hybrid measurement method is introduced here to keep the IV matrix $\mathbf{G}$, constructed from an initial source location estimate, and the data matrix $\mathbf{A}$ strongly correlated as there is a high probability that $\mathbf{G}$ and $\mathbf{A}$ lose correlation when the measurement noise is large [18]. The principle of selective hybrid measurements is to decide which rows of $\mathbf{G}$ should remain identical to those of $\mathbf{A}$ based on a measure of difference between them.

Consider the difference between the first N rows of $\mathbf{A}$ and $\mathbf{G}$ corresponding to the AOA measurements:

$$\mathbf{A}_{\theta_i} - \mathbf{G}_{\theta_i} = \left[\sin\tilde{\theta}_i - \sin\hat{\theta}_i, \quad \cos\tilde{\theta}_i - \cos\hat{\theta}_i\right]$$

$$= 2\sin\left(\frac{\tilde{\theta}_i - \hat{\theta}_i}{2}\right)\left[\cos\left(\frac{\tilde{\theta}_i+\hat{\theta}_i}{2}\right), \quad -\sin\left(\frac{\tilde{\theta}_i+\hat{\theta}_i}{2}\right)\right]. \tag{58}$$

The common factor $\sin\left((\tilde{\theta}_i - \hat{\theta}_i)/2\right)$ suggests that it will be appropriate to use the angle difference $|\tilde{\theta}_i - \hat{\theta}_i|$ as a measure of row difference [18], which leads to the following criterion for using $\hat{\theta}_i$, instead of $\tilde{\theta}_i$, in the ith row of the IV matrix $\mathbf{G}$:

$$|\tilde{\theta}_i - \hat{\theta}_i| \leq \lambda_1. \tag{59}$$

The recommended range of values for the threshold is $5\sigma_{\theta_i} \leq \lambda_1 \leq 20\sigma_{\theta_i}$, $i = 1,\ldots, N$. Following extensive simulation studies, we have concluded that selecting λ_1 in this range achieves the intended effect of making the IV matrix strongly correlated with the data matrix. In general, the larger the angle noise, the larger λ_1 should be.

The row difference between $\mathbf{A}$ and $\mathbf{G}$ for the DRSS measurements is

$$\mathbf{A}_{p_{i,1}} - \mathbf{G}_{p_{i,1}} = \left(10^{\frac{\tilde{p}_{i,1}}{-10\gamma}}\cos\tilde{\alpha}_{2i} + \cos\tilde{\alpha}_{1i}\right)r_i^T - \left(10^{\frac{\hat{p}_{i,1}}{-10\gamma}}\cos\hat{\alpha}_{2i} + \cos\hat{\alpha}_{1i}\right)r_i^T$$

$$= \left(\left(10^{\frac{\tilde{p}_{i,1}}{-10\gamma}} - 10^{\frac{\hat{p}_{i,1}}{-10\gamma}}\right)(\cos\tilde{\alpha}_{2i} - \cos\hat{\alpha}_{2i}) + \left(10^{\frac{\tilde{p}_{i,1}}{-10\gamma}} - 10^{\frac{\hat{p}_{i,1}}{-10\gamma}}\right)\cos\hat{\alpha}_{2i}\right. \tag{60}$$

$$\left. + 10^{\frac{\hat{p}_{i,1}}{-10\gamma}}(\cos\tilde{\alpha}_{2i} - \cos\hat{\alpha}_{2i}) + (\cos\tilde{\alpha}_{1i} - \cos\hat{\alpha}_{1i})\right)r_i^T$$

where the following terms determine the magnitude of difference

$$\left|10^{\frac{\tilde{p}_{i,1}}{-10\gamma}} - 10^{\frac{\hat{p}_{i,1}}{-10\gamma}}\right| \tag{61a}$$

$$|\cos\tilde{\alpha}_{1i} - \cos\hat{\alpha}_{1i}| = \left|2\sin\left(\frac{\tilde{\alpha}_{1i} + \hat{\alpha}_{1i}}{2}\right)\sin\left(\frac{\tilde{\theta}_i - \hat{\theta}_i}{2}\right)\right| \tag{61b}$$

$$|\cos\tilde{\alpha}_{2i} - \cos\hat{\alpha}_{2i}| = \left|2\sin\left(\frac{\tilde{\alpha}_{2i} + \hat{\alpha}_{2i}}{2}\right)\sin\left(\frac{\tilde{\theta}_1 - \hat{\theta}_1}{2}\right)\right|. \tag{61c}$$

From (61a) we obtain the following criterion for using $\hat{p}_{i,1}$, instead of $\tilde{p}_{i,1}$, in **G**:

$$|\tilde{p}_{i,1} - \hat{p}_{i,1}| \leq \lambda_2 \tag{62}$$

where the recommended range for λ_2 is $5\sigma_{p_{i,1}} \leq \lambda_2 \leq 20\sigma_{p_{i,1}}$ with $\sigma_{p_{i,1}} = \sqrt{\sigma_{p_1}^2 + \sigma_{p_i}^2}$. This range was confirmed to yield satisfactory estimation performance through extensive simulation studies.

Equations (61b) and (61c) result in the same criterion as (59). Thus, applying the difference measures in (59) and (62) to (60) yields the hybrid measurement selection criterion:

$$|\tilde{p}_{i,1} - \hat{p}_{i,1}||\tilde{\theta}_1 - \hat{\theta}_1| + |\tilde{p}_{i,1} - \hat{p}_{i,1}| + |\tilde{\theta}_1 - \hat{\theta}_1| + |\tilde{\theta}_i - \hat{\theta}_i| \leq \lambda_1\lambda_2 + \lambda_2 + 2\lambda_1. \tag{63}$$

We refer to the WIV estimate incorporating (59) and (63) in the construction of the IV matrix **G** as the *selective hybrid measurement WIV (SHM-WIV)* algorithm.

6. Simulation Results

6.1. Simulation Set-Up

The RMSE and bias performance of the MLE, LS, WLS, WIV and SHM-WIV algorithms is compared using Monte Carlo simulations. The simulated network topology has ten sensor nodes at fixed locations and a source, all contained within a 60 m × 60 m region. The path loss exponent is assumed to be $\gamma = 4$. The range of AOA and DRSS measurement noise is indicated by a noise index given in Table 1. The average SNR values for AOA and DRSS measurements are also included. AOA and DRSS measurements have different SNR values because the AOA noise power is the variance of the additive thermal (Gaussian) noise and the DRSS measurements are corrupted by the shadowing log-normal noise. The AOA measurements are obtained from an antenna array with $m = 10$ elements, using [48]

$$\text{SNR}_i = \frac{6}{m^3\sigma_{\theta_i}^2}, \quad i = 1, \ldots, N \tag{64}$$

which assumes the Cramer–Rao lower bound is achieved. The DRSS SNR values are for a source with transmit power of 40 dBm (10 W). The MLE uses the iterative Gauss–Newton method with initialization obtained from the LS estimate. The SHM-WIV threshold values are given in Table 2.

Table 1. Noise index for AOA/DRSS measurements.

Index	1	2	3	4	5	6	7
σ_{θ_i} (degrees)	0.1	0.2	0.3	0.4	0.5	0.6	0.7
$\sigma_{p_{i,1}}$ (dBm)	1	1.5	2	2.5	3	4	5
AOA SNR (dB)	32.95	26.92	23.40	20.90	18.96	17.38	16.04
DRSS SNR (dB)	-22.70	-23.20	-23.70	-24.20	-24.70	-25.70	-26.70

Table 2. λ_1 and λ_2 for SHM-WIV.

Noise Index	1	2	3	4	5	6	7
AOA λ_1	6.5σ	6.5σ	6.5σ	6.5σ	6.5σ	18σ	20σ
DRSS λ_2	6.5σ	6.5σ	6.5σ	6.5σ	6.5σ	18σ	20σ

6.2. Fixed Source Location

We start with a fixed network topology simulation where the source is stationary at a fixed location $p = [10, 56]^T$ as shown in Figure 3. The simulations consist of 10,000 Monte Carlo runs. Figures 4 and 5 present the RMSE and bias results versus noise index.

The MLE achieves the CRLB at small noise (noise index 1 and 2), but starts to diverge for large noise. The LS exhibits significant bias and poor RMSE compared to the other estimates for all noise levels. The WLS has a better bias and RMSE performance than the LS, but still shows a large bias and deviates from the CRLB for large noise. The WIV attains the CRLB when the noise index is below 6, but its RMSE rapidly deviates from the CRLB at noise index 7. The SHM-WIV exhibits the best overall RMSE and bias performance for the entire noise range.

For noise index 1 and 4, individual location estimates along with mean locations for the simulated algorithms are shown in Figures 6 and 7, respectively, to demonstrate the spread of estimates. The standard deviations of Monte Carlo simulation results that led to the bias and RMSE values plotted in Figures 4 and 5 are listed in Table 3. The standard deviation is left blank for algorithms that exhibit divergence.

The total run times of the simulated algorithms are listed in Table 4. We observe that the LS runs the fastest; however, it has a poor performance. The WLS is approximately three times slower than the LS due to weighting matrix computation. The MLE and WLS have comparable run times, even though the Gauss–Newton iterations can take longer time depending on initialization. The WIV is roughly five times slower than the LS because of computational overheads associated with weighting matrix and IV matrix computations. The SHM-WIV is slightly slower than the WIV method because of the additional SHM step.

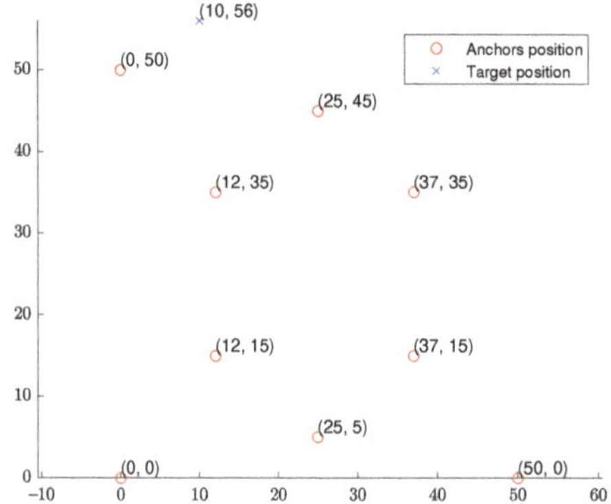

Figure 3. DRSS-AOA geometry with fixed source location.

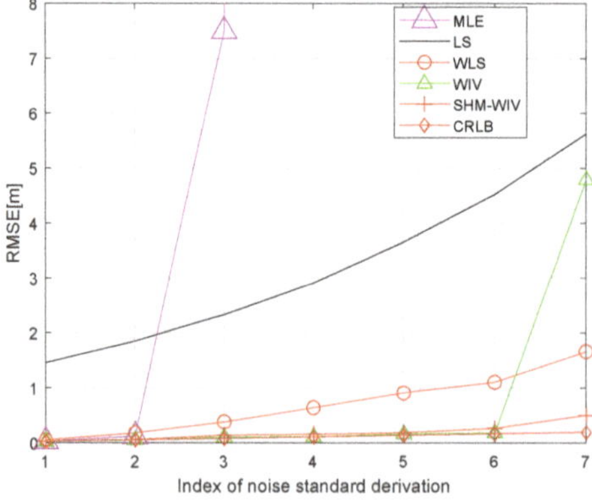

Figure 4. RMSE versus noise with fixed source location.

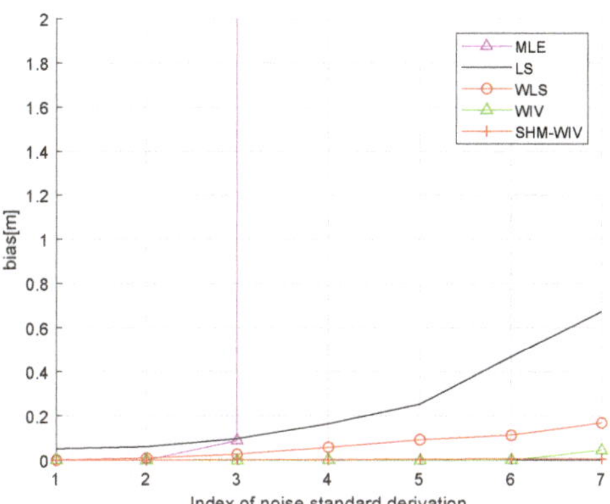

Figure 5. Bias versus noise with fixed source location.

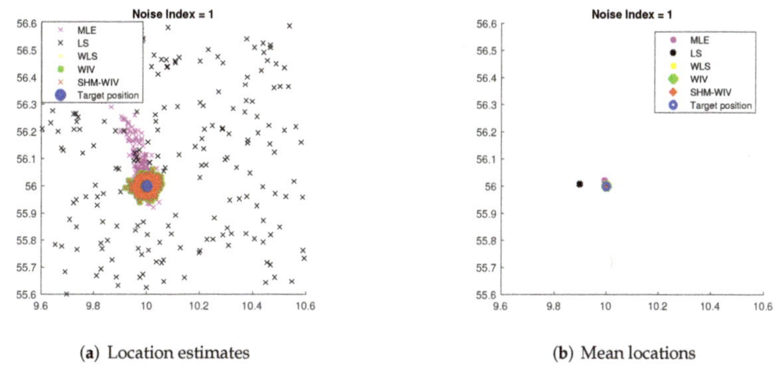

(**a**) Location estimates

(**b**) Mean locations

Figure 6. (**a**) Plot of individual location estimates for noise index 1; (**b**) Plot of mean location estimates for noise index 1.

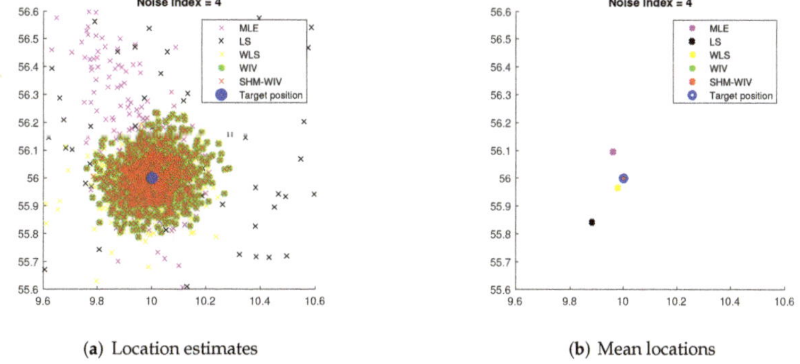

(**a**) Location estimates

(**b**) Mean locations

Figure 7. (**a**) Plot of individual location estimates for noise index 4; (**b**) Plot of mean location estimates for noise index 4.

Table 3. Standard deviations of Monte Carlo results for bias/RMSE values.

Index	bias/RMSE	MLE	LS	WLS	WIV	SHM-WIV
1	Bias	0.0007	0.0069	0.0004	0.0002	0.0001
	RMSE	0.0024	0.0098	0.0183	0.0147	0.0044
2	Bias	0.0013	0.0095	0.0013	0.0002	0.0003
	RMSE	0.0037	0.0115	0.0268	0.0019	0.0110
3	Bias	0.0017	0.0113	0.0029	0.0012	0.0008
	RMSE	0.0053	0.0123	0.0351	0.1219	0.0203
4	Bias	0.0033	0.0170	0.0045	0.0018	0.0012
	RMSE	0.1676	0.0159	0.0417	0.1596	0.0237
5	Bias		0.0184	0.0062	0.3103	0.0020
	RMSE		0.0206	0.0503	0.4246	0.0453
6	Bias		0.0233	0.0070	0.0044	0.0041
	RMSE		0.0294	0.0641	0.4337	0.4022
7	Bias		0.0288	0.0125	56.8201	0.0048
	RMSE		0.0302	0.0771	5682	0.4105

Table 4. Total simulation run time in MATLAB.

	MLE	LS	WLS	WIV	SHM-WIV
Time (s)	15.7860	5.5127	14.5313	24.3188	25.7447

6.3. Randomized Source Location

In these simulations 100 source locations are generated randomly in the 60 m $\times$ 60 m region, and, for each source location, RMSE and bias are evaluated using 10,000 Monte Carlo runs. The RMSE and bias results are shown in Figures 8 and 9, respectively. The MLE has a divergence problem across the whole noise range. Among the remaining algorithms, the LS has the largest bias and RMSE. The WLS shows improved performance compared to the LS. The WIV and SHM-WIV have the best RMSE and bias performance with the SHM-WIV slightly outperforming the WIV at large noise levels.

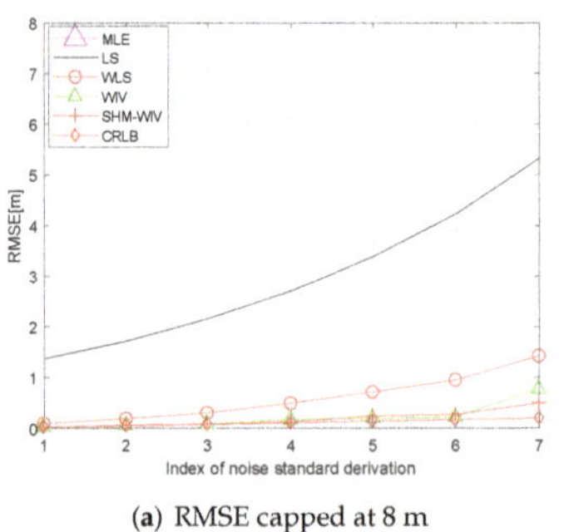

(a) RMSE capped at 8 m

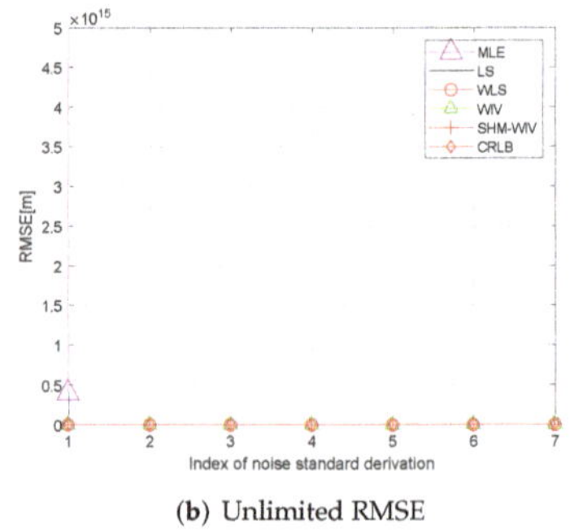

(b) Unlimited RMSE

Figure 8. (**a**) RMSE versus noise with randomized source location (RMSE is capped at 8 m); (**b**) RMSE versus noise with randomized source location (MLE is missing in (**a**) as it diverges for entire noise range).

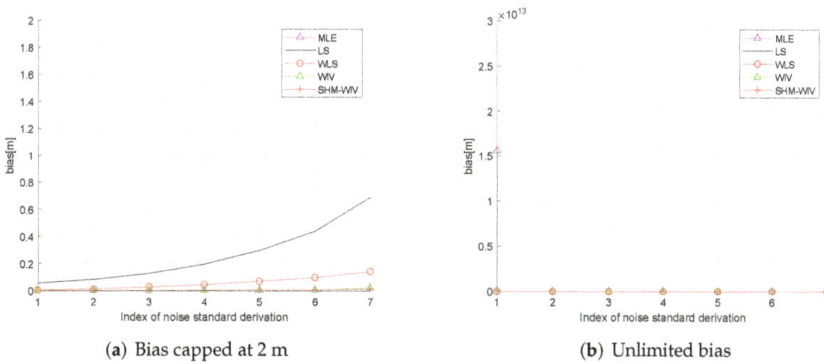

(a) Bias capped at 2 m   (b) Unlimited bias

Figure 9. (**a**) Bias versus noise with randomized source location (bias is capped at 2 m); (**b**) Bias versus noise with randomized source location (MLE is missing in (**a**) as it diverges for entire noise range).

7. Conclusions

A new pseudolinear hybrid DRSS-AOA localization method, free of nuisance parameter (squared source range from the reference sensor), was developed by exploiting the geometric relationship between AOA and DRSS measurements. To solve the resulting linear matrix equation for the source location, several variants of the pseudolinear estimator were proposed. These estimators are closed-form, leading to fewer computational steps than the MLE. However, the LS and WLS solutions have severe bias problems as verified by the simulations. The WIV estimator, on the other hand, was seen to be capable of alleviating the bias problem, achieving approximately zero bias for a large number of sensor measurements and small noise. The SHM-WIV was developed to guarantee a strong correlation between the IV matrix and the linearized data matrix for the WIV method as this correlation can be weakened when the noise is large. Simulation studies were carried out to compare the performance of the proposed estimators in fixed and randomized localization geometries. It was observed that the MLE has severe stability issues and cannot be considered an optimal solution at large noise. In the simulation studies the SHM-WIV outperformed all the estimators with an RMSE close to the CRLB and bias approaching zero.

Author Contributions: Conceptualization, J.L. and K.D.; methodology, J.L., K.D. and H.H.; software, J.L.; validation, J.L., K.D. and H.H.; formal analysis, J.L., K.D. and H.H.; investigation, J.L. and K.D.; resources, K.D.; data curation, J.L.; writing—original draft preparation, J.L., K.D. and H.H.; writing—review and editing, J.L., K.D. and H.H.; visualization, J.L. and K.D.; supervision, K.D. and H.H.; project administration, K.D.; funding acquisition, K.D. All authors have read and agreed to the published version of the manuscript.

Funding: This research received no external funding.

Institutional Review Board Statement: Not applicable.

Informed Consent Statement: Not applicable.

Data Availability Statement: No new data were created or analyzed in this study. Data sharing is not applicable to this article.

Conflicts of Interest: The authors declare no conflict of interest.

Appendix A. Derivation of Linearized DRSS Equation Noise

From (29), we have

$$
\begin{aligned}
e_{p_{i,1}} &= \mathbf{A}_{p_{i,1}} \boldsymbol{p} - b_{p_{i,1}} \\
&= \left(10^{\frac{\tilde{p}_{i,1}}{-10\gamma}} \cos(\alpha_{2i} - n_i) + \cos(\alpha_{1i} + n_1) \right) \boldsymbol{r}_i^T \boldsymbol{p} - \|\boldsymbol{r}_i\|^2 \cos(\alpha_{1i} + n_1).
\end{aligned}
\tag{A1}
$$

Expanding (A1), we obtain

$$
\begin{aligned}
e_{p_{i,1}} =\; & 10^{\frac{\tilde{p}_{i,1}}{-10\gamma}} \|r_i\|^2 \left(\cos\alpha_{2i}\cos n_i + \sin\alpha_{2i}\sin n_i\right) \\
& - 10^{\frac{\tilde{p}_{i,1}}{-10\gamma}} \|r_i\|\|d_i\| \left(\cos^2\alpha_{2i}\cos n_i + \sin\alpha_{2i}\cos\alpha_{2i}\sin n_i\right) \\
& + \|r_i\|\|p\| \left(\cos^2\alpha_{1i}\cos n_1 - \sin\alpha_{1i}\cos\alpha_{1i}\sin n_1\right) \\
& - \|r_i\|^2 \left(\cos\alpha_{1i}\cos n_1 - \sin\alpha_{1i}\sin n_1\right).
\end{aligned}
\tag{A2}
$$

When the measurement noise is sufficiently small, we have

$$
\cos n_i \approx 1 - \frac{n_i^2}{2},
\tag{A3a}
$$

$$
\sin n_i \approx n_i,
\tag{A3b}
$$

and

$$
\begin{aligned}
10^{\frac{p_{i,1}}{-10\gamma}} &= 10^{\frac{\tilde{p}_{i,1}}{-10\gamma}} 10^{\frac{\epsilon_{i,1}}{-10\gamma}} \\
&\approx 10^{\frac{\tilde{p}_{i,1}}{-10\gamma}} \left(1 - \frac{\epsilon_{i,1}}{10\gamma}\ln 10 + \frac{\left(\frac{\epsilon_{i,1}}{-10\gamma}\ln(10)\right)^2}{2}\right) \\
&\approx 10^{\frac{\tilde{p}_{i,1}}{-10\gamma}} - \frac{\epsilon_{i,1}}{10\gamma} 10^{\frac{\tilde{p}_{i,1}}{-10\gamma}}\ln 10 + 10^{\frac{\tilde{p}_{i,1}}{-10\gamma}} \frac{\epsilon_{i,1}^2 \ln^2(10)}{200\gamma^2}.
\end{aligned}
\tag{A3c}
$$

Substituting (A3) into (A2) yields

$$
\begin{aligned}
e_{p_{i,1}} \approx\; & -C_{1i}\|r_i\|^2 \frac{n_i^2}{2}\cos\alpha_{2i} - C_{2i}\|r_i\|^2\epsilon_{i,1}\cos\alpha_{2i} \\
& + C_{2i}\|r_i\|^2\epsilon_{i,1}\frac{n_i^2}{2}\cos\alpha_{2i} + C_{3i}\|r_i\|^2\epsilon_{i,1}^2\cos\alpha_{2i} \\
& - C_{3i}\|r_i\|^2\epsilon_{i,1}^2\frac{n_i^2}{2}\cos\alpha_{2i} + C_{1i}\|r_i\|^2 n_i\sin\alpha_{2i} \\
& - C_{2i}\|r_i\|^2\epsilon_{i,1}n_i\sin\alpha_{2i} + C_{3i}\|r_i\|^2\epsilon_{i,1}^2 n_i\sin\alpha_{2i} \\
& + C_{1i}\|d_i\|\|r_i\|\frac{n_i^2}{2}\cos^2\alpha_{2i} + C_{2i}\|d_i\|\|r_i\|\epsilon_{i,1}\left(1 - \frac{n_i^2}{2}\right)\cos^2\alpha_{2i} \\
& - C_{3i}\|d_i\|\|r_i\|\epsilon_{i,1}^2\left(1 - \frac{n_i^2}{2}\right)\cos^2\alpha_{2i} - C_{1i}\|d_i\|\|r_i\| n_i\cos\alpha_{2i}\sin\alpha_{2i} \\
& + C_{2i}\|d_i\|\|r_i\|\epsilon_{i,1}n_i\cos\alpha_{2i}\sin\alpha_{2i} - C_{3i}\|d_i\|\|r_i\|\epsilon_{i,1}^2 n_i\cos\alpha_{2i}\sin\alpha_{2i} \\
& + \|r_i\|^2 \frac{n_1^2}{2}\cos\alpha_{1i} + \|r_i\|^2 n_1\sin\alpha_{1i} - \|p\|\|r_i\|\frac{n_1^2}{2}\cos^2\alpha_{1i} \\
& - \|p\|\|r_i\|\cos\alpha_{1i}\sin\alpha_{1i}n_1
\end{aligned}
\tag{A4}
$$

where $C_{1i} = 10^{-\frac{\tilde{p}_{i,1}}{10\gamma}}$, $C_{2i} = \frac{\ln 10}{10\gamma}10^{-\frac{\tilde{p}_{i,1}}{10\gamma}}$ and $C_{3i} = \frac{\ln^2 10}{200\gamma^2}10^{-\frac{\tilde{p}_{i,1}}{10\gamma}}$. Neglecting the second and higher-order noise terms in (A4), $e_{p_{i,1}}$ can be further simplified:

$$
\begin{aligned}
e_{p_{i,1}} \approx\; & -C_{2i}\|r_i\|^2\epsilon_{i,1}\cos\alpha_{2i} + C_{1i}\|r_i\|^2 n_i\sin\alpha_{2i} \\
& + C_{2i}\|r_i\|\|d_i\|\epsilon_{i,1}\cos^2\alpha_{2i} - C_{1i}\|r_i\|\|d_i\| n_i\sin\alpha_{2i}\cos\alpha_{2i} \\
& + \|r_i\|^2 n_1\sin\alpha_{1i} - \|r_i\|\|p\| n_1\sin\alpha_{1i}\cos\alpha_{1i}.
\end{aligned}
\tag{A5}
$$

References

1. Stansfield, R.G. Statistical theory of d.f. fixing. *J. Inst. Electr. Eng. Part IIIA Radiocommun.* **1947**, *94*, 762–770. [CrossRef]
2. Torrieri, D.J. Statistical Theory of Passive Location Systems. *IEEE Trans. Aerosp. Electron. Syst.* **1984**, *AES-20*, 183–198. [CrossRef]
3. Nardone, S.; Lindgren, A.; Gong, K. Fundamental properties and performance of conventional bearings-only target motion analysis. *IEEE Trans. Autom. Control* **1984**, *29*, 775–787. [CrossRef]
4. Ljung, L.; Söderström, T. *Theory and Practice of Recursive Identification*; MIT Press: Cambridge, MA, USA, 1983.
5. Athley, F. Threshold region performance of maximum likelihood direction of arrival estimators. *IEEE Trans. Signal Process.* **2005**, *53*, 1359–1373. [CrossRef]
6. Lingren, A.G.; Gong, K.F. Position and velocity estimation via bearing observations. *IEEE Trans. Aerosp. Electron. Syst.* **1978**, *AES-14*, 564–577. [CrossRef]
7. Pages-Zamora, A.; Vidal, J.; Brooks, D. Closed-form solution for positioning based on angle of arrival measurements. In Proceedings of the 13th IEEE International Symposium on Personal, Indoor and Mobile Radio Communications, Lisboa, Portugal, 15–18 September 2002; Volume 4, pp. 1522–1526. [CrossRef]
8. Aidala, V.J.; Nardone, S.C. Biased Estimation Properties of the Pseudolinear Tracking Filter. *IEEE Trans. Aerosp. Electron. Syst.* **1982**, *AES-18*, 432–441. [CrossRef]
9. Chan, Y.T.; Rudnicki, S.W. Bearings-only and Doppler-bearing tracking using instrumental variables. *IEEE Trans. Aerosp. Electron. Syst.* **1992**, *28*, 1076–1083. [CrossRef]
10. Holtsberg, A.; Holst, J.H. A nearly unbiased inherently stable bearings-only tracker. *IEEE J. Ocean. Eng.* **1993**, *18*, 138–141. [CrossRef]
11. Pham, D.T. Some quick and efficient methods for bearing-only target motion analysis. *IEEE Trans. Signal Process.* **1993**, *41*, 2737–2751. [CrossRef]
12. Nardone, S.C.; Graham, M.L. A closed-form solution to bearings-only target motion analysis. *IEEE J. Ocean. Eng.* **1997**, *22*, 168–178. [CrossRef]
13. Doğançay, K. Bias compensation for the bearings-only pseudolinear target track estimator. *IEEE Trans. Signal Process.* **2005**, *54*, 59–68. [CrossRef]
14. Ho, K.; Chan, Y. An asymptotically unbiased estimator for bearings-only and Doppler-bearing target motion analysis. *IEEE Trans. Signal Process.* **2006**, *54*, 809–822. [CrossRef]
15. Ho, K.C.; Chan, Y.T. Geometric-Polar Tracking From Bearings-Only and Doppler-Bearing Measurements. *IEEE Trans. Signal Process.* **2008**, *56*, 5540–5554. [CrossRef]
16. Gu, G. A novel power-bearing approach and asymptotically optimum estimator for target motion analysis. *IEEE Trans. Signal Process.* **2011**, *59*, 912–922. [CrossRef]
17. Zhang, Y.; Xu, G.Z. Bearings-only target motion analysis via instrumental variable estimation. *IEEE Trans. Signal Process.* **2010**, *58*, 5523–5533. [CrossRef]
18. Dogancay, K. 3D Pseudolinear Target Motion Analysis From Angle Measurements. *IEEE Trans. Signal Process.* **2015**, *63*, 1570–1580. [CrossRef]
19. Pang, F.; Wen, X. A Novel Closed-Form Estimator for AOA Target Localization Without Prior Knowledge of Noise Variances. *Circ. Syst. Signal Process.* **2021**, *40*, 3573–3591. [CrossRef]
20. Shao, H.J.; Zhang, X.P.; Wang, Z. Efficient Closed-Form Algorithms for AOA Based Self-Localization of Sensor Nodes Using Auxiliary Variables. *IEEE Trans. Signal Process.* **2014**, *62*, 2580–2594. [CrossRef]
21. Dogancay, K. Self-localization from landmark bearings using pseudolinear estimation techniques. *IEEE Trans. Aerosp. Electron. Syst.* **2014**, *50*, 2361–2368. [CrossRef]
22. Lee, J.H.; Buehrer, R.M. Location estimation using differential RSS with spatially correlated shadowing. In Proceedings of the Global Telecommunications Conference, 2009 GLOBECOM 2009, IEEE, Honolulu, HI, USA, 30 November–4 December 2009; pp. 1–6.
23. Salman, N.; Kemp, A.H.; Ghogho, M. Low complexity joint estimation of location and path-loss exponent. *IEEE Wirel. Commun. Lett.* **2012**, *1*, 364–367. [CrossRef]
24. Hu, Y.; Leus, G. Robust Differential Received Signal Strength-Based Localization. *IEEE Trans. Signal Process.* **2017**, *65*, 3261–3276. [CrossRef]
25. Ogilvy, C.S. *Excursions in Geometry*; Dover Publications: Mineola, NY, USA, 1990.
26. Lin, L.; So, H.C.; Chan, Y.T. Accurate and Simple Source Localization Using Differential Received Signal Strength. *Digit. Signal Process.* **2013**, *23*, 736–743. [CrossRef]
27. Sun, Y.; Li, X.; Huang, Z.; Tian, J. An Improved Closed-Form Solution for Differential RSS-based Localization. In Proceedings of the 2020 IEEE Radar Conference (RadarConf20), Florence, Italy, 21–25 September 2020; pp. 1–5. [CrossRef]
28. Vaghefi, R.M.; Gholami, M.R.; Ström, E.G. RSS-based sensor localization with unknown transmit power. In Proceedings of the 2011 IEEE International Conference on Acoustics, Speech and Signal Processing (ICASSP), Prague, Czech Republic, 22–27 May 2011; pp. 2480–2483. [CrossRef]
29. Li, J.; Doğançay, K.; Nguyen, N.H.; Law, Y.W. Reducing the bias in DRSS-based localization: An instrumental variable approach. In Proceedings of the 2019 27th European Signal Processing Conference (EUSIPCO), IEEE, Coruña, Spain, 2–6 September 2019; pp. 1–5.

30. Li, J.; Doğançay, K.; Nguyen, N.H.; Law, Y.W. DRSS-Based Localisation Using Weighted Instrumental Variables and Selective Power Measurement. In Proceedings of the ICASSP 2020—2020 IEEE International Conference on Acoustics, Speech and Signal Processing (ICASSP), IEEE, Barcelona, Spain, 4–8 May 2020; pp. 4876–4880.

31. Lee, J.H.; Buehrer, R.M. *Handbook of Position Location: Theory, Practice, and Advances*; Chapter Fundamentals of Received Signal Strength-Based Position Location; Wiley: New York, NY, USA, 2012.

32. Wang, S.; Jackson, B.R.; Inkol, R. Hybrid RSS/AOA emitter location estimation based on least squares and maximum likelihood criteria. In Proceedings of the 2012 26th Biennial Symposium on Communications (QBSC), IEEE, Kingston, ON, Canada, 28–29 May 2012; pp. 24–29.

33. Chan, Y.T.; Chan, F.; Read, W.; Jackson, B.R.; Lee, B.H. Hybrid localization of an emitter by combining angle-of-arrival and received signal strength measurements. In Proceedings of the 2014 IEEE 27th Canadian Conference on Electrical and Computer Engineering (CCECE), IEEE, Toronto, ON, Canada, 5–8 May 2014; pp. 1–5.

34. Tomic, S.; Marikj, M.; Beko, M.; Dinis, R.; Órfão, N. Hybrid RSS-AoA technique for 3-D node localization in wireless sensor networks. In Proceedings of the 2015 International Wireless Communications and Mobile Computing Conference (IWCMC), IEEE, Dubrovnik, Croatia, 24–28 August 2015; pp. 1277–1282.

35. Tomic, S.; Beko, M.; Dinis, R.; Montezuma, P. A closed-form solution for RSS/AoA target localization by spherical coordinates conversion. *IEEE Wirel. Commun. Lett.* **2016**, *5*, 680–683. [CrossRef]

36. Beko, M.; Tomic, S.; Dinis, R.; Carvalho, P. Apparatus and Method for RSS/AoA Target 3-D Localization in Wireless Networks. U.S. Patent Number 10338193, 2 July 2019.

37. Nguyen, T.; D Vy, T.; Shin, Y. An efficient hybrid RSS-AoA localization for 3D wireless sensor networks. *Sensors* **2019**, *19*, 2121. [CrossRef] [PubMed]

38. Liu, R.; Wang, D.; Yin, J.; Wu, Y. Constrained total least squares localization using angle of arrival and time difference of arrival measurements in the presence of synchronization clock bias and sensor position errors. *Int. J. Distrib. Sens. Networks* **2019**, *15*, 1550147719858591. [CrossRef]

39. Deng, B.; Sun, Z.B.; He, Q. Efficient closed-form estimator for moving source localization using TDOA-FDOA-AOA measurements. In Proceedings of the 6th International Conference on Information Engineering, Liaoning, China, 17–18 August 2017; p. 20.

40. Nguyen, N.H.; Doğançay, K. Multistatic pseudolinear target motion analysis using hybrid measurements. *Signal Process.* **2017**, *130*, 22–36. [CrossRef]

41. Costa, M.S.; Tomic, S.; Beko, M. An SOCP Estimator for Hybrid RSS and AOA Target Localization in Sensor Networks. *Sensors* **2021**, *21*, 1731. [CrossRef]

42. Yang, S.; Wang, G.; Hu, Y.; Chen, H. Robust Differential Received Signal Strength Based Localization With Model Parameter Errors. *IEEE Signal Process. Lett.* **2018**, *25*, 1740–1744. [CrossRef]

43. Kay, S.M. *Fundamentals of Statistical Signal Processing: Estimation Theory*; PTR Prentice-Hall: Englewood Cliffs, NJ, USA, 1993.

44. Zekavat, R.; Buehrer, R.M. *Handbook of Position Location: Theory, Practice and Advances*; John Wiley & Sons: Hoboken, NJ, USA, 2011; Volume 27.

45. Mendel, J.M. *Lessons in Estimation Theory for Signal Processing, Communications, and Control*; Prentice-Hall: Englewood Cliffs, NJ, USA, 1995.

46. Ljung, L. *System Identification: Theory for The User*, 2nd ed.; Prentice-Hall: Saddle River, NJ, USA, 1999.

47. Doğançay, K. Passive emitter localization using weighted instrumental variables. *Signal Process.* **2004**, *84*, 487–497. [CrossRef]

48. Stoica, P.; Nehorai, A. MUSIC, maximum likelihood, and Cramer-Rao bound. *IEEE Trans. Acoust. Speech, Signal Process.* **1989**, *37*, 720–741. [CrossRef]

Article

A New Coarse Gating Strategy Driven Multidimensional Assignment for Two-Stage MHT of Bearings-Only Multisensor-Multitarget Tracking

Zheng Wei [1], Zhansheng Duan [1,*], Yina Han [2] and Mahendra Mallick [3]

1   Center for Information Engineering Science Research, Xi'an Jiaotong University, Xi'an 710049, China; weizheng179@stu.xjtu.edu.cn
2   School of Marine Science and Technology, Northwestern Polytechnical University, Xi'an 710072, China; yina.han@nwpu.edu.cn
3   Independent Researcher, Anacortes, WA 98221, USA; mmallick.us@gmail.com
*   Correspondence: zsduan@mail.xjtu.edu.cn

Abstract: The problem of two-dimensional bearings-only multisensor-multitarget tracking is addressed in this work. For this type of target tracking problem, the multidimensional assignment (MDA) is crucial for identifying measurements originating from the same targets. However, the computation of the assignment cost of all possible associations is extremely high. To reduce the computational complexity of MDA, a new coarse gating strategy is proposed. This is realized by comparing the Mahalanobis distance between the current estimate and initial estimate in an iterative process for the maximum likelihood estimation of the target position with a certain threshold to eliminate potential infeasible associations. When the Mahalanobis distance is less than the threshold, the iteration will exit in advance so as to avoid the expensive computational costs caused by invalid iteration. Furthermore, the proposed strategy is combined with the two-stage multiple hypothesis tracking framework for bearings-only multisensor-multitarget tracking. Numerical experimental results verify its effectiveness.

Keywords: bearings-only multisensor-multitarget tracking; multidimensional assignment (MDA); coarse gating; Mahalanobis distance; maximum likelihood estimation; multiple hypothesis tracking

Citation: Wei, Z.; Duan, Z.; Han, Y.; Mallick, M. A New Coarse Gating Strategy Driven Multidimensional Assignment for Two-Stage MHT of Bearings-Only Multisensor-Multitarget Tracking. *Sensors* **2022**, *22*, 1802. https://doi.org/10.3390/s22051802

Academic Editor: Andrzej Stateczny

Received: 9 January 2022
Accepted: 21 February 2022
Published: 24 February 2022

Publisher's Note: MDPI stays neutral with regard to jurisdictional claims in published maps and institutional affiliations.

1. Introduction

Multitarget tracking (MTT) refers to jointly estimating the number of targets and their states in the presence of false alarms and missed detections using single or multiple sensors [1]. It has been widely used in many fields such as surveillance and tracking of ground moving targets [2], maritime surveillance [3], sonar tracking of submarines [4], simultaneous localization and mapping [5], unmanned air vehicles [6], etc. For different application scenarios, tracked targets can be considered as point targets or extended targets [7]. If the distance between the sensor and target is large enough as in radar-based air surveillance applications, the target can be treated as a point target. In this case, it is usually assumed that a target can give rise to at most one measurement in a scan [8]. However, if multiple resolution cells of the sensor are occupied by a target, for example, in vehicle tracking using automotive radar, the target is regarded as an extended target [9]. In such a case, each target can give rise to multiple measurements [10]. Only point targets will be discussed below.

Multitarget tracking has been studied for decades and many effective algorithms are available. The earliest and simplest MTT algorithm is the global nearest neighbor (GNN) algorithm [11], which attempts to search for the single most likely hypothesis for track update and new track initiation [12]. Although the GNN algorithm is intuitively attractive and easy to implement, it is prone to track loss in scenarios with closely spaced targets and high false alarm density [13]. The joint probabilistic data association (JPDA) algorithm is an extension of the probabilistic data association (PDA) algorithm to the multitarget case [14]. The standard JPDA algorithm evaluates the association probabilities of measurement-to-track and combines them to obtain the state estimate of the target [15], which means that one observation may contribute to updating multiple tracks [16]. Many variants of the JPDA algorithm are abundant, such as the joint integrated PDA (JIPDA) algorithm [17] and multiscan JPDA (MS-JPDA) algorithm [18]. Multiple hypothesis tracking (MHT) is a deferred decision algorithm for MTT. It handles uncertainty of measurement-to-track associations by considering all possible association hypotheses in subsequent multiple scans [19]. Compared with GNN and JPDA algorithms that rely on the current scan, the MHT algorithm is computationally expensive, but it has significantly better tracking performance [20]. There are two different implementations of MHT algorithm, namely hypothesis-oriented MHT [21] and track-oriented MHT [22]. Between them, the track-oriented MHT algorithm, which uses the score function to evaluate the quality of tracks, is considered a more effective alternative to a hypothesis-oriented MHT [21]. Among the above three data association-based MTT algorithms, i.e., GNN, JPDA, and MHT, MHT is considered as a leading algorithm in high false alarm density and dense target scenarios [23].

The random finite set (RFS) algorithm [24] represents the multitarget state and measurements as a random finite set, which allows multitarget tracking to be cast in a Bayesian framework to obtain an optimal multitarget Bayes filter. Due to the high computational complexity of a multitarget Bayes filter [25], many approximate filters have been developed, such as probability hypothesis density (PHD) [26], cardinalized PHD (CPHD) [27], second-order PHD [28], and multitarget multi-Bernoulli (MeMBer) [29] filters. It should be note that none of these filters can obtain distinguishable target tracks. The generalized labeled multi-Bernoulli (GLMB) [20] is the RFS based MTT algorithm that produces tracks. In recent years, the GLMB filter has been widely studied, and fruitful achievements have been achieved in both theory and application [30]. In addition, the GLMB filter has been used to develop an MTT algorithm with structures similar to MHT [19].

Multisensor-multitarget tracking (MSMTT) has two basic architectures: centralized and distributed tracking [7]. In centralized MSMTT, the raw measurements from all sensors are sent to the fusion center (FC) where data association is followed by filtering, while in distributed MSMTT, each sensor first processes its own measurements and then sends the results to FC for further processing. Both frameworks have their own advantages and disadvantages in terms of communication requirements, computational complexity, performance, robustness, etc. In general, the centralized MSMTT framework has higher accuracy [31]. However, in practical applications, due to network bandwidth limitations, it is often not feasible to communicate all measurements to FC. Comparatively, the distributed MSMTT framework can reduce communication cost and has better flexibility and reliability, but it is more challenging.

For distributed MSMTT based on data association, one approach is that each sensor sends the local track estimates to the FC, which performs track-to-track association and fusion [32]. Another type of approach is to perform measurement space tracking at individual local sensors to suppress clutter and then send the associated measurements to the FC where the measurement-to-track association is performed [33]. In addition, distributed MSMTT based on RFS has also been widely studied in recent years [34].

Depending on the types of sensors used, target tracking can be split into two classes: active and passive tracking [35]. The sensors used for active tracking first transmit signals (such as acoustic waves, electromagnetic waves) into the environment and then obtains range, bearing, elevation, and other measurements of the target of interest from the received echo [36]. Passive sensors sense the signal from the target of interest to acquire bearing, elevation, and other measurements. In comparison, passive tracking has the advantages of strong anti-interference and good concealment [37].

Passive tracking also involves a unique set of challenges. One of the key challenges in bearings-only tracking is that the range between the passive sensor and the target is unavailable. This results in an unobservability of the target state [38]. A basic observable condition is that the sensor performs a higher order maneuver than all targets [39]. An alternative approach is to use multiple spatially separated sensors for triangulation, that is, the passive MSMTT [40]. But for this approach, the attendant problem is the well-known ghosting. In order to reduce the number of ghosts, three or more sensors should be used [41]. In this case, multidimensional assignment (MDA) can be used to associate the measurements from different sensors to identify common targets, which also makes this approach computationally costly for a large number of measurements. One of the main reasons is that in MDA, most of the time (at least up to 80%), is spent in calculating the association cost [42]. To reduce calculation times, many fast MDA methods have been proposed. Among them, it was proposed in [43] to cluster the measurements of different sensors before forming possible association hypotheses, thus reducing the requirement for calculating the association cost. In addition, two improved MDA methods using prior track information were proposed in [44].

A new coarse gating strategy is studied for the passive MSMTT. First, in order to reduce the computational complexity of MDA, a new coarse gating strategy is proposed. Second, the proposed strategy is combined with a two-stage MHT (TS-MHT) framework for distributed MSMTT. The remainder of the paper is organized as follows. Section 2 formulates the problems of bearings-only MSMTT. Section 3 briefly summarizes MDA for measurement-to-measurement association. In Section 4, a new coarse gating is proposed. Section 5 presents the combination of the proposed new coarse gating driven MDA with the TS-MHT framework. Section 6 provides numerical examples to illustrate the effectiveness of the proposed coarse gating strategy. Section 7 concludes the paper.

2. Problem Formulation and Notations

The two-dimensional (2D) bearings-only MSMTT is considered. The bearing measurement is shown in Figure 1.

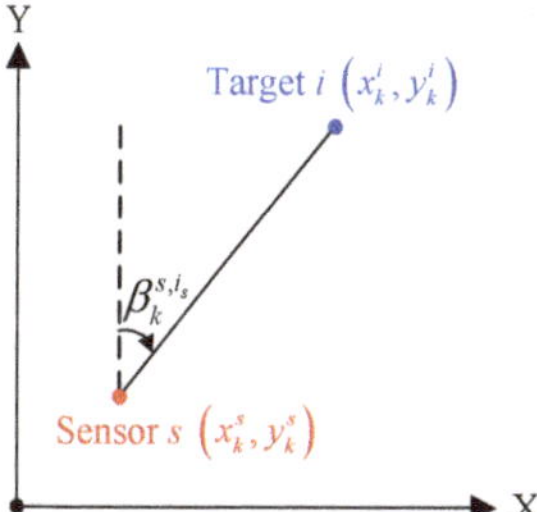

Figure 1. Illustration of two-dimensional bearing measurement.

Assume that there are S synchronous passive sensors and sensor s, $s \in \{1, 2, \cdots, S\}$, can acquire N_s bearing measurements $\{z_k^{s,j_s}\}_{j_s=1}^{N_s}$ at time k. Here, N_s may not be equal to the number of true targets due to false alarms and nonunity detection probability P_{D_s} of sensor s. For the sake of simplicity, each target is assumed to move with nearly constant velocity (NCV) in the XY-plane. Then, the discrete-time dynamic system can be written as follows:

$$\mathbf{x}_k^i = \mathbf{F}_{k-1}\mathbf{x}_{k-1}^i + \mathbf{w}_{k-1}^i, \tag{1}$$

$$z_k^{s,j_s} = \begin{cases} h\left(\mathbf{x}_k^i, \mathbf{p}_k^s\right) + v_k^{s,j_s} & \text{if } z_k^{s,j_s} \text{ originates from target } i \\ \tilde{z}_k^{j_s} & \text{otherwise} \end{cases}, \tag{2}$$

where $\mathbf{x}_k^i$ is the state vector consisting of the target position $[x_k^i \ y_k^i]'$ and velocity $[\dot{x}_k^i \ \dot{y}_k^i]'$, i.e., $\mathbf{x}_k^i = [x_k^i \ \dot{x}_k^i \ y_k^i \ \dot{y}_k^i]'$, $\mathbf{F}_{k-1}$ is the state transition matrix for NCV motion model, $\langle \mathbf{w}_{k-1}^i \rangle$ is a sequence of zero-mean white Gaussian process noise, $\mathbf{p}_k^s = [x_k^s \ y_k^s]'$ is the position of the sensor s, $\langle v_k^{s,j_s} \rangle$ is a sequence of zero-mean white Gaussian bearing measurement noise with variance σ_s^2, and the measurement noises across sensors are independent; h is a nonlinear function. The nonlinear relationship among $\beta_k^{i,s}$, $\mathbf{x}_k^i$ and $\mathbf{p}_k^s$ is given by the following:

$$\beta_k^{s,is} = h(\mathbf{x}_k^i, \mathbf{p}_k^s) = \tan^{-1}(x_k^i - x_k^s, y_k^i - y_k^s), \tag{3}$$

where $\tan^{-1}$ refers to the four-quadrant inverse tangent function [45].

The purpose is to estimate the number of targets and their corresponding states in real time. A list of nomenclatures is provided in Nomenclatures.

3. Measurement-to-Measurement Association

A brief description of measurement-to-measurement association is required to illustrate the proposed strategy more clearly. For a single passive sensor, the range measurement between target and sensor is not available, which makes the target state unobservable. During target tracking, especially for track initiation, at least two passive sensors are needed to obtain the full position of the potential target. It should be noted that, in a two-dimensional multitarget tracking scenario with only two sensors, one of the major problems is the occurrence of false intersections or ghosts . For example, as shown in Figure 2, the dashed lines of different colors indicate bearing measurements originating from target 1, and the solid lines of different colors indicate bearing measurements originating from target 2. Obviously, the correct association pair cannot be identified with only two bearings-only sensors.

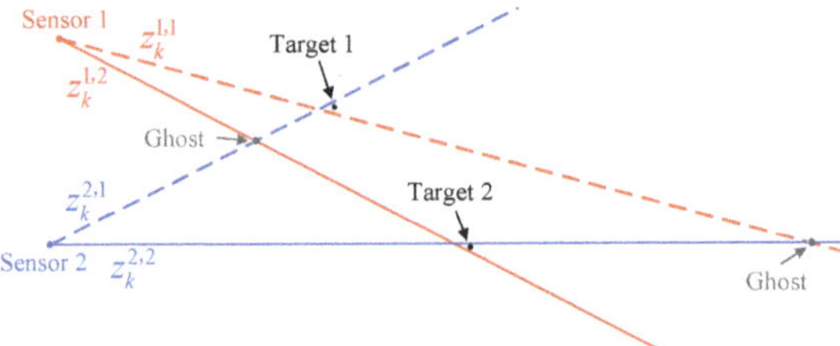

Figure 2. A scenario with 2 passive sensors and 2 targets.

Therefore, it is necessary to use three or more sensors if possible. However, the consequent problem is that this also makes it computationally expensive for a large number of measurements. Taking Figure 3 as an example, it shows the situation of two targets observed by three passive sensors with measurement errors.

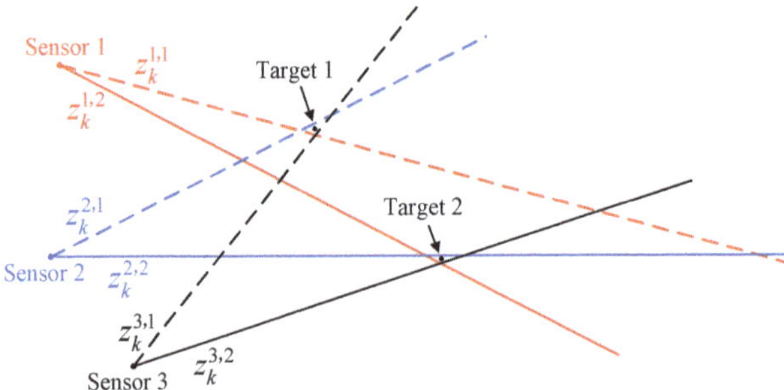

Figure 3. A scenario with 3 passive sensors and 2 targets.

As shown in the above figure, the sets of measurements obtained by different sensors originating from the targets can be denoted by $\{z_k^{1,1}, z_k^{1,2}\}$, $\{z_k^{2,1}, z_k^{2,2}\}$, and $\{z_k^{3,1}, z_k^{3,2}\}$, respectively. For measurement-to-measurement associations, each candidate association, consisting one measurement from each sensor, is denoted as the S-tuple of measurements $\mathbf{z}_k^{j_1 j_2 j_3}$. Even in the case where there are no false alarms or missed detections, the number of S-tuples is as follows:

$$c = \begin{pmatrix} 2 \\ 1 \end{pmatrix} \begin{pmatrix} 2 \\ 1 \end{pmatrix} \begin{pmatrix} 2 \\ 1 \end{pmatrix} = 2 \times 2 \times 2 = 8, \tag{4}$$

where $\begin{pmatrix} m \\ n \end{pmatrix}$ denotes the number of combinations of selecting n choices from m choices. The corresponding geometric relationship is shown in Figure 4.

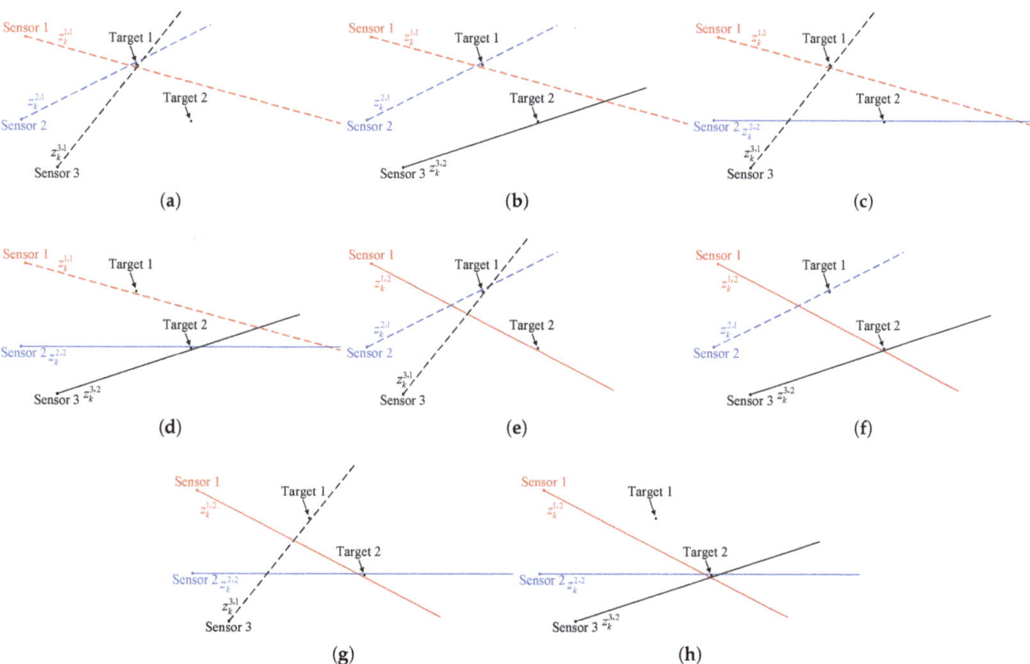

Figure 4. Geometric relationship between sensors and S-tuples of measurements (a) $\mathbf{Z}_k^{111}$. (b) $\mathbf{Z}_k^{112}$. (c) $\mathbf{Z}_k^{121}$. (d) $\mathbf{Z}_k^{122}$. (e) $\mathbf{Z}_k^{211}$. (f) $\mathbf{Z}_k^{212}$. (g) $\mathbf{Z}_k^{221}$. (h) $\mathbf{Z}_k^{222}$.

Each S-tuple of measurements is an association hypothesis. Obviously, only S-tuples $\mathbf{Z}_k^{111}$ and $\mathbf{Z}_k^{222}$ (as in Figure 4a,h) originate from the targets, and the others are spurious association hypotheses. Note that when there are false alarms or missed detections, and the number of S-tuples that can be formed will increase.

The process of associating the S-tuples of measurements to targets is the well-known measurement-to-measurement association problem. MDA based on likelihood ratio is widely considered to be the most efficient method to deal with this problem, which formulates the association between measurements from different sensors as a discrete optimization problem given by the following:

$$\min_{\rho_k^{j_1 j_2 \cdots j_S}} \sum_{j_1=0}^{N_1} \sum_{j_2=0}^{N_2} \cdots \sum_{j_S=0}^{N_S} c_k^{j_1 j_2 \cdots j_S} \rho_k^{j_1 j_2 \cdots j_S} \tag{5}$$

subject to

$$\sum_{j_2=0}^{N_2} \sum_{j_3=0}^{N_3} \cdots \sum_{j_S=0}^{N_S} \rho_k^{j_1 j_2 \cdots j_S} = 1, \quad j_1 = 1, 2, \cdots, N_1$$

$$\sum_{j_1=0}^{N_1} \sum_{j_3=0}^{N_3} \cdots \sum_{j_S=0}^{N_S} \rho_k^{j_1 j_2 \cdots j_S} = 1, \quad j_2 = 1, 2, \cdots, N_2 \tag{6}$$

$$\vdots$$

$$\sum_{j_1=0}^{N_1} \sum_{j_2=0}^{N_2} \cdots \sum_{j_{S-1}=0}^{N_{S-1}} \rho_k^{j_1 j_2 \cdots j_S} = 1, \quad j_S = 1, 2, \cdots, N_S$$

where $j_s = 0$ is the index of dummy measurement to indicate sensor s's missed detection, $c_k^{j_1 j_2 \cdots j_S}$ is the cost of associating the S-tuple of measurements $\mathbf{Z}_k^{j_1 j_2 \cdots j_S}$ to a target, and $\rho_k^{j_1 j_2 \cdots j_S}$ is a binary decision variable such that the following is the case.

$$\rho_k^{j_1 j_2 \cdots j_S} = \begin{cases} 1 & \text{if } \mathbf{Z}_k^{j_1 j_2 \cdots j_S} \text{ is associated with a candidate target} \\ 0 & \text{otherwise} \end{cases}. \tag{7}$$

The equality constraints in Equation (6) are to ensure that each measurement is associated with a unique target, or declared false, and that each target is assigned to at most one measurement from each sensor. In Equation (5), cost $c_k^{j_1 j_2 \cdots j_S}$ is defined as the following negative log-likelihood ratio:

$$c_k^{j_1 j_2 \cdots j_S} = -\ln \frac{p\left(\mathbf{Z}_k^{j_1 j_2 \cdots j_S} \mid \mathbf{p}_k^i\right)}{p\left(\mathbf{Z}_k^{j_1 j_2 \cdots j_S} \mid \mathbf{p}_k^i = \varnothing\right)}, \tag{8}$$

where $p\left(\mathbf{Z}_k^{j_1 j_2 \cdots j_S} \mid \mathbf{p}_k^i = \varnothing\right)$ is the likelihood that measurements in S-tuple $\mathbf{Z}_k^{j_1 j_2 \cdots j_S}$ are all spurious, and $p\left(\mathbf{Z}_k^{j_1 j_2 \cdots j_S} \mid \mathbf{p}_k^i\right)$ is the likelihood that these measurements originate from a common target at position $\mathbf{p}_k^i = [\zeta_k^i \ \eta_k^i]'$. They can be calculated as follows, respectively:

$$p\left(\mathbf{Z}_k^{j_1 j_2 \cdots j_S} \mid \mathbf{p}_k^i = \varnothing\right) = \prod_{s=1}^{S} \left[\frac{1}{\psi_s}\right]^{u(j_s)}, \tag{9}$$

$$p\left(\mathbf{Z}_k^{j_1 j_2 \cdots j_S} \mid \mathbf{p}_k^i\right) = \prod_{s=1}^{S} (1 - P_{D_s})^{1 - u(j_s)} \left[P_{D_s} p\left(z_k^{s, j_s} \mid \mathbf{p}_k^i\right)\right]^{u(j_s)}, \tag{10}$$

where ψ_s is the volume of the field of view of sensor s, and $u(j_s)$ is a binary indicator function.

$$u(j_s) = \begin{cases} 1 & \text{if } j_s \neq 0 \ (\text{an actual measurement of sensor } s) \\ 0 & \text{if } j_s = 0 \ (\text{a dummy measurement}) \end{cases}. \tag{11}$$

It should be noted that, in Equation (10), $\mathbf{p}_k^i$ is unknown. Therefore, in order to calculate likelihood $p\left(z_k^{s, j_s} \mid \mathbf{p}_k^i\right)$, the corresponding $\mathbf{Z}_k^{j_1 j_2 \cdots j_S}$ is used to obtain the maximum likelihood estimation (MLE) of the target position.

$$\hat{\mathbf{p}}_k^i = \arg\max_{\mathbf{p}_k^i} p\left(\mathbf{Z}_k^{j_1 j_2 \cdots j_S} \mid \mathbf{p}_k^i\right) \tag{12}$$

Substituting Equations (9), (10) and (12) into Equation (8), required cost $c_k^{j_1 j_2 \cdots j_S}$ can be calculated. Note that the optimization problem given by Equations (5) and (6) is NP-hard for $S \geq 3$. However, a number of efficient methods to obtain sub-optimal solution have been proposed [46–49].

4. A New Coarse Gating Strategy for MDA

The MLE $\hat{\mathbf{p}}_k^i$ of the position of potential target in Equation (12) is a nonlinear optimization problem. In this section, a new coarse gating strategy is proposed to eliminate infeasible association hypotheses by comparing the Mahalanobis distance between the current estimate and initial estimate in an iterative process for the MLE of the target position.

Each S-tuple of measurements $\mathbf{Z}_k^{j_1 j_2 \cdots j_S}$ can form a corresponding stacked measurement vector denoted by $\mathbf{z}_k^{j_1 j_2 \cdots j_S} = [z_k^{1,j_1}, z_k^{2,j_2}, \cdots z_k^{S,j_S}]'$. The relationship between the stacked measurement vector and position of the corresponding target can be written as follows:

$$\mathbf{z}_k^{j_1 j_2 \cdots j_S} = \begin{bmatrix} z_k^{1,j_1} \\ z_k^{2,j_2} \\ \vdots \\ z_k^{S,j_S} \end{bmatrix} = \begin{bmatrix} h(\mathbf{p}_k^i, \mathbf{p}_k^1) \\ h(\mathbf{p}_k^i, \mathbf{p}_k^2) \\ \vdots \\ h(\mathbf{p}_k^i, \mathbf{p}_k^S) \end{bmatrix} + \mathbf{w}_k = \mathbf{h}(\mathbf{p}_k^i, \mathbf{p}_k^s) + \mathbf{w}_k, \quad s = 1, 2, \cdots, S \quad (13)$$

where $\mathbf{p}_k^i = [\xi_k^i \ \eta_k^i]'$ is the position of target in XY-plane, $\mathbf{p}_k^s = [\xi_k^s \ \eta_k^s]'$ is the position of sensor s, and $\mathbf{w}_k$ is the stacked measurement vector of measurement noises with covariance $\mathbf{R}_k = \mathrm{diag}(\sigma_1^2, \sigma_2^2, \cdots, \sigma_S^2)$.

The MLE $\hat{\mathbf{p}}_k^i$ of the target position can be solved by iteration, and the iterative process can be denoted [50] by the following:

$$\hat{\mathbf{p}}_k^{i,l+1} = \hat{\mathbf{p}}_k^{i,l} + \left((\mathbf{J}_k^l)' \mathbf{R}_k^{-1} \mathbf{J}_k^l \right)^{-1} (\mathbf{J}_k^l)' \mathbf{R}_k^{-1} [\mathbf{z}_k^{j_1 j_2 \cdots j_S} - \mathbf{h}(\hat{\mathbf{p}}_k^{i,l}, \mathbf{p}_k^s)] \quad (14)$$

where the following is the Jacobian matrix:

$$\mathbf{J}_k^l = \left. \frac{\partial \mathbf{h}(\mathbf{p}_k^i, \mathbf{p}_k^s)}{\partial \mathbf{p}_k^i} \right|_{\mathbf{p}_k^i = \hat{\mathbf{p}}_k^{i,l}} \quad (15)$$

and $\hat{\mathbf{p}}_k^{i,l}$ is the position estimation of target after iteration l. The initial estimate $\hat{\mathbf{p}}_k^{i,0}$ can be obtained from the intersection of the bearing measurements of any two of all sensors.

The mean square error of final target position estimate can be calculated by the following.

$$\mathbf{R}_k^{i,l+1} \triangleq E\left[(\hat{\mathbf{p}}_k^{i,l+1} - \mathbf{p}_k^i)(\hat{\mathbf{p}}_k^{i,l+1} - \mathbf{p}_k^i)' \right] = \left((\mathbf{J}_k^{i,l})' \mathbf{R}_k^{-1} \mathbf{J}_k^{i,l} \right)^{-1}. \quad (16)$$

For stacked measurement vectors formed by incorrect associations, their elements do not originate from common targets. Therefore, in this case, it is irrational to solve the position estimation given in Equation (12). A natural idea is to analyze the differences of different measurement vectors in the iterative process so as to roughly delete some infeasible associations.

In the iteration, the initial position estimate $\hat{\mathbf{p}}_k^{i,0}$ can be obtained from the intersection of any two bearing components of stacked measurement vector. Moreover, the corresponding covariance $\mathbf{R}_k^{i,0}$ can be computed by Equation (16). Note that the initial estimate $(\hat{\mathbf{p}}_k^{i,0}, \mathbf{R}_k^{i,0})$ is determined by the measurements of only two sensors, while the estimate $(\hat{\mathbf{p}}_k^{i,l}, \mathbf{R}_k^{i,l})$ after l iterations, $l \geq 1$, is determined by the measurements of all sensors together. That is, these two estimates are not generated by the same measurements. If these measurements are not originated from a common target, the position estimate $\hat{\mathbf{p}}_k^{i,l}$ will deviate from the initial estimate $\hat{\mathbf{p}}_k^{i,0}$ in the iterative process. This will easily result in inconsistencies between these two estimates. Here, the inconsistency between two estimates refers to the fact that the difference between their means is greater than what can be expected based on their respective error covariance estimates [51].

Taking Figure 4c in Section 3 as an example, the stacked measurement vector formed by the S-tuple of measurements $\mathbf{Z}_k^{121}$ is $\mathbf{z}_k^{121} = [z_k^{1,1}, z_k^{2,2}, z_k^{3,1}]'$. Suppose that, in the iterative process, the initial position estimate $(\hat{\mathbf{p}}_k^{i,0}, \mathbf{R}_k^{i,0})$ is obtained by the bearing measurements of sensors 1 and 3. If the initial estimate $(\hat{\mathbf{p}}_k^{i,0}, \mathbf{R}_k^{i,0})$ and the estimate $(\hat{\mathbf{p}}_k^{i,l}, \mathbf{R}_k^{i,l})$ after l iterations, $l \geq 1$, are as shown in Figure 5, it means that the two estimates are inconsistent with each other. It should be noted that Figure 5 is only a schematic diagram and not a real experimental result. Numerical experiments will be presented in Section 6.

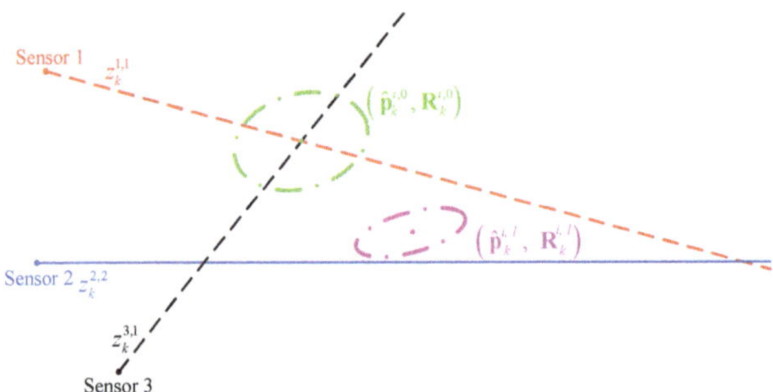

Figure 5. An illustration of the inconsistency between the two estimates. $(\hat{\mathbf{p}}_k^{i,0}, \mathbf{R}_k^{i,0})$ is the initial estimate and $(\hat{\mathbf{p}}_k^{i,l}, \mathbf{R}_k^{i,l})$ is the estimate after l iterations.

Therefore, it is necessary to quantitatively analyze the difference between the two estimates. One mechanism for detecting statistically significant deviations between estimates is to calculate the Mahalanobis distance [52]. The Mahalanobis distance between estimates $(\hat{\mathbf{p}}_k^{i,0}, \mathbf{R}_k^{i,0})$ and $(\hat{\mathbf{p}}_k^{i,l}, \mathbf{R}_k^{i,l})$ is defined as follows.

$$d_k^{i,l} = \left(\hat{\mathbf{p}}_k^{i,0} - \hat{\mathbf{p}}_k^{i,l}\right)'\left(\mathbf{R}_k^{i,0} + \mathbf{R}_k^{i,l}\right)^{-1}\left(\hat{\mathbf{p}}_k^{i,0} - \hat{\mathbf{p}}_k^{i,l}\right). \tag{17}$$

It can be roughly interpreted to mean that $\hat{\mathbf{p}}_k^{i,l}$ lies within an ellipsoid centered around $\hat{\mathbf{p}}_k^{i,0}$ [53]. A larger Mahalanobis distance tends to indicate that the two estimates are inconsistent; that is, the components in the corresponding stacked measurement vector do not originate from the common target [51]. Therefore, it is necessary to set an appropriate threshold T according to the measurement accuracy of the sensors. When $d_k^{i,l} \leq T$, it means that the components may originate from the common target. In this case, iteration (14) will be repeated until $l > N_{\max}$ or the following occurs:

$$\Delta p \triangleq \|\hat{\mathbf{p}}_k^{i,l+1} - \mathbf{p}_k^{i,l}\| < \varepsilon \tag{18}$$

where $N_{\max}$ is preset maximum number of iterations, $\|\cdot\|$ is the norm of a vector, ε is a sufficiently small positive real number. Final position estimate $\hat{\mathbf{p}}_k^{i,l}$ will be used to calculate assignment cost $c_k^{j_1 j_2 \cdots j_S}$. When $d_k^{i,l} > T$, this means that measurements in the vector originate from different targets. Therefore, the iteration will be terminated and the corresponding association cost will be assigned to infinity.

A threshold T is required to detect inconsistencies between the two estimates $(\hat{\mathbf{p}}_k^{i,0}, \mathbf{R}_k^{i,0})$ and $(\hat{\mathbf{p}}_k^{i,l}, \mathbf{R}_k^{i,l})$, $l \geq 1$, which decides whether it is necessary to further calculate the association cost $c_k^{j_1 j_2 \cdots j_S}$ for MDA. The choice of the threshold T is inherently problem dependent [54]. In bearings-only MSMTT, it is closely related to the position and measurement accuracy of the passive sensors. In order to avoid deleting incorrect associations, the threshold should not be too small. For a small number of remaining incorrect associations, the subsequent MDA can be used for further identification. In practical applications, an a priori threshold can be determined in advance with the help of cooperative targets.

For some infeasible associations, terminating the iterations when the Mahalanobis distance between the initial estimate of the iterative estimate is greater than a set threshold T can effectively save computational cost. The proposed strategy is denoted by coarse gating in iterations (CGI). The CGI-driven MDA is summarized in Algorithm 1.

Algorithm 1: CGI driven MDA

Input: position $\mathbf{p}_k^s = [\xi_k^s \eta_k^s]'$ of each sensor, stacked measurement vectors, minimum threshold ε of iterations, maximum number $N_{\max}$ of iterations, threshold T

Output: the binary decision variable $\rho_k^{j_1 j_2 \cdots j_S}$ in Equation (5)

1   **foreach** *stacked measurement vector* $\mathbf{z}_k^{j_1 j_2 \cdots j_S}$ **do**

2   calculate the initial position estimate $\hat{\mathbf{p}}_k^{i,0}$ and covariance $\mathbf{R}_k^{i,0}$ using any two non-dummy measurements;

3     $l \leftarrow 1$;

4   **while** $(\Delta p > \varepsilon \ and \ l < N_{\max})$ **do**

5   calculate the position estimation $\hat{\mathbf{p}}_k^{i,l}$ after l iterations and corresponding covariance matrix $\mathbf{R}_k^{i,l}$ via Equations (14) and (16), respectively;

6   calculate Mahalanobis distance d_k^i via Equation (17);

7   only if;

8   **if** $d_k^{i,l} > T$ **then**

9         $c_k^{j_1 j_2 \cdots j_S} \leftarrow +\infty$;

10   **break**;

11   **end**

12   calculate Δp via Equation (18);

13       $l \leftarrow l + 1$;

14   **end**

15   calculate the assignment cost $c_k^{j_1 j_2 \cdots j_S}$ via Equation (8);

16 **end**

17 solve the optimization problem in Equation (5)

---

5. Two-Stage MSMTT

In this section, the CGI-driven MDA is combined with a TS-MHT framework to perform bearings-only MSMTT. The framework is given in Figure 6.

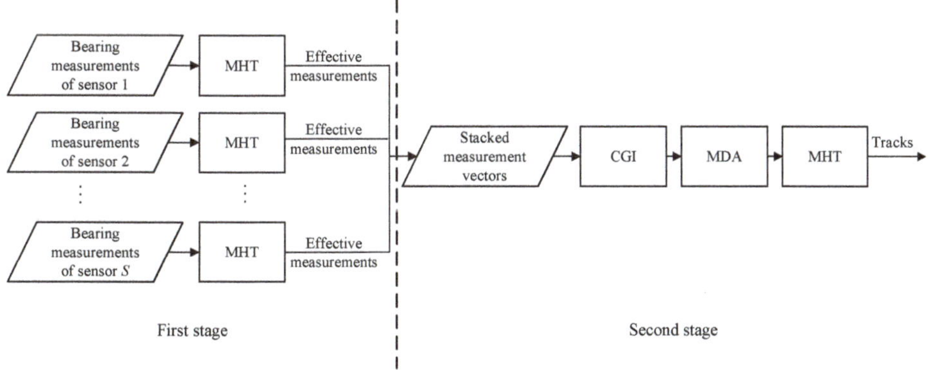

Figure 6. Framework of TS-MHT.

First, MHT is performed at each sensor, and only the measurements used to update the tracks are sent to the FC. Here, these measurements are referred to as "effective measurements." Second, the effective measurements from different sensors are combined and augmented to form stacked measurement vectors. Note that each measurement vector is a potential association hypothesis. The proposed CGI is then used to eliminate infeasible association hypotheses. After this, the measurement-to-measurement association is performed using the MDA algorithm. Finally, target tracks are obtained by using the second stage MHT.

The advantages of the above framework are mainly in the following aspects. In the framework shown in Figure 6, using the first stage MHT can eliminate most of the false measurements obtained by individual sensors, thus reducing the number of stacked measurement vectors. This further reduces the computational requirement of associations, and it also helps improve the accuracy of MDA. In turn, accurate data association facilitates track initialization in the second stage MHT and avoids infeasible hypothesis generation.

5.1. First Stage MHT

For the first stage, bearings-only multitarget tracking needs to be performed at each local passive sensor. Many existing methods are available [23,33,55]. Since this part is not the focus of this work, only one of the methods is considered.

The method proposed in [33] is to define the target state in Cartesian coordinates, thus performing single sensor state-space tracking. It should be noted that in [33], the target moves in three-dimensional space, and frequency information is available. In order to use the strategy for two-dimensional bearings-only MSMTT, it is simplified so that the dynamical system of the target can be described by Equations (1) and (2).

First, the one-point initialization approach is performed by combining the detection range of the sensor and all measurements at the initial time. Suppose that the detection range of sensor s is within the interval $[r_{min}^s, r_{max}^s]$. Correspondingly, the initial range between the target and the sensor and the corresponding variance can be calculated [33] as follows.

$$r^s = \frac{r_{min}^s + r_{max}^s}{2}, \quad \sigma_r^2 = \frac{(r_{max}^s - r_{min}^s)^2}{12}. \tag{19}$$

Then, the estimate of the initial state vector and the associated covariance are the following:

$$\hat{x}_{0|0}^{j_s} = \begin{bmatrix} x_0^{j_s} \\ \dot{x}_0^{j_s} \\ y_0^{j_s} \\ \dot{y}_0^{j_s} \end{bmatrix} = \begin{bmatrix} r^s \sin(z_0^{s,j_s}) + x_0^s \\ 0 \\ r^s \cos(z_0^{s,j_s}) + y_0^s \\ 0 \end{bmatrix}, \tag{20}$$

$$\mathbf{P}_{0|0}^{j_s} = \mathbf{J}'\mathbf{R}\mathbf{J}, \tag{21}$$

where the following is the case:

$$\mathbf{R} = \text{diag}\left(\sigma_r^2, \sigma_s^2, \sigma_{\dot{x}}^2, \sigma_{\dot{y}}^2\right), \tag{22}$$

$$\mathbf{J} = \frac{\partial z_0^{s,j_s}}{\partial \hat{x}_{0|0}^{j_s}}, \tag{23}$$

and σ_s^2 represents the measurement noise variance of sensor s, and $\sigma_{\dot{x}}^2$ and $\sigma_{\dot{y}}^2$ are the velocity variances based on their a priori maximum values.

It should be noted that, for this method, parameter r^s is only used for track initiation. That is to say that only bearing measurements are used to update tracks during the course of track maintenance. In addition, the measurements used for updating will be sent to the second stage.

5.2. Second Stage MHT

After the first stage MHT, most false measurements from each local sensor are eliminated, and the effective measurements are sent to the FC. Considering that the tracking performance of single passive sensor is quite limited in the first stage, these effective measurements can be divided into three categories: measurements originated from the target, false measurements due to false association, and dummy measurements due to missed detection. Therefore, in the second stage, the measurement-to-measurement association still needs to be performed.

First, all effective measurements from different sensors are combined and augmented to form stacked measurement vectors. Each stacked measurement vector is a potential association hypothesis. Then, the proposed CGI is used to delete infeasible associations. For each stacked measurement vector, in the iterative process of obtaining the MLE of target position, if the Mahalanobis distance $d_k^{i,l}$ between the initial estimate $(\hat{\mathbf{p}}_k^{i,0}, \mathbf{R}_k^{i,0})$ and the iterative estimate $(\hat{\mathbf{p}}_k^{i,l}, \mathbf{R}_k^{i,l})$ is greater than threshold T, then the association is determined as infeasible and deleted. When $d_k^{i,l} \leq T$, the estimate from the final iteration is naturally regarded as the MLE of the target position in the XY-plane, i.e., the solution of Equation (12). At the same time, it can be used for subsequent MDA. Finally, target tracks are obtained through the second stage MHT.

6. Illustrative Examples

In this section, five illustrative examples are presented. First, a scenario with three stationary targets (Scenario 1) is used to illustrate that, for incorrect associations, the initial estimation and iterative estimation generated in the iterative process are often inconsistent so as to verify the rationality and feasibility of the proposed strategy CGI. Second, a scenario with 18 stationary targets (Scenario 2) is used to compare the performance difference of three methods, MDA, CGI-driven MDA, and clustering-based MDA [43], to verify the effectiveness of the proposed strategy. Finally, a single-target tracking scenario (scenario 3) and multi-target tracking scenarios (scenarios 4 and 5) are used to further validate the performance of the framework shown in Figure 6.

6.1. Verification of Inconsistency

This subsection uses a numerical example about stationary targets to illustrate the difference in Mahalanobis distance between the current and initial estimates in an iterative process for the MLE of different target positions so as to verify the feasibility of the CGI proposed in Section 4.

Suppose there are three fixed passive sensors located at (0 m, 0 m), (1000 m, 600 m), and (3000 m, 0 m) in the XY-plane. At time k, sensor $s, s \in \{1,2,3\}$, acquires bearing measurements $\{z_k^{s,1}, z_k^{s,2}\}$, where $z_k^{s,1}$ and $z_k^{s,2}$ represents the measurements originated from the targets 1 and 2, respectively. The positions of these two targets in the XY-plane are (1500 m, 200 m) and (1800 m, 500 m). The standard deviations of the measurement errors of these three sensors are $\sigma_s = 17.5$ mrad, $s \in \{1,2,3\}$.

In the absence of false alarms and missed detections, eight stacked measurement vectors, i.e., association hypotheses, can be obtained. Figure 7 shows the bearing measurements of each sensor in one of the Monte Carlo runs, where the dashed lines represent the measurements originated from target 1, and the solid lines represent the measurements originated from target 2. Figures 8–10 show initial estimate $(\hat{\mathbf{p}}_k^{i,0}, \mathbf{R}_k^{i,0})$ and iterative estimate $(\hat{\mathbf{p}}_k^{i,l}, \mathbf{R}_k^{i,l})$, $l = N_{max}$ obtained using these stacked measurement vectors. Note that the only condition for iteration termination in this scenario is $l > N_{max}$. The uncertainty of the position estimates in the XY-plane is represented by the 95% probability ellipses.

From Figure 8a,f, when all components of the stacked measurement vector originate from the same target, the uncertainty ellipse of the iterative estimate is smaller than that of the initial estimate, and these two estimates are consistent. From Figure 8b,d,e, it can be observed that these two estimates obtained by z_k^{121}, z_k^{211}, and z_k^{221} are inconsistent. For the other two stacked measurement vectors z_k^{112} and z_k^{212}, since the initial and iterative estimates are too far away from each other, they are shown in the subfigures of Figures 9 and 10, respectively. It can be observed that the uncertainty ellipses of the iterative estimates are extremely large. For this two cases, the initial and iterative estimates are also obviously inconsistent.

It can be demonstrated through the above experiments that for many infeasible associations, the two estimates, $(\hat{\mathbf{p}}_k^{i,0}, \mathbf{R}_k^{i,0})$ and $(\hat{\mathbf{p}}_k^{i,l}, \mathbf{R}_k^{i,l})$, obtained in the iterations are often inconsistent.

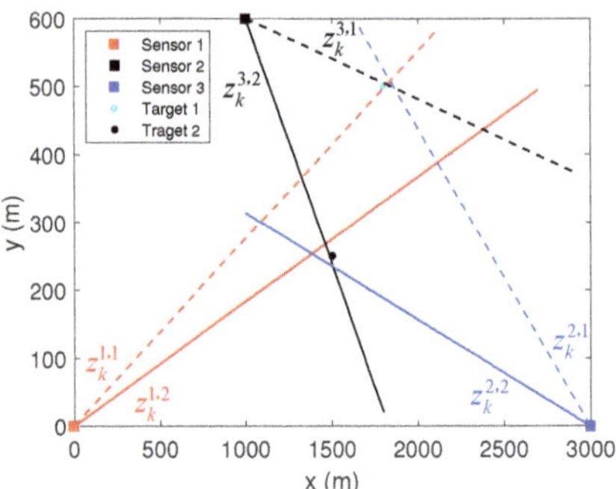

Figure 7. Scenario 1 with 2 stationary targets and 3 passive sensors.

Furthermore, for each stacked measurement vector, the Mahalanobis distances between initial estimate $(\hat{\mathbf{p}}_k^{i,0}, \mathbf{R}_k^{i,0})$ and all iterative estimates $(\hat{\mathbf{p}}_k^{i,l}, \mathbf{R}_k^{i,l})$, $l = \{1, 2, \cdots, N_{\max}\}$ are calculated. Table 1 presents the minimum and maximum Mahalanobis distances obtained in the iterative process with the different stacked measurement vectors. It is the statistic obtained from 2000 Monte Carlo runs.

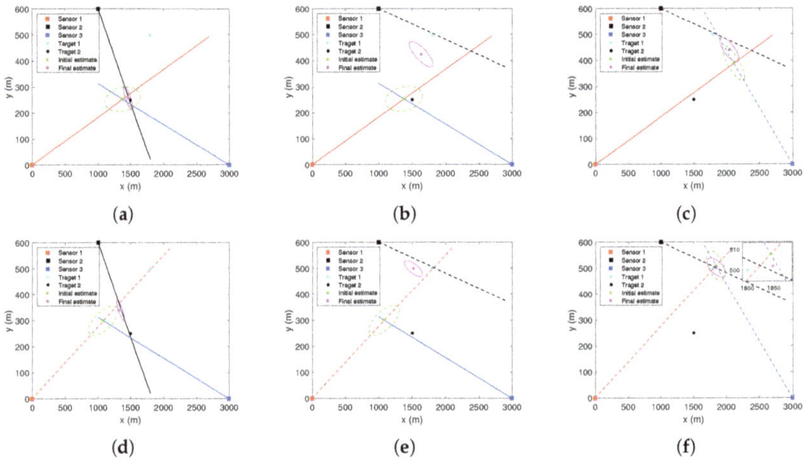

Figure 8. The initial estimate and the estimate after l iterations obtained using different stacked measurement vectors. (**a**) $\mathbf{z}_k^{111}$. (**b**) $\mathbf{z}_k^{121}$. (**c**) $\mathbf{z}_k^{122}$. (**d**) $\mathbf{z}_k^{211}$. (**e**) $\mathbf{z}_k^{221}$. (**f**) $\mathbf{z}_k^{222}$.

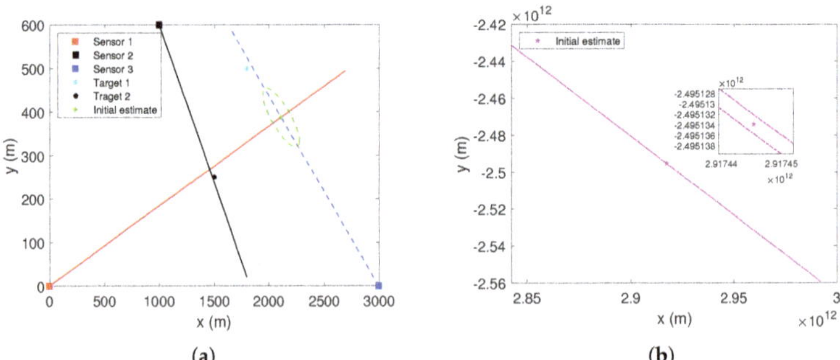

Figure 9. The initial estimate and the estimate after l iterations obtained using the stacked measurement vector $\mathbf{z}_k^{112}$. (**a**) Initial estimate. (**b**) Estimate after l iterations.

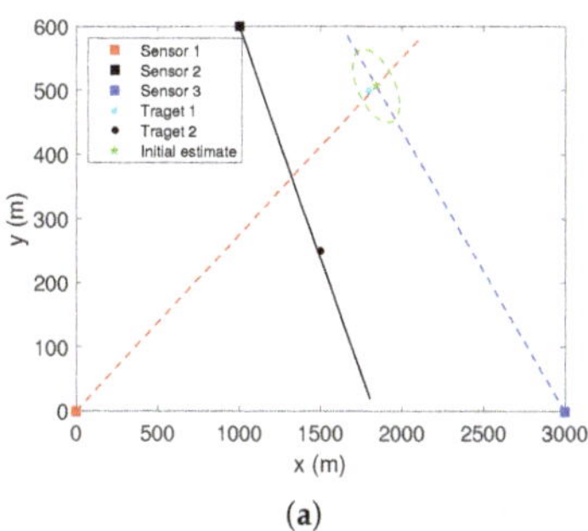

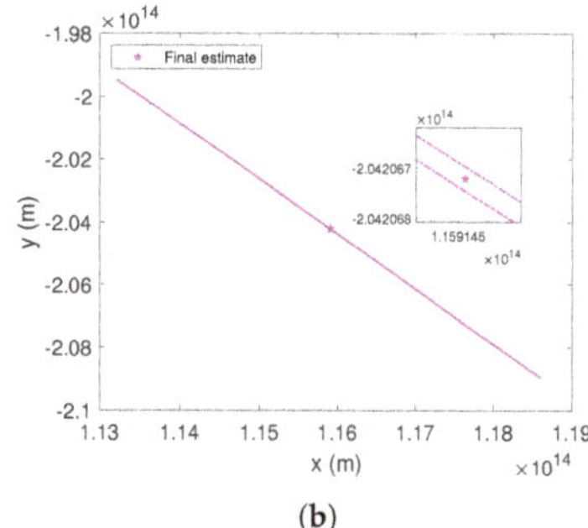

(a)	(b)

Figure 10. The initial estimate and the estimate after l iterations obtained using the stacked measurement vector $\mathbf{z}_k^{212}$. (**a**) Initial estimate. (**b**) Estimate after l iterations.

Table 1. Mahalanobis distances between the initial estimate and the iterative estimates.

	$\mathbf{z}_k^{111}$	$\mathbf{z}_k^{112}$	$\mathbf{z}_k^{121}$	$\mathbf{z}_k^{122}$	$\mathbf{z}_k^{211}$	$\mathbf{z}_k^{212}$	$\mathbf{z}_k^{221}$	$\mathbf{z}_k^{222}$
Minimum Mahalanobis distances	0.8336	80.9022	17.9394	3.8289	4.7382	102.5766	15.8057	0.4970
Maximum Mahalanobis distances	1.1744	5.1467×10^4	45.7706	4.7942	7.0418	2.1429×10^4	56.0292	0.5112

It can be observed that, throughout the iterative process, the Mahalanobis distances obtained by using correctly associated vectors $\mathbf{z}_k^{111}$ and $\mathbf{z}_k^{222}$ are significantly smaller. Therefore, infeasible associations can be effectively eliminated by setting an appropriate threshold T.

6.2. CGI Driven MDA for Stationary Targets

In this subsection, the impact of the proposed CGI on the performance of MDA will be analyzed. This scenario, as illustrated in Figure 11, consists of 3 bearing-only passive sensors, 1 cooperative target, and 18 unknown non-cooperative targets.

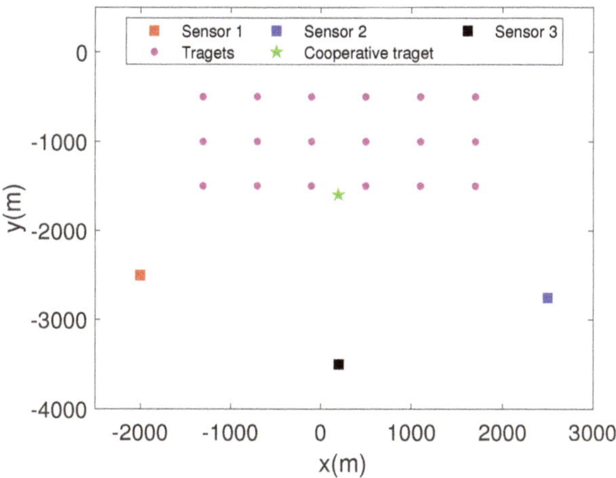

Figure 11. Scenario 2 with 18 stationary targets and 3 passive sensors.

The positions of three passive sensors in the XY-plane are $(-2000\ \text{m}, -2500\ \text{m})$, $(2500\ \text{m}, -2750\ \text{m})$, and $(200\ \text{m}, -3500\ \text{m})$, respectively. The standard deviations of the measurement errors for all sensors are $\sigma_s = 1$ mrad, $s \in \{1, 2, 3\}$. The position of the cooperative target is $(600\ \text{m}, -1600\ \text{m})$. The positions of other unknown non-cooperative targets are shown in Table 2. For the sake of simplicity, it is assumed that all sensors have unity detection probability for each target and there are no false measurements. It is also supposed that, at some point before these unknown targets are detected, three passive sensors can only acquire the bearing measurements originating from the cooperative target.

Table 2. Positions of all targets in XY-plane.

$(-1500\text{ m}, -500\text{ m})$	$(-900\text{ m}, -500\text{ m})$	$(-300\text{ m}, -500\text{ m})$	$(900\text{ m}, -500\text{ m})$	$(1500\text{ m}, -500\text{ m})$	$(-1500\text{ m}, -500\text{ m})$
$(-1500\text{ m}, -1000\text{ m})$	$(-900\text{ m}, -1000\text{ m})$	$(-300\text{ m}, -1000\text{ m})$	$(900\text{ m}, -1000\text{ m})$	$(1500\text{ m}, -1000\text{ m})$	$(-1500\text{ m}, -1000\text{ m})$
$(-1500\text{ m}, -1500\text{ m})$	$(-900\text{ m}, -1500\text{ m})$	$(-300\text{ m}, -1500\text{ m})$	$(900\text{ m}, -1500\text{ m})$	$(1500\text{ m}, -1500\text{ m})$	$(-1500\text{ m}, -1500\text{ m})$

In order to set a reasonable threshold T for the proposed CGI, the bearing measurements originating from the cooperative target are used iteration of Equations (14) and (16). The maximum Mahalanobis distance between the initial estimate $(\hat{\mathbf{p}}_k^{i,0}, \mathbf{R}_k^{i,0})$ and each iterative estimate $(\hat{\mathbf{p}}_k^{i,l}, \mathbf{R}_k^{i,l})$, $l \in \{1, 2, \cdots N_{\max}\}$ was $d_{\max} = 11.6977$ over 2000 Monte Carlo runs. Considering that the Mahalanobis distance is closely related to the geometric structure between the sensors and the cooperative target, threshold T should not be less than $d_{\max}$. In order to avoid deleting the correct association, the threshold in this scenario is set to $T = 12$.

The Lagrangian relaxation method in [48] is used to obtain suboptimal solutions of the MDA problem in Equation (5). Table 3 presents the performance comparison of three different methods, MDA, CGI driven MDA, and clustering-based MDA [43], based on a 2000-run Monte Carlo average. The experimental results are obtained on MATLAB R2020b with Intel(R) Core(TM) i5-9500 CPU @3.00GHz and RAM of 8 GB.

Table 3. The performance comparison of different methods.

	MDA	Clustering-Based MDA	CGI Driven MDA
Number of all S-tuples	5832	5832	5832
Number of S-tuples after coarse gating	-	103.74	83.82
Number of identified targets	19.58	18.97	18.02
Percent correct association	33.35%	81.67%	99.61%
Execution time to calculate assignment costs	3.9069 s	0.3917 s	0.3625 s
Execution time to obtain suboptimal solution	0.9163 s	0.1457 s	0.3543 s

It can be observed from Table 3 that the number of S-tuples reduced from 5832 to 83.82 after CGI. Moreover, the correct association rate of CGI-driven MDA is 99.61%, much higher than that of the other two methods. This means that CGI can effectively eliminate a large number of infeasible associations and can significantly improve the correct association rate of MDA. In addition, for MDA, the execution time to calculate the assignment costs of all S-tuples is 3.9069 s. This takes about 81% of the total execution time. For CGI-driven MDA, the execution time for calculating all assignment costs is only 3 s, which accounts for about 50% of the total execution time. Obviously, the proposed CGI driven MDA has a significant improvement in both computational efficiency and correct association probability. For the clustering-based MDA method, the execution time of obtaining the suboptimal solution of Equation (5) by the Lagrangian relaxation algorithm is less than that of the proposed CGI-driven MDA method. This is due to the fact that the clustering method decomposes the entire assignment problem into smaller subproblems, thus improving computational efficiency.

Table 4 illustrates the impact of five different thresholds on the performance of the proposed CGI-driven MDA. It can be observed that the larger T is, the larger the number of S-tuples obtained after CGI, and the more execution time is required. The correct association rate when $T = 1$ is 78.14%, which is less than the correct association rate when $T = 12$. Therefore, it is not the case that the smaller the threshold is, the better. Smaller thresholds may result in the removal of some correct associations. Moreover, it can be found that when

$T = 24$, the correct association rate is significantly smaller than the other two groups. This is mainly due to the large number of retained S-tuples, which results in the degradation of the performance of the Lagrangian relaxation algorithm used.

Table 4. The effect of different threshold T on the performance of CGI driven MDA.

	$T = 1$	$T = 6$	$T = 12$	$T = 16$	$T = 24$
Number of all S-tuples	5832	5832	5832	5832	5832
Number of S-tuples after CGI	29.43	61.03	83.82	95.89	114.93
Number of identified targets	18.27	18.01	18.02	18.37	19.68
Percent correct association	78.14%	99.33%	99.61%	87.61%	69.39%
Execution time to calculate assignment costs	0.3597	0.3397 s	0.3625 s	0.3472 s	0.3439 s
Execution time to obtain suboptimal solution	0.0164	0.0964 s	0.3543 s	0.8761 s	1.3270 s

6.3. TS-MHT for Single Target Tracking in Clutter

In this subsection, a single target tracking scenario is considered to verify the performance of the TS-MHT framework shown in Figure 6.

There are four passive sensors located at (1000 m, 2250 m), (1000 m, −2250 m), (6000 m, 2250 m), and (6000 m, −2250 m), respectively. Each sensor can only measure the bearing to the target, and the sampling interval is 10 s. Their measurement errors are modeled as zero-mean Gaussian white noises with same standard deviations $\sigma_s = 17.5$ mrad, $s \in \{1, 2, 3, 4\}$. The maximum detection range of each sensor is 5 km and the detection probability is $P_{D_s} = 0.9$, $s \in \{1, 2, 3, 4\}$. False measurements are uniformly distributed over the detection range and their number is Poisson distributed with an average of 4 false measurements per sensor per scan. Target moves in two dimensions with NCV, and its initial position and velocity are [3500 m, −4000 m] and [0 m/s, 7.2 m/s], respectively. The process noise covariance is $Q = 0.01^2 I$. The true trajectory of the target motion and the sensor positions are shown in Figure 12, where the detection range of each sensor is indicated by dashed lines of different colors.

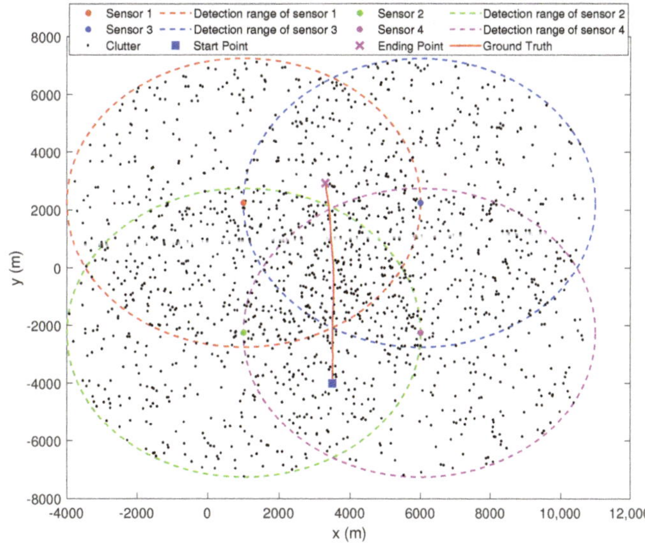

Figure 12. Scenario 3 for single target tracking with 4 passive sensors.

Figure 13 shows the tracking results of each sensor at the first stage. Here, threshold T is set to 16. Superficially, there are significant differences between the tracking results of each sensor and the true track of the target. This is due to the unobservability of target state for single passive sensor. Although, in track initiation, the initial position estimate of the target can be obtained by the detection range of the sensor, it is also inaccurate.

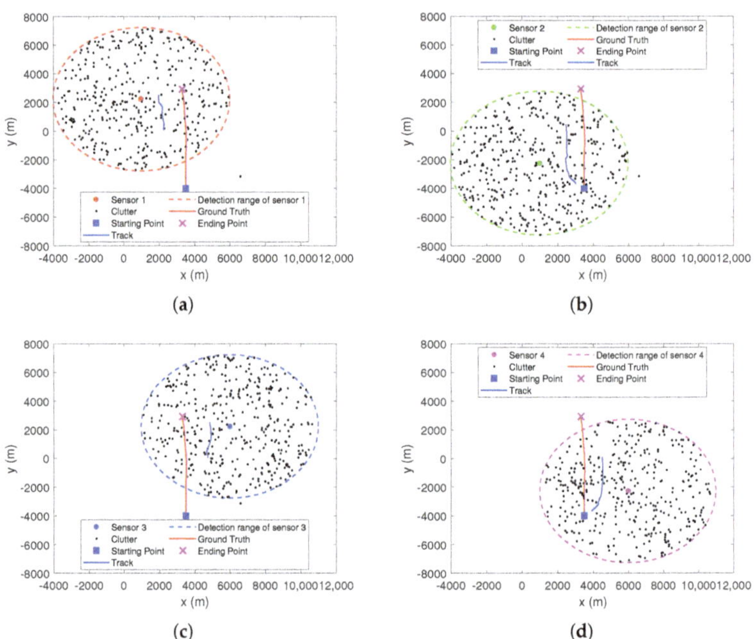

Figure 13. Tracking results of first stage MHT in each local sensor. (**a**) Sensor 1. (**b**) Sensor 2. (**c**) Sensor 3. (**d**) Sensor 4.

It should be noted that, for the first stage of the TS-MHT framework shown in Figure 6, its main purpose is to eliminate as many false measurements as possible by using preliminary tracking. Moreover, only the measurements used to update these tracks are sent to the second stage in real time. That is to say that it is more interested in whether the measurements sent to the second stage originate from the true target than in the accuracy of target state estimation. From Figure 13, it can be observed that the number of tracks obtained by sensors are all one. These estimated numbers of tracks are close to the number of true target. In addition, the estimated tracks by each sensor and the true track of the target are on the same side of the corresponding sensor, and their orientations with respect to the sensors are roughly the same. This means that the tracking results of the first stage may not be that bad, although the tracking results still need to be further improved by the measurements from other sensors during the second stage.

Figure 14 shows the tracking result of the second stage. It can be observed that the second stage MHT can effectively track the target in clutter. At the same time, this in turn shows that CGI-driven MDA can effectively delete infeasible associations.

The execution time of each stage is calculated over 2000 Monte Carlo runs. For the first stage, the average execution time per frame of the MHT algorithm in each sensor is approximately equal, and it is about 1.0741 s. For the second stage, the average execution time of MHT algorithm is 0.1782 s per frame. Obviously, the execution time of the second stage MHT is significantly smaller than that of the first stage MHT. This is due to the fact that the first stage can effectively eliminate a large number of false measurements, thus effectively reducing the number of feasible assumptions in the second stage. It should be noted that the effective measurements in the first stage are sent to the second stage in real time.

In addition, to verify the effect of different thresholds T on tracking performance, the root mean square error (RMSE) is used to measure the performance of target tracking, as shown in Figure 15. It can be observed that when $T = 16$, tracking performance is significantly better than the other two groups. Combined with the experimental results of Scenario 2, it can be further demonstrated that when preset threshold T is too large or too small, and it may result in a decrease in the correct association rate of MDA, which further affects tracking performance.

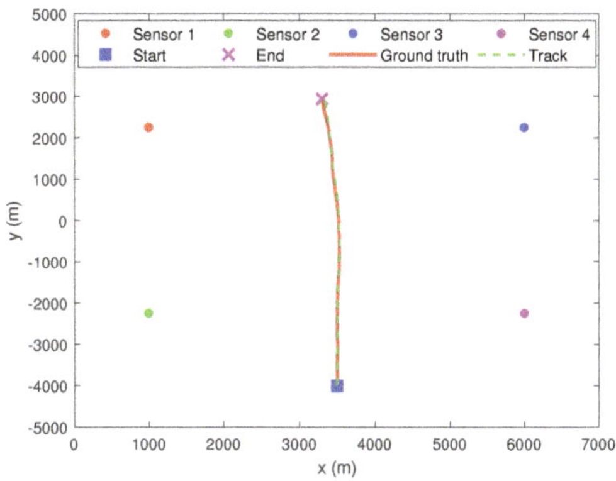

Figure 14. Tracking results of second stage MHT.

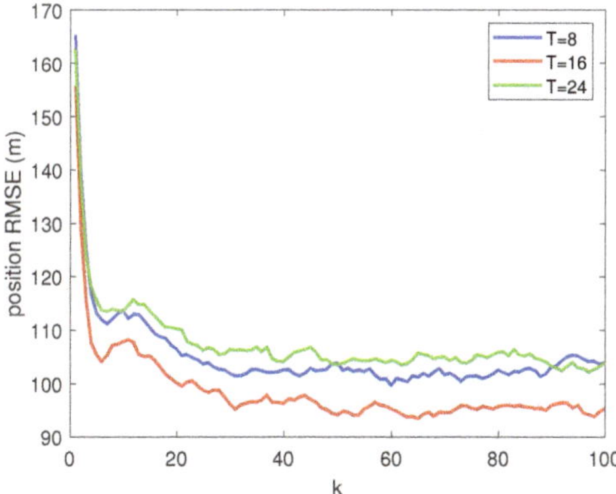

Figure 15. Position RMSE of different threshold T.

6.4. TS-MHT for Multitarget Tracking in Clutter

Consider two multitarget tracking scenarios with four sensors. For scenario 4, as shown in Figure 16a, the two targets move simultaneously along the Y direction with a nearly constant speed of 6.2 m/s, and their initial positions are [3500 m, −3500 m] and [6500 m, −3500 m], respectively. For scenario 5 as shown in Figure 16b, the initial positions of the two targets are [5000 m, −3500 m] and [8500 m, 0 m], and their initial velocities are [0 m/s, 7.2 m/s] and [−5.2 m/s, 0 m/s], respectively. The other parameters are the same as in Scenario 2.

By comparing Figures 16 and 17, it can be observed that the proposed strategy can effectively tackle MSMTT.

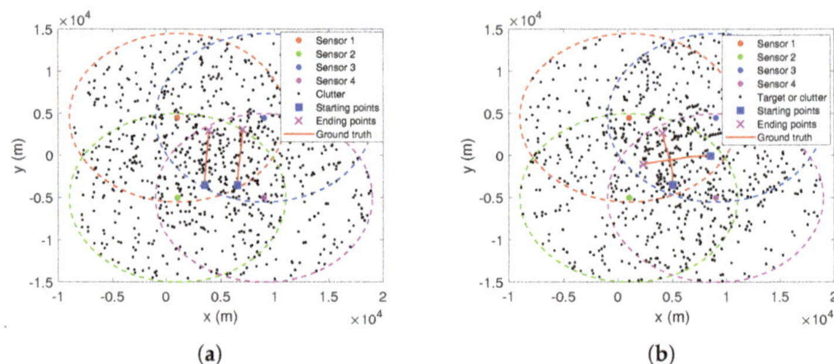

Figure 16. Multitarget tracking scenarios with 4 passive sensors. (**a**) Scenarios 4. (**b**) Scenarios 5.

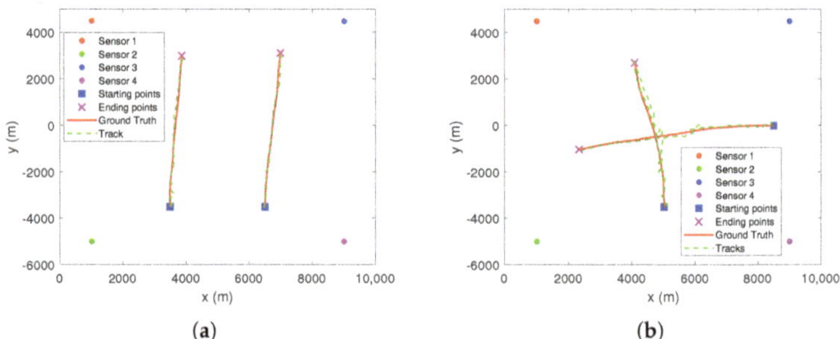

Figure 17. Tracking results. (**a**) Scenarios 4. (**b**) Scenarios 5.

7. Conclusions

The bearings-only multitarget tracking problem is investigated for synchronous passive sensors. In the target tracking process, especially for track initiation, MDA can be used to identify the measurements originating from common targets. In order to reduce the computational cost of the multidimensional assignment and improve its correct association rate, a new coarse gating strategy, the CGI, has been proposed first. For MDA, iterative processes can be used to obtain the MLE of target position corresponding to each possible association and, thus, further calculate the assignment cost of that association. Since the initial estimate and the iterative estimate are not obtained by the same measurements, it has been proposed to eliminate infeasible associations by using the Mahalanobis distance between the initial estimate and the iterative estimate as a measure. The feasibility and effectiveness of the proposed CGI is verified by two scenarios, i.e., scenarios 1 and 2, respectively. In addition, MDA driven by this strategy is combined with the TS-MHT framework for distributed MSMTT. Numerical examples have verified the performance of the proposed strategy. Moreover, the effectiveness of the proposed strategy in the tracking process is further verified by two scenarios of single target and multitarget in clutter.

Author Contributions: Conceptualization, Z.W., Z.D. and M.M.; methodology, Z.W., Z.D. and M.M.; software, Z.W.; validation, Z.W., Z.D. and Y.H.; writing—original draft preparation, Z.W., Z.D. and M.M.; writing—review and editing, Z.D., Y.H. and M.M. All authors have read and agreed to the published version of the manuscript.

Funding: This work is supported in part by National Key Research and Development Plan under Grants 2021YFC2202600 and 2021YFC2202603, and National Natural Science Foundation of China through Grants 62033010, 61773147 and 61673317.

Institutional Review Board Statement: Not applicable.

Informed Consent Statement: Not applicable.

Data Availability Statement: Not applicable.

Abbreviations

The following abbreviations are used in this manuscript:

MSMTT	Multisensor-multitarget tracking;
MDA	Multidimensional assignment;
MLE	Maximum likelihood estimation;
TS-MHT	Two-stage multiple hypothesis tracking;
MTT	Multitarget tracking;
GNN	Global nearest neighbor;
PDA	Probabilistic data association;
JPDA	Joint probabilistic data association;
JIPDA	Joint integrated probabilistic data association;
MS-JPDA	Multiscan joint probabilistic data association;
MHT	Multiple hypothesis tracking;
RFS	Random finite set;
PHD	Probability hypothesis density;
CPHD	Cardinalized probability hypothesis density;
GLMB	Generalized labeled multi-Bernoulli;
FC	Fusion center;
CGI	Coarse gating in iterations;
2D	Two-dimensional;
RMSE	Root mean square error;
NCV	Nearly constant velocity.

Nomenclatures

Notations	Definitions
S	Number of sensors
s	Sensor index, $s = 1, 2, \cdots, S$
i	Target index
k	Time index
j_s	Bearing measurement index acquired by sensor s
N_s	Number of bearing measurements acquired by sensor s
z_k^{s,j_s}	The j_sth bearing measurement acquired by sensor s at time k
β_k^{s,i_s}	True bearing between target i and sensor s
$\tilde{z}_k^{j_s}$	False measurements
$\mathbf{x}_k^i$	State vector of target i at time k
$\mathbf{p}_k^s$	Position vector of sensor s
F_k	State transition matrix at time k
$\mathbf{w}_k^i$	Process noise vector of target i at time k
v_k^{s,j_s}	Measurement noise of sensor s at time k
σ_s^2	Measurement noise variance of sensor s
$\mathbf{Z}_k^{j_1 j_2 \cdots j_s}$	An S-tuple of measurements, one from each sensor
$c_k^{j_1 j_2 \cdots j_s}$	Cost of associating S-tuple of measurements with a target
$\rho_k^{j_1 j_2 \cdots j_s}$	Binary decision variables
P_{D_s}	Detection probability of sensor s
ψ_s	Volume of the field of view of sensor s
$u(j_s)$	Binary indicator function
l	iteration index
$\mathbf{z}_k^{j_1 j_2 \cdots j_s}$	Stacked measurement vector corresponding to S-tuple $\mathbf{Z}_k^{j_1 j_2 \cdots j_s}$
$\mathbf{p}_k^i$	Position vector of target i
$\hat{\mathbf{p}}_k^{i,0}$	Initial position estimate of target i

$\hat{\mathbf{p}}_k^{i,l}$	Position estimation of target i after iteration l
$\mathbf{R}_k^{i,l}$	Covariance matrix corresponding to the position estimate $\hat{\mathbf{p}}_k^{i,l}$
$\mathbf{J}_k^l$	Jacobian matrix
$d_k^{i,l}$	Mahalanobis distance
T	Threshold

References

1. Mallick, M.; Vo, B.N.; Kirubarajan, T.; Arulampalam, S. Introduction to the issue on multitarget tracking. *IEEE J. Sel. Top. Signal Process.* **2013**, *7*, 373–375. [CrossRef]
2. Mallick, M.; Bar-Shalom, Y.; Kirubarajan, T.; Moreland, M. An improved single-point track initiation using GMTI measurements. *IEEE Trans. Aerosp. Electron. Syst.* **2015**, *51*, 2697–2714. [CrossRef]
3. Lima, K.M.D.; Costa, R.R. Cooperative-PHD Tracking Based on Distributed Sensors for Naval Surveillance Area. *Sensors* **2022**, *22*, 729. [CrossRef] [PubMed]
4. Luo, J.; Han, Y.; Fan, L. Underwater Acoustic Target Tracking: A Review. *Sensors* **2018**, *18*, 112. [CrossRef] [PubMed]
5. Bahraini, M.S.; Rad, A.B.; Bozorg, M. SLAM in Dynamic Environments: A Deep Learning Approach for Moving Object Tracking Using ML-RANSAC Algorithm. *Sensors* **2019**, *19*, 3699. [CrossRef]
6. Panicker, S.; Gostar, A.K.; Bab-Hadiashar, A.; Hoseinnezhad, R. Recent Advances in Stochastic Sensor Control for Multi-Object Tracking. *Sensors* **2019**, *19*, 3790. [CrossRef]
7. Mallick, M.; Krishnamurthy, V.; Vo, B.N. Multitarget Tracking Using Multiple Hypothesis Tracking. In *Integrated Tracking, Classification, and Sensor Management: Theory and Applications*; Wiley: Piscataway, NJ, USA, 2012; pp. 163–202. [CrossRef]
8. Lundquist, C.; Granström, K.; Orguner, U. An Extended Target CPHD Filter and a Gamma Gaussian Inverse Wishart Implementation. *IEEE J. Sel. Top. Signal Process.* **2013**, *7*, 472–483. [CrossRef]
9. Tang, X.; Li, M.; Tharmarasa, R.; Kirubarajan, T. Seamless Tracking of Apparent Point and Extended Targets Using Gaussian Process PMHT. *IEEE Trans. Signal Process.* **2019**, *67*, 4825–4838. [CrossRef]
10. Hoher, P.; Wirtensohn, S.; Baur, T.; Reuter, J.; Govaers, F.; Koch, W. Extended Target Tracking with a Lidar Sensor Using Random Matrices and a Virtual Measurement Model. *IEEE Trans. Signal Process.* **2022**, *70*, 228–239. doi: 10.1109/TSP.2021.3138006. [CrossRef]
11. Smith, J.; Particke, F.; Hiller, M.; Thielecke, J. Systematic Analysis of the PMBM, PHD, JPDA and GNN Multi-Target Tracking Filters. In Proceedings of the 2019 International Conference on Information Fusion, Ottawa, ON, Canada, 2–5 July 2019.
12. Ishtiaq, S.; Wang, X.; Hassan, S. Multi-Target Tracking Algorithm Based on 2-D Velocity Measurements Using Dual-Frequency Interferometric Radar. *Electronics* **2021**, *10*, 1969. [CrossRef]
13. Blackman, S.S.; Popoli, R. *Design and Analysis of Modern Tracking Systems*; Radar Library: Norwood, MA, USA, 1999.
14. Bar-Shalom, B.Y.; Willett, P.K.; Tian, A.X. *Tracking and Data Fusion: A Handbook of Algorithms*; YBS Publishing: Storrs, CT, USA, 2011.
15. He, S.; Shin, H.S.; Tsourdos, A. Joint Probabilistic Data Association Filter with Unknown Detection Probability and Clutter Rate. *Sensors* **2018**, *18*, 269. [CrossRef]
16. Blackman, S. Multiple hypothesis tracking for multiple target tracking. *IEEE Aerosp. Electron. Syst. Mag.* **2004**, *19*, 5–18. [CrossRef]
17. Musicki, D.; Evans, R. Joint integrated probabilistic data association: JIPDA. *IEEE Trans. Aerosp. Electron. Syst.* **2004**, *40*, 1093–1099. [CrossRef]
18. Roecker, J. Multiple scan joint probabilistic data association. *IEEE Trans. Aerosp. Electron. Syst.* **1995**, *31*, 1204–1210. [CrossRef]
19. Chong, C.Y.; Mori, S.; Reid, D.B. Forty Years of Multiple Hypothesis Tracking. *J. Adv. Inf. Fusion* **2019**, *14*, 131–153.
20. Vo, B.N.; Mallick, M.; bar Shalom, Y.; Coraluppi, S., III; Osborne, R.; Mahler, R.; Vo, B.T. *Multitarget Tracking*. Wiley: Hoboken, NJ, USA, 2015; doi: 10.1002/047134608X.W8275. [CrossRef]
21. Reid, D. An algorithm for tracking multiple targets. *IEEE Trans. Autom. Control.* **1979**, *24*, 843–854. [CrossRef]
22. Bar-Shalom, Y.; Blackman, S.S.; Fitzgerald, R.J. Dimensionless score function for multiple hypothesis tracking. *IEEE Trans. Aerosp. Electron. Syst.* **2007**, *43*, 392–400. [CrossRef]
23. Coraluppi, S.; Rago, C.; Carthel, C.; Bale, B. Distributed MHT with Passive Sensors. In Proceedings of 2021 International Conference on Information Fusion, Sun City, South Africa, 1–4 November 2021.
24. Mahler, R.P.S. *Advances in Statistical Multisource-Multitarget Information Fusion*; Artech House: Norwood, MA, USA, 2014.
25. Moratuwage, D.; Adams, M.; Inostroza, F. δ-Generalized Labeled Multi-Bernoulli Simultaneous Localization and Mapping with an Optimal Kernel-Based Particle Filtering Approach. *Sensors* **2019**, *19*, 2290. [CrossRef]
26. Mahler, R. Multitarget Bayes filtering via first-order multitarget moments. *IEEE Trans. Aerosp. Electron. Syst.* **2003**, *39*, 1152–1178. [CrossRef]
27. Mahler, R. PHD filters of higher order in target number. *IEEE Trans. Aerosp. Electron. Syst.* **2007**, *43*, 1523–1543. [CrossRef]
28. Schlangen, I.; Delande, E.D.; Houssineau, J.; Clark, D.E. A Second-Order PHD Filter With Mean and Variance in Target Number. *IEEE Trans. Signal Process.* **2018**, *66*, 48–63. [CrossRef]
29. Vo, B.T.; Vo, B.N.; Cantoni, A. The Cardinality Balanced Multi-Target Multi-Bernoulli Filter and Its Implementations. *IEEE Trans. Signal Process.* **2009**, *57*, 409–423. [CrossRef]

30. Beard, M.; Vo, B.T.; Vo, B.N. A Solution for Large-Scale Multi-Object Tracking. *IEEE Trans. Signal Process.* **2020**, *68*, 2754–2769. [CrossRef]

31. Chen, H.; Kirubarajan, T.; Bar-Shalom, Y. Performance limits of track-to-track fusion versus centralized estimation: Theory and application [sensor fusion]. *IEEE Trans. Aerosp. Electron. Syst.* **2003**, *39*, 386–400. [CrossRef]

32. Yu, Y.; Hou, Q.; Zhang, W.; Zhang, J. A Sequential Two-Stage Track-to-Track Association Method in Asynchronous Bearings-Only Sensor Networks for Aerial Targets Surveillance. *Sensors* **2019**, *19*, 3185. [CrossRef] [PubMed]

33. Lexa, M.; Coraluppi, S.; Carthel, C.; Willett, P. Distributed MHT and ML-PMHT Approaches to Multi-Sensor Passive Sonar Tracking. In Proceedings of the 2020 IEEE Aerospace Conference, Big Sky, MT, USA, 7–14 March 2020. [CrossRef]

34. Shen, K.; Dong, P.; Jing, Z.; Leung, H. Consensus-Based Labeled Multi-Bernoulli Filter for Multitarget Tracking in Distributed Sensor Network. *IEEE Trans. Cybern.* **2021**, 1–12. doi: 10.1109/TCYB.2021.3087521. [CrossRef]

35. Kazimierski, W.; Zaniewicz, G. Determination of Process Noise for Underwater Target Tracking with Forward Looking Sonar. *Remote Sens.* **2021**, *13*, 1014. [CrossRef]

36. Wang, M.; Qiu, B.; Zhu, Z.; Xue, H.; Zhou, C. Study on Active Tracking of Underwater Acoustic Target Based on Deep Convolution Neural Network. *Appl. Sci.* **2021**, *11*, 7530. [CrossRef]

37. Li, X.; Lu, B.; Ali, W.; Jin, H. Passive Tracking of Multiple Underwater Targets in Incomplete Detection and Clutter Environment. *Entropy* **2021**, *23*, 1082. [CrossRef] [PubMed]

38. Zhang, Y.; Lan, J.; Mallick, M.; Li, X.R. Bearings-Only Filtering Using Uncorrelated Conversion Based Filters. *IEEE Trans. Aerosp. Electron. Syst.* **2021**, *57*, 882–896. [CrossRef]

39. Mušicki, D. Bearings only single-sensor target tracking using Gaussian mixtures. *Automatica* **2009**, *45*, 2088–2092. [CrossRef]

40. Do, C.T.; Nguyen, T.T.D.; Nguyen, H.V. Robust multi-sensor generalized labeled multi-Bernoulli filter. *Signal Process.* **2022**, *192*, 108368. [CrossRef]

41. Bar-Shalom, Y.; Li, X. *Multitarget-Multisensor Tracking: Principles and Techniques*; YBS Publishing: Storrs, CT, USA, 1995.

42. Deb, S.; Pattipati, K.; Bar-Shalom, Y. A multisensor-multitarget data association algorithm for heterogeneous sensors. *IEEE Trans. Aerosp. Electron. Syst.* **1993**, *29*, 560–568. [CrossRef]

43. Chummun, M.; Kirubarajan, T.; Pattipati, K.; Bar-Shalom, Y. Fast data association using multidimensional assignment with clustering. *IEEE Trans. Aerosp. Electron. Syst.* **2001**, *37*, 898–913. [CrossRef]

44. Sathyan, T.; Sinha, A.; Kirubarajan, T.; Mcdonald, M.; Lang, T. MDA-Based Data Association with Prior Track Information for Passive Multitarget Tracking. *IEEE Trans. Aerosp. Electron. Syst.* **2011**, *47*, 539–556. [CrossRef]

45. Mallick, M. A Note on Bearing Measurement Model. *Mach. Eng.* **2018**. doi: 10.13140/RG.2.2.13441.35681. [CrossRef]

46. Leung, H. Neural network data association with application to multiple-target tracking. *Opt. Eng.* **1996**, *35*, 693–700. [CrossRef]

47. Carrier, J.Y.; Litva, J.; Leung, H.; Lo, T.K.Y. Genetic algorithm for multiple-target-tracking data association. In Proceedings of the SPIE Conference on Acquisition, Tracking, Pointing, Orlando, FL, USA, 7 June 1996; Volume 2739. [CrossRef]

48. Deb, S.; Yeddanapudi, M.; Pattipati, K.; Bar-Shalom, Y. A generalized S-D assignment algorithm for multisensor-multitarget state estimation. *IEEE Trans. Aerosp. Electron. Syst.* **1997**, *33*, 523–538. [CrossRef]

49. Poore, A.B.; Robertson, A.J. A New Lagrangian Relaxation Based Algorithm for a Class of Multidimensional Assignment Problems. *Comput. Optim. Appl.* **1997**, *8*, 129–150. [CrossRef]

50. Bar-Shalom, Y.; Kirubarajan, T.; Li, X.R. *Estimation with Applications to Tracking and Navigation*; Wiley: New York, NY, USA, 2001.

51. Uhlmann, J.K. Covariance consistency methods for fault-tolerant distributed data fusion. *Inf. Fusion* **2003**, *4*, 201–215. [CrossRef]

52. Uhlmann, J., Introduction to the Algorithmics of Data Association in Multiple-Target Tracking. In *Handbook of Multisensor Data Fusion*; CRC Press: Boca Raton, FL, USA, 2008; Chapter 3. [CrossRef]

53. Collins, J.; Uhlmann, J. Efficient gating in data association with multivariate Gaussian distributed states. *IEEE Trans. Aerosp. Electron. Syst.* **1992**, *28*, 909–916. [CrossRef]

54. Klingner, J.; Ahmed, N.; Correll, N. Fault-tolerant Covariance Intersection for localizing robot swarms. *Robot. Auton. Syst.* **2019**, *122*, 103306. [CrossRef]

55. Coraluppi, S.; Carthel, C.; Coon, A. An MHT Approach to Multi-Sensor Passive Sonar Tracking. In Proceedings of the 2018 International Conference on Information Fusion, Cambridge, UK, 10–13 July 2018. [CrossRef]

Article

A Total Lp-Norm Optimization for Bearing-Only Source Localization in Impulsive Noise with $S\alpha S$ Distribution

Ji-An Luo [1], Chang-Cheng Xue [1], Ying-Jiao Rong [2,*] and Shen-Tu Han [1]

[1] School of Automation, Hangzhou Dianzi University, Hangzhou 310018, China; luojian@hdu.edu.cn (J.-A.L.); xcc@hdu.edu.cn (C.-C.X.); hanshentu@hdu.edu.cn (S.-T.H.)
[2] Science and Technology on Near-Surface Detection Laboratory, Wuxi 214035, China
* Correspondence: ryingjiao@hdu.edu.cn

Abstract: This paper considers the problem of robust bearing-only source localization in impulsive noise with symmetric α-stable distribution based on the Lp-norm minimization criterion. The existing Iteratively Reweighted Pseudolinear Least-Squares (IRPLS) method can be used to solve the least LP-norm optimization problem. However, the IRPLS algorithm cannot reduce the bias attributed to the correlation between system matrices and noise vectors. To reduce this kind of bias, a Total Lp-norm Optimization (TLPO) method is proposed by minimizing the errors in all elements of system matrix and data vector based on the minimum dispersion criterion. Subsequently, an equivalent form of TLPO is obtained, and two algorithms are developed to solve the TLPO problem by using Iterative Generalized Eigenvalue Decomposition (IGED) and Generalized Lagrange Multiplier (GLM), respectively. Numerical examples demonstrate the performance advantage of the IGED and GLM algorithms over the IRPLS algorithm.

Keywords: bearing-only; source localization; robust estimation; least Lp-norm; total Lp-norm optimization

Citation: Luo, J.-A.; Xue, C.-C.; Rong, Y.-J.; Han, S.-T. A Total Lp-Norm Optimization for Bearing-Only Source Localization in Impulsive Noise with $S\alpha S$ Distribution. *Sensors* **2021**, *21*, 6471. https://doi.org/10.3390/s21196471

Academic Editors: Mahendra Mallick and Ratnasingham Tharmarasa

Received: 30 August 2021
Accepted: 23 September 2021
Published: 28 September 2021

Publisher's Note: MDPI stays neutral with regard to jurisdictional claims in published maps and institutional affiliations.

1. Introduction

Bearing-Only Source Localization (BOSL) using spatially distributed sensors can be widely applied in network localization [1], vehicle localization [2], gunshot localization [3], animal behavior monitoring [4] and rigid body localization [5], to name but a few. The problem of BOSL is to estimate the source location from a set of noise-corrupted bearing measurements where its main challenge originates in the highly nonlinear nature of the angle observations with the true source position. Under the assumption of Gaussian measurement noise, many estimation approaches have been proposed to handle the nonlinearity, including the grid search method, the pseudolinear estimators [6,7], the iterative maximum likelihood methods [8] and the subspace approaches [9]. However, in the complex field environment, the sensor is vulnerable to external interference, enemy attack or node failure. The bearing measurements may suffer impulse noise [10–13] and those outlier data can degrade the localization performance dramatically. Therefore, it is necessary to develop new estimators that are robust to impulsive noise.

In fact, non-Gaussian models corresponding to impulsive noise have been extensively studied in the literature [13–15]. These studies have demonstrated that the Symmetric α-Stable ($S\alpha S$) distribution is more suitable to model impulsive noise than the Gaussian distribution. The family of stable distribution is a generalization of the Gaussian distribution under the stable law, including a large range of distributions with mutable values of impulsiveness (α), skewness (β) and dispersion (γ). In particular, two special cases of α-Stable distribution can be obtained by letting impulsiveness parameter α take values of 1 and 2. One is Cauchy distribution ($\alpha = 1$), and the other is Gaussian distribution ($\alpha = 2$).

In the present study, we focus on the problem of robust BOSL with impulsive noise modeled as $S\alpha S$ distribution. In this situation, the methods derived using L2-norm optimization exhibit unreliable estimates because they are not robust to outliers. Therefore, robust statistics need to be used to improve the localization performance. Under the assumption of $S\alpha S$ measurement noise, a Maximum Likelihood Estimator (MLE) was proposed in [16] to produce optimal estimates. Unfortunately, no closed-form expressions exist for the general likelihood function in the cases of $1 < \alpha < 2$ and therefore no explicit solution for MLE is possible. Although the $S\alpha S$ distribution does not have finite variance, it has finite Fractional Lower Order Moments (FLOM) [13], which can be calculated by its dispersion γ and its characteristic exponent α. Moreover, minimizing the FLOM of estimation errors is equivalent to minimizing the dispersion. It is well known as the minimum dispersion criterion, which minimizes the Lp-norm of the estimation residuals. Unlike L2-norm minimization, the least Lp-norm estimator ($1 < p < 2$) does not have a closed-form solution and consequently needs to be solved in an iterative manner [17].

The least Lp-norm estimator belongs to the M-estimator. The main idea of M-estimate is to replace the sum of squares of least squares residuals with cost functions more robust to impulsive noise, so as to reduce the sensitivity of estimators with respect to model errors. These functions include Huber [18], Bi-square [19], the negative log-likelihood of the Cauchy distribution [20], Wilcoxon [21], L1-norm [22,23], Lp-norm [24] and L∞-norm [25], etc. Ref. [19] presented a distributed robust localization algorithm based on energy information for sensor networks. The algorithm uses Bi-square function as the cost function of M-estimate. A distributed incremental least mean square algorithm based on Wilcoxon norm was proposed in [21] for parameter estimation of sensor networks. Ref. [26] proposed a robust structure total least-squares algorithm for BOSL by using the improved Danish weight function to suppress the impact of outlier data on the localization performance. In addition to the M-estimator, there are other algorithms that can handle outlier data, such as outlier detection [27], clustering [28] and game theoretic techniques [29]. The outlier detection method [27] is to detect suspected outlier data first, and separate it from the original data set. The clustering based techniques [28] can be used to classify normal and abnormal sensors. In game theory [29], the defense strategies could be employed to detect the outlier data and adaptive threshold selection can be also introduced.

Recently, an Iteratively Reweighted Pseudolinear Least-Squares (IRPLS) method was proposed in [14] to reduce biases attributed to the impulsive noise. However, IRPLS still suffers from a major bias problem caused by the correlation between system matrix and the noise vector. This bias can be reduced by exploiting the use of an Instrumental-Variable (IV) matrix [14] that is approximately uncorrelated with the noise vector. Inconsistent with the IV method, we present a robust total least-squares method using Lp-norm minimization that can reduce biases by minimizing the errors in the system matrix and the data vector simultaneously. We first formulate the problem of BOSL subjected to impulsive noise modeled as $S\alpha S$ distribution and review the pseudo-linear measurement model for BOSL. Next, we present the Total LP-norm Optimization (TLPO) method for BOSL to minimize the the errors in the system matrix and the data vector under the minimum dispersion criterion. Two algorithms, named Iterative Generalized Eigenvalue Decomposition (IGED) and Generalized Lagrange Multiplier (GLM), are designed to solve the TLPO problem. The performance advantage of the proposed algorithms is demonstrated by numerical simulations in terms of both bias and Root-Mean-Square-Error (RMSE) performance. The main contributions of the proposed method can be summarized as follows:

- Development of a new bias reduced estimator based on TLPO for BOSL when the measurement noise is modeled as $S\alpha S$ distribution;

- Development of two algorithms for TLPO optimization using the IGED approach and the GLM method, respectively.

The rest of this paper is organized below. Section 2 briefly summarizes the $S\alpha S$ distribution, presents the measurement model for BOSL and discusses the nonlinear least Lp-norm for BOSL. In Section 3, the pseudolinear estimator and the iteratively reweighted pseudolinear least-squares algorithm are reviewed. In Section 4, a new TLPO method is presented, and two iterative algorithms are developed to solve the TLPO problem. Section 5 presents the Cramér-Rao Lower Bound and Section 6 illustrates the bias and RMSE performance of PLE, TLS, LAR and TLAR using various numerical examples. Lastly, conclusions are drawn in Section 7.

2. Lp-Norm Optimization for Robust BOSL

2.1. Symmetric Alpha-Stable Distribution

The impulsive noise is more likely to exhibit outliers than normal noise. Studies [13] have shown that $S\alpha S$ distribution can model the impulsive noise better than Gaussian distribution due to the reason that the $S\alpha S$ densities have heavier tails than the Gaussian density. The characteristic function of $S\alpha S$ distribution is described as:

$$\chi(\tau) = \exp\{i\delta\tau - \gamma|\tau|^{\alpha}\} \tag{1}$$

where α $(0 < \alpha \leq 2)$ denotes the characteristic exponent, γ indicates the dispersion parameter and δ stands for the location parameter. The $S\alpha S$ distribution is completely determined by these three parameters. The value of α indicates the heaviness of the tails of the density function. A small positive value of α implies high impulsiveness, while a value of α close to 2 shows a type of Gaussian-like shape. The dispersion γ performs like the variance and δ controls the location of the density function.

A $S\alpha S$ distribution is called standard if $\delta = 0$, $\gamma = 1$. Let x be a random variable that follows the standard $S\alpha S$ distribution. By taking the inverse Fourier transform of its characteristic function, the density function of x is of the form:

$$f_{\alpha}(x) = \frac{1}{2\pi} \int_{-\infty}^{+\infty} \exp(-ix\tau - \gamma|\tau|^{\alpha})d\tau \tag{2}$$

It is known that a $S\alpha S$ distribution with characteristic exponent α only has finite moments for orders less than α, which are called the Fractional Lower Order Moments (FLOM). In particular, the FLOM of a standard $S\alpha S$ random variable x is given by

$$\mathbb{E}\{|x|^{p}\} = C(p,\alpha)\gamma^{p/\alpha}, \quad 0 < p < \alpha \tag{3}$$

where $\mathbb{E}\{\cdot\}$ denotes expectation operator,

$$C(p,\alpha) = \frac{2^{p+1}\Gamma\left(\frac{p+1}{2}\right)\Gamma\left(-\frac{p}{\alpha}\right)}{\alpha\sqrt{\pi}\Gamma\left(-\frac{p}{2}\right)} \tag{4}$$

is a constant that depends only on q and α and $\Gamma(a) = \int_{0}^{+\infty} y^{(a-1)} \exp(-y)dy$ is the Gamma function. It is worth mentioning that the linear space of a $S\alpha S$ process is a Banach space for $\alpha \in [1,2)$, and it is only a metric space for $\alpha \in (0,1)$ [30]. Therefore, the tools of Hilbert space are not applicable when one solves a linear estimation problem with the $S\alpha S$ process. In the present study, we only focus on the case of $\alpha \in (1,2)$.

2.2. Measurement Model

The problem of robust BOSL is depicted in Figure 1, where $t = [t_x, t_y]^T$ denotes the unknown target location vector, $r_m = [r_{x,m}, r_{y,m}]^T$ represents the mth sensor location vector and θ_m is the true bearing at sensor m. The nonlinear relationship between θ_m, t and r_m is given by [31]

$$\theta_m = \tan^{-1}(t_y - r_{y,m}, t_x - r_{x,m}), \quad \theta_m \in [0, 2\pi) \tag{5}$$

where $\tan^{-1}$ denotes the two-argument inverse tangent function and $m = 1, 2, \ldots, M$.

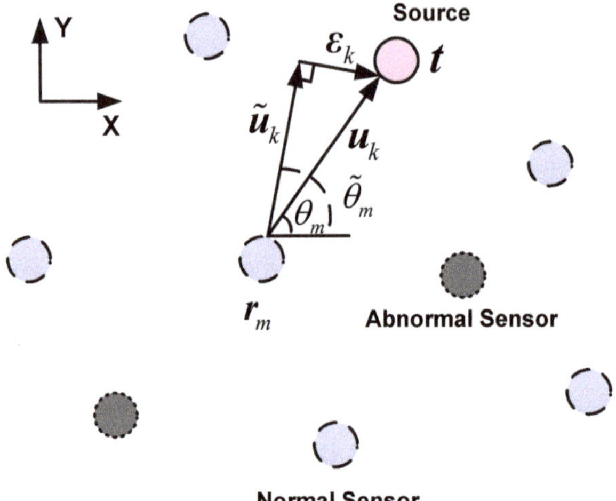

Figure 1. Illustration of the BOSL system.

The bearing measurement taken at sensor m is given by

$$\tilde{\theta}_m = \theta_m + n_m \tag{6}$$

where n_m follows independent zero mean $S\alpha S$ distribution with γ_m representing the dispersion parameter and α accounting for the characteristic exponent. The dispersion γ_m can vary with m. If the noise dispersion γ_m is known *a priori*, then the noise term n_m can be normalized. Let $e_m = \gamma_m^{-1/\alpha} n_m$ denote the normalized noise that has unit dispersion. The normalized measurement equation can be written as

$$\tilde{\phi}_m = \phi_m + e_m \tag{7}$$

where $\tilde{\phi}_m = \gamma_m^{-1/\alpha} \tilde{\theta}_m$ is the normalized bearing measurement and $\phi_m = \gamma_m^{-1/\alpha} \theta_m$ is the normalized true bearing. Stacking (7) in the vector form yields

$$\tilde{\phi} = \phi + e \tag{8}$$

where $\tilde{\phi} = [\tilde{\phi}_1, \tilde{\phi}_2, \ldots, \tilde{\phi}_M]^T$ denotes the normalized measurement vector, $\phi = [\phi_1, \phi_2, \ldots, \phi_M]^T$ represents the true bearing vector and $e = [e_1, e_2, \ldots, e_M]^T$ indicates the measurement noise vector.

2.3. Minimum Dispersion Criterion

The main difficulty for parameter estimation with the α-stable process is that all non-Gaussian α-stable distributions have infinite variance. As such, the traditional nonlinear least-squares techniques [32], which rely on the second order moments, are not suitable for solving BOSL problems with impulsive noise. Fortunately, we can use the minimum dispersion criterion instead of minimizing the variance. To do this, we first define the norm of the $S\alpha S$ random variable x as

$$\|x\|_\alpha = \gamma^{\frac{1}{\alpha}}, \quad \alpha \in [1, 2) \tag{9}$$

Thus, a suitable measure of the dispersion could be $\gamma = \|x\|_\alpha^\alpha$. Combining (3) and (9), one can easily obtain

$$\left(\frac{\mathbb{E}\{|x|^p\}}{C(p,\alpha)}\right)^{\frac{1}{p}} \equiv \|x\|_\alpha, \quad p \in (0,\alpha), \quad \alpha \in [1,2) \tag{10}$$

Therefore, we may estimate the source location by solving the following nonlinear Lp-norm optimization problem:

$$\hat{t} = \arg\min_t J(t), \quad J(t) = \|\tilde{\phi} - \phi\|_p^p = \sum_{m=1}^{M} |\tilde{\phi}_m - \phi_m(t)|^p \tag{11}$$

where $p \in (1,\alpha)$. The definition of the Lp-norm of the vector ζ is

$$\|\zeta\|_p = \left(\sum_k |\zeta_k|^p\right)^{1/p} \tag{12}$$

In the present study, we do not consider the case of $0 < p \le 1$, since the corresponding Lp-norm cost function is not differentiable. For non-normalized bearing measurements, the Lp-norm objective function in (11) becomes

$$J(t) = \sum_{m=1}^{M} |\gamma_m^{-1/\alpha}(\tilde{\theta}_m - \theta_m(t))|^p \tag{13}$$

The optimization problem listed in (11) can be solved numerically for a given 2D space of interest. To be more specific, we can perform a grid search over that 2D space. The global solution of maximum likelihood (ML) estimator is guaranteed for the given set of data as long as the spacing between grids is small enough. However, if the range of the parameter of interest is not confined to a relatively small interval or the dimension of unknown parameter vector is high, the grid search approach is computationally infeasible. Instead, one may resort to iterative optimization methods, such as gradient decent, and Gauss–Newton [33], etc. In particular, the Gauss–Newton algorithm is

$$\hat{t}^{(i+1)} = \hat{t}^{(i)} - \left(\nabla J(t)^T \nabla J(t)\right)^{-1} \nabla J(t)^T \left(\tilde{\phi} - \phi(\hat{t}^{(i)})\right) \tag{14}$$

where $\nabla J(t)$ denotes the Jacobian matrix,

$$\nabla J(t) = p \sum_{m=1}^{M} \frac{\partial \phi_m(t)}{\partial t} |e_m|^{p-1} \mathrm{sign}(e_m) \tag{15}$$

If $e_m > 0$, $\mathrm{sign}(e_m) = 1$ and -1 otherwise. The statistical properties of the nonlinear Lp-norm minimizer have been studied in the literature [24]. The theoretical covariance B is related to the value of p, and it is given by [24]:

$$B \approx \frac{C(2p-2,\alpha)}{(p-1)^2 C^2(p-2,\alpha)} (\nabla\phi^T \nabla\phi)^{-1} \tag{16}$$

where $\nabla\phi$ denotes the Jacobian matrix of ϕ with respect to t. By minimizing the scalar term of B, we obtain the optimal choice of p

$$p^o = \arg\min_p \frac{C(2p-2,\alpha)}{(p-1)^2 C^2(p-2,\alpha)} \tag{17}$$

where p^o denotes the optimal value of p.

3. Pseudolinear Lp-Norm Minimization

3.1. Pseudolinear Estimator

The robust BOSL problem is nontrivial because the measurement Equation in (7) is nonlinearly related to the unknown source location. An attractive solution is to set up a pseudolinear equation by lumping the nonlinearities into the noise term. As illustrated in Figure 1, an orthogonal vector sum between the measured angle vector and the true angle vector can be geometrically described by

$$u_m = \tilde{u}_m + \varepsilon_m = t - r_m \tag{18}$$

where u_m is the true angle vector between r_m and t, $\tilde{u}_m$ is the measured angle vector starting from r_m and produces the noisy bearing $\tilde{\theta}_m$ with respect to the horizontal direction and ε_m is the error vector. Let $\mu_m = [\cos\tilde{\theta}_m, \sin\tilde{\theta}_m]^T$ and $v_m = [\sin\tilde{\theta}_m, -\cos\tilde{\theta}_m]^T$ denote two orthogonal unit trigonometric vectors. Then, $\tilde{u}_m$ and ε_m are given by

$$\tilde{u}_m = \|u_m\| \cos n_m \cdot \mu_m \tag{19}$$

$$\varepsilon_m = \|u_m\| \sin n_m \cdot v_m \tag{20}$$

where $\|\cdot\|$ denotes Euclidean norm. Note from (19) that $\tilde{u}_m^T v_m = 0$. Substituting (19) and (20) into (18) and multiplying (18) with v_m^T yields

$$\xi_m = v_m^T t - v_m^T r_m \tag{21}$$

where $\xi_m = \|u_m\| \sin n_m$ is a nonlinear transformed measurement error. Collecting the pseudolinear equation errors as a vector $\xi = [\xi_1, \xi_2, \ldots, \xi_K]^T$, we obtain

$$\xi = At - b \tag{22}$$

where $A = [v_1, v_2, \ldots, v_M]^T$, $b = [v_1^T r_1, \ldots, v_M^T r_M]^T$ are the measurement matrix and vector, respectively.

The PLE requires that t be estimated by minimizing $\|At - b\|_2^2$ with respect to t in an L2-norm optimization sense, and the solution is

$$\hat{t}_{\text{PLE}} = (A^T A)^{-1} A^T b \tag{23}$$

The above PLE solution has a large bias because the abnormal nodes can produce significantly large bearing measurement errors. As a typical robust estimation algorithm, Lp-norm minimization has been widely used for parameter estimation using measurements corrupted by impulse noise. In this paper, we aim to develop robust estimators using Lp-norm optimization to reduce the bias as much as possible.

3.2. Iteratively Reweighted Pseudolinear Least-Squares Algorithm

This subsection reviews an Iteratively Reweighted Pseudolinear Least-Squares (IRPLS) algorithm, which has been proposed in [14]. The IRPLS algorithm is derived from the following Lp-norm optimization problem:

$$\min_t \|G(At - b)\|_p^p = \min_t \sum_{m=1}^{M} |g_m((At)_m - b_m)|^p \tag{24}$$

where $G = \text{diag}(g_1, g_2, \ldots, g_M)$, $g_m = (\gamma_m^{1/\alpha} \|t - r_m\|)^{-1}$, $(At)_m$ and b_m are the m-th element of At and b, respectively. When the measurement noise is small, we have $\sin n_m \approx n_m$. Then, $|g_m((At)_m - b_m)|^p$ can be expressed approximately as

$$|g_m((At)_m - b_m)|^p \approx \gamma_m^{-p/\alpha} \|t - r_m\|^{-2} |\tilde{\theta}_m - \theta_m|^{(p-2)} |(Ap)_m - b_m|^2 \tag{25}$$

The above interpretation suggests an iterative reweighted pseudolinear least-squares with the weight matrix for the i-th iteration given by

$$\mathbf{W}^{(i)} = \text{diag}\left(\gamma_1^{-\frac{p}{2\alpha}}\|\hat{\mathbf{t}}^{(i-1)} - \mathbf{r}_1\|^{-1}\left|\tilde{\theta}_1 - \hat{\theta}_1^{(i-1)}\right|^{\frac{p-2}{2}}, \ldots, \gamma_M^{-\frac{p}{2\alpha}}\|\hat{\mathbf{t}}^{(i-1)} - \mathbf{r}_M\|^{-1}\left|\tilde{\theta}_M - \hat{\theta}_M^{(i-1)}\right|^{\frac{p-2}{2}}\right) \tag{26}$$

where $\hat{\mathbf{t}}^{(i-1)}$ and $\hat{\theta}_m^{(i-1)}$ denote the estimated source location and the mth bearing obtained from the previous iteration $i - 1$. Using the weight matrix, the weight error for the k-th iteration is given by

$$\left(\mathbf{W}^{(i)}\boldsymbol{\xi}^{(i)}\right)^T\mathbf{W}^{(i)}\boldsymbol{\xi}^{(i)} = \sum_{m=1}^{M} |g_m((A\mathbf{t})_m - b_m)|^p \tag{27}$$

The homotopy method can be applied in the iterations by starting with a value for p equals to 2 (the weight matrix is unit matrix) and decreasing it each iteration until it reaches the designated value. Thus,

$$p^{(i)} = \max(p, \kappa p^{(i-1)}) \tag{28}$$

where $\kappa < 1$ is the homotopy parameter. The source location can be estimated by performing least-squares estimation:

$$\hat{\mathbf{t}}^{(i)} = \left((\mathbf{W}^{(i)}A)^T\mathbf{W}^{(i)}A\right)^{-1}(\mathbf{W}^{(i)}A)^T\mathbf{W}^{(i)}\mathbf{b} \tag{29}$$

The IRPLS algorithm is stopped when $\|\hat{\mathbf{t}}^{(i-1)} - \hat{\mathbf{t}}^{(i)}\| \leq \epsilon$, where ϵ is a tolerance parameter.

Finally, the whole process of the IRPLS algorithm is summarized in Algorithm 1.

Algorithm 1 The IRPLS algorithm.

1: **Initialization:** Set $\hat{\mathbf{t}}^{(0)} = \hat{\mathbf{t}}_{\text{PLE}}$.
2: **for** $i = 0; i + + $ **do**
3:   Compute the weight matrix using (31).
4:   Determine the value of p for each iteration using (28).
5:   Perform weighted least squares estimation using (29).
6:   if $\|\hat{\mathbf{t}}^{(i-1)} - \hat{\mathbf{t}}^{(i)}\| \leq \epsilon$, we obtain the final solution, and stop the iteration. Otherwise $i = i + 1$, go to step 3.
7: **end for**

The Lp-norm criterion ensures that the cost function in (24) gives less weight to large deviations, and therefore reduces the bias formed by pseudolinear errors with large residuals. However, the IRPLS algorithm implicitly assumes that only $\mathbf{b}$ is subjected to errors. This is not the case, since the system matrix A is corrupted with measurement noises as well. The correlation between A and $\mathbf{b}$ causes the IRPLS estimator to be inconsistent. In this subsection, we analyze such bias attributed to the correlation between A and $\mathbf{b}$.

Let $\hat{\mathbf{t}}^*$ denote the final solution of the IRPLS algorithm. This solution satisfies

$$\hat{\mathbf{t}}^* = \left(A^T\mathbf{W}(\hat{\mathbf{t}}^*)A\right)^{-1}A^T\mathbf{W}(\hat{\mathbf{t}}^*)\mathbf{b} \tag{30}$$

where

$$\mathbf{W}(\hat{\mathbf{t}}^*) = \text{diag}\left(\gamma_1^{-\frac{p}{\alpha}}\|\hat{\mathbf{t}}^* - \mathbf{r}_1\|^{-2}|\tilde{\theta}_1 - \hat{\theta}_1^*|^{(p-2)}, \ldots, \gamma_M^{-\frac{p}{\alpha}}\|\hat{\mathbf{t}}^* - \mathbf{r}_M\|^{-2}|\tilde{\theta}_M - \hat{\theta}_M^*|^{(p-2)}\right) \tag{31}$$

and $\hat{\theta}_m^*$ denotes the mth bearing obtained from (5) by using $\hat{\mathbf{t}}^*$.

According to (22) and (30), we can get

$$\Delta t = \hat{t}^* - t = -\left(A^T \mathcal{W}(\hat{t}^*)A\right)^{-1} A^T \mathcal{W}(\hat{t}^*)\xi \tag{32}$$

The analytical bias of IRPLS is given by [14]

$$\mathbb{E}\{\Delta t\} = -\frac{1}{p-1}\mathbb{E}\left\{\frac{A^T \mathcal{W}(t)A}{M}\right\}^{-1}\mathbb{E}\left\{\frac{A^T \mathcal{W}(t)\xi}{M}\right\} \tag{33}$$

where the weighting matrix $\mathcal{W}(t)$ is computed from (31) using true source location parameter t.

4. Total Lp-Norm Optimization

In this section, we present a method of TLPO for BOSL. The IRPLS algorithm only considers the deviation of the system vector b. But in fact, the system matrix A also has a measurement residual. The IRPLS algorithm will inevitably cause a large bias, because the system matrix A and the system vector b are statistically correlated.

4.1. Method Description

In order to improve the performance of the IRPLS algorithm and reduce the bias caused by the correlation between A and b, we develop the TLPO method in this subsection. Unlike the IRPLS method, the TLPO method uses the correction matrix ΔA and the correction vector Δb to compensate the system matrix A and system vector b, respectively. The normalized equation can be written as

$$G(A + \Delta A)t = G(b + \Delta b) \tag{34}$$

where ΔA is the perturbation matrix of A and Δb is the perturbation vector of b. Let $K = G[A, b]$ and $v = [t, -1]^T$ be the augmented matrix and vector, and Equation (34) can be rewritten as

$$(K + \Omega)v = 0 \tag{35}$$

where $\Omega = G[\Delta A, \Delta b]$ is the error matrix.

In order to reduce the bias of IRPLS, we use the TLPO approach to minimize the perturbation matrix ΔA and perturbation vector Δb simultaneously, and the TLPO problem for BOTL can be formulated as

$$\min_{E,t} \|\Omega\|_p, \quad \text{s. t. } (K + \Omega)v = 0 \tag{36}$$

for $1 < p < 2$. Note that if $p = 1$, (36) becomes the total least absolute residual method, and if $p = 2$, it is the well-known TLS method. To solve the TLPO problem conveniently, it is necessary to develop an equivalent form of (36). The following proposition holds.

Proposition 1. *Given the TLPO problem defined in* (36), *the estimation of t from the minimization of $\|\Omega\|_p$ subject to $(K + \Omega)v = 0$ is equivalent to*

$$\min_{v} \|Kv\|_p, \quad \text{s. t. } \|v\|_q - 1 = 0 \tag{37}$$

where $v = [t, -1]^T$, $\frac{1}{p} + \frac{1}{q} = 1$ and $1 < p < \alpha$.

Proof. From the constraint $(K + \Omega)v = 0$, we have $\|Kv\|_p = \|\Omega v\|_p$. Let ω_m^T denote the mth row of Ω. $\|\Omega v\|_p^p$ becomes

$$\|\Omega v\|_p^p = \sum_{m=1}^{M} \left| \omega_m^T v \right|^p$$

$$\leq \sum_{m=1}^{M} \|\omega_m\|_p^p \|v\|_q^p \tag{38}$$

for p and q satisfy the equation $\frac{1}{p} + \frac{1}{q} = 1$ and $1 < p < \alpha$. To derive (38), we used the properties of hölder's inequality [34]. If the Lq-norm of v satisfies $\|v\|_q = 1$, then the inequality (38) becomes $\|\Omega v\|_p \leq \|\Omega\|_p$. Therefore, we can conclude that the minimization of $\|\Omega\|_p$ subject to $(K + \Omega)v = 0$ is equivalent to (37). $\quad\square$

Proposition 1 provides a facilitated way to solve the TLPO problem. In the following two subsections, we concentrate on deriving the solution on this particular problem and developing the IGED and GLM algorithms.

4.2. The IGED Algorithm

The optimization problem in Proposition 1 is equivalent to

$$\min_{v} \|Kv\|_p^p, \quad \text{s. t. } \|v\|_q^q - 1 = 0 \tag{39}$$

To solve (39), we first transform it into an unconstrained optimization problem by using the Lagrange multiplier method. The appropriate Lagrangian function of (39) can be written as

$$\mathcal{L}(v, \lambda) = \|Kv\|_p^p + \lambda(1 - \|v\|_q^q)$$
$$= v^T K^T DKv + \lambda(1 - v^T Cv) \tag{40}$$

where $D = \text{diag}(|(Kv)_m|^{p-2})$, $(Kv)_m$ represents the mth element of vector Kv for $m = 1, 2, \ldots, M$, $C = \text{diag}(|v_1^{q-2}|, |v_2^{q-2}|, |v_3^{q-2}|)$ and λ is the Lagrangian multiplier. Taking the partial derivative of $\mathcal{L}(v, \lambda)$ with respect to v and setting it to zero yields

$$K^T DKv = \lambda Cv \tag{41}$$

After multiplying both sides of (41) with v^T, it can be found that only $\lambda = v^T K^T DKv$ is the cost to be minimized. So, the optimal solution v^* is the eigenvector of the smallest generalized eigenvalue of $(K^T DK, C)$.

Figure 2 gives a flowchart to depict the whole process of the IGED algorithm. The algorithm starts from the initialization procedure, where v_0 is set as $[\hat{t}_{\text{PLE}}^T, -1]^T$ and $\hat{t}_{\text{PLE}}$ is obtained from (23). Given p and q, matrices C and D can be computed, respectively. After performing generalized eigenvalue decomposition from the pair $(K^T DK, C)$, the solution of v in (41) is the generalized eigenvector of $(K^T DK, C)$ that gives the minimum generalized eigenvalue. Since IGED is an iterative algorithm, it requires the convergence check. Let λ_i and λ_{i+1} be the minimum generalized eigenvalues at the ith and $i + 1$th iterations, respectively. If there exists ε such that

$$\left| \frac{\lambda_{i+1} - \lambda_i}{\lambda_{i+1}} \right| \leq \varepsilon \tag{42}$$

then we would get the final solution v^* and the estimate for source location is

$$\hat{t} = \frac{v^*(1:2)}{v^*(3)}. \tag{43}$$

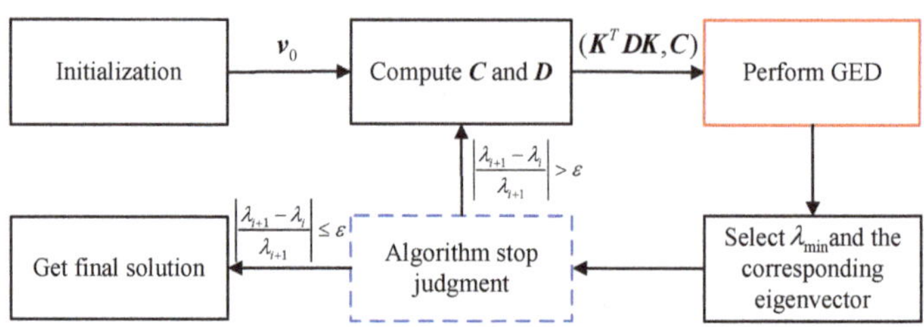

Figure 2. The flowchart of the IGED algorithm.

In summary, the IGED Algorithm 2 can be implemented by the following procedures.

Algorithm 2 The IGED algorithm.

1: **Initialization:** $v_0 = [\hat{t}_{\mathrm{PLE}}^T, -1]^T$.
2: **for** $i = 0; i + +$ **do**
3:     Compute C and D.
4:     Perform generalized eigenvalue decomposition of the pair $(K^T DK, C)$.
5:     Select the smallest generalized eigenvalue and its corresponding generalized eigen-
    vector.
6:     Set λ_i and λ_{i+1} as the minimum generalized eigenvalues for iterations i and $i + 1$.
    If $|(\lambda_{i+1} - \lambda_i)/\lambda_{i+1}| > \varepsilon$, go to step 3. Otherwise, obtain the final solution v^* and
    stop.
7:     Get the source location estimate $\hat{t}$ using (43).
8: **end for**

4.3. The GLM Algorithm

The GLM method [35] is widely used for constrained optimization. Its basic principle is to add a penalty term to the Lagrangian function to form an augmented Lagrangian function, which can impose a larger penalty on infeasible points. Thus, the constrained optimization problem (37) would be transformed into a new unconstrained optimization problem that can be solved efficiently.

To solve (37) with the GLM method, we formulate an augmented Lagrangian function as follows:

$$\mathcal{L}(v, \lambda, s) = g(v) + \frac{s}{2}\|h(v)\|_2^2 - \lambda h(v) \tag{44}$$

where $g(v) = \|Kv\|_p$ and $h(v) = 1 - \|v\|_q$. Compared with the Lagrangian function, the GLM cost function has an additional term $s/2\|h(v)\|_2^2$. This item is a punishment for violation of constraint $h(v) = 0$. The punishment parameter s determines the degree of punishment, and it is generally sufficiently large. λ is a positive scalar, and the term $\lambda h(v)$ is to ensure that the optimal solution is a strict local minimum point of $F(v) = g(v) + \frac{s}{2}\|h(v)\|_2^2$ under the condition of obeying the constraint $h(v) = 0$.

Figure 3 shows the flow diagram of the GLM algorithm. For the initialization step, we set $v_0 = [\hat{t}_{\mathrm{PLE}}^T, -1]^T$. Given initial multiplier λ, penalty factor s, admissible error ε, scaling parameter $a > 1$ and parameter $\rho \in (0, 1)$, the problem (44) can be solved by using the existing unconstrained optimization methods, such as Newton, Interior point method, simplex method, etc. We use fminsearch function in MATLAB to estimate v from (44). Then, we check whether the termination criteria are satisfied. If $\|h(v_i)\| \le \delta$, the iteration terminates and v_i is the near optimal solution of (37). Otherwise, go to the next step which is to control the convergence speed. If there exists $\rho \in (0, 1)$ such that

$$\frac{\|h(v_i)\|}{\|h(v_{i-1})\|} \ge \rho \tag{45}$$

we can increase the penalty factor s by setting $s = a \cdot s$ $(a > 1)$. If not, then we leave the value of s unchanged. The penalty factor is used for multiplier update $\lambda = \lambda + s\|h(v_i)\|$. The updated λ and s are used to solve the unconstrained optimization problem (44) again.

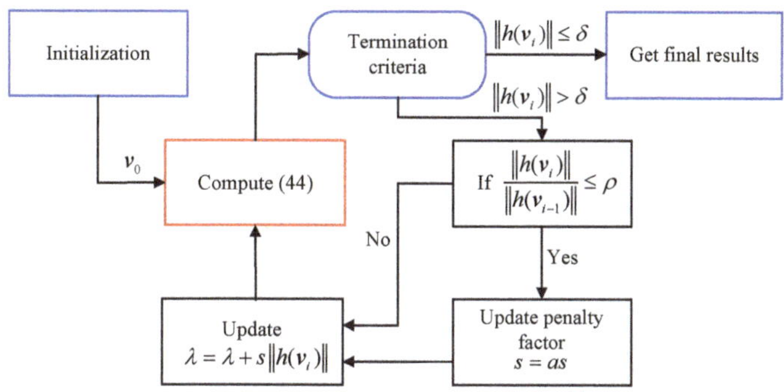

Figure 3. The flow diagram of the GLM algorithm.

Finally, the implementation process of the GLM algorithm is summarized in Algorithm 3.

Algorithm 3 The GLM algorithm.

1: **Initialization:** $v_0 = [\hat{t}_{\text{PLE}}^T, -1]^T, \delta, \lambda, s, a > 1, \rho \in (0,1)$.
2: **for** $i = 0; i++$ **do**
3:　　Using v_i as the initial point, find the minimum value of $\mathcal{L}(v, \lambda, s)$.
4:　　If $\|h(v_i)\| \le \delta$ holds, stop the iterations and get the final solution v^*, and go to step 7. Otherwise, go to step 5.
5:　　If $\|h(v_i)\| / \|h(v_{i-1})\| \ge \rho$, update the penalty coefficient $s = a \cdot s$, and go to step 6. Otherwise, go to step 6 directly.
6:　　Update $\lambda = \lambda - \rho\|h(v_i)\|$ and go to step 3.
7:　　Find the target position $\hat{t} = v^*(1:2)/v^*(3)$.
8: **end for**

4.4. Computational Complexity Analysis

In this subsection, we analyze the computational complexities of the proposed algorithms and compare with those of PLE and IRPLS. We only consider the asymptotic computational cost. For each algorithm, we separate the complexity calculation to small steps, e.g., multiplying two matrices of size $n \times m$ and $m \times p$ costs $\mathcal{O}(nmp)$. Let L_1, L_2 and L_3 denote the number of iterations with regarding to IRPLS, IGED and GLM, respectively. The iterative implementation of IRPLS, IGED and GLM, however, could be very sensitive to initialization. In this paper, we use the solution of PLE for initialization.

The computational costs are listed in Table 1. Table 1 summarizes the computational complexities of the PLE and IRPLS, IGED and GLM algorithms. PLE requires the least amounts of computation. IRPLS and IGED exhibit the similar computational costs. For the GLM algorithm, we employ the Broyden–Fletcher–Goldfarb–Shanno (BFGS) [33] algorithm to solve the unconstrained problem presented in (44). BFGS is an iterative method for solving unconstrained nonlinear optimization problems by providing an approximation to the Hessian matrix. We assume that the number of iterations for BFGS is N_i for the ith iteration of GLM. For each iteration, we assume the order of computational complexity of BFGS is $O_i(M)$. The GLM algorithm needs to estimate v which is computationally demanding. The estimation procedure needs to repeat L_3 times since it is iterative.

Table 1. Computational complexity.

Method	Operation	Cost
PLE	$A^T A, A^T b$	$\mathcal{O}(4M) + \mathcal{O}(2M)$
	$(A^T A)^{-1}$	$\mathcal{O}(2^3)$
	$(A^T A)^{-1} A^T b$	$\mathcal{O}(4M)$
IRPLS	Compute $W^{(i)} A$ L_1 times	$\mathcal{O}(2L_1 M)$
	Compute $W^{(i)} b$ L_1 times	$\mathcal{O}(L_1 M)$
	Compute (29) L_1 times	$\mathcal{O}(L_1(10M + 8))$
IGED	Compute DK L_2 times	$\mathcal{O}(3L_2 M)$
	Perform GED L_2 times	$\mathcal{O}(3^3 L_2)$
GLM	L_3 evaluations of BFGS	$\mathcal{O}(\sum_{i=1}^{L_3} N_i O_i(M))$

5. Performance Bound

In this section, we present the Cramér-Rao Lower Bound (CRLB) for robust BOSL when the measurement noise is modeled as $S\alpha S$ distribution. We then discuss how the CRLB is calculated. The CRLB for any unbiased estimator of t is derived by the inverse of the Fisher Information Matrix (FIM) [36], which is computed as

$$\text{CRLB}(t) = J^{-1} \tag{46}$$

where J is the FIM given by the 2×2 matrix

$$J = \mathbb{E}\left\{ \begin{bmatrix} \left(\frac{\partial f(\tilde{\phi}|t)}{\partial t_x}\right)^2 & \frac{\partial f(\tilde{\phi}|t)}{\partial t_x}\frac{\partial f(\tilde{\phi}|t)}{\partial t_y} \\ \frac{\partial f(\tilde{\phi}|t)}{\partial t_x}\frac{\partial f(\tilde{\phi}|t)}{\partial t_y} & \left(\frac{\partial f(\tilde{\phi}|t)}{\partial t_y}\right)^2 \end{bmatrix} \right\} \tag{47}$$

where $f(\tilde{\phi}|t)$ denotes the Probability Density Function (PDF) of $\tilde{\phi}$. Since $e_1, e_2, \ldots, e_M$ have the same standard dispersion, the PDF of $\tilde{\phi}_m$ is

$$f_\alpha(e) = f_\alpha(\tilde{\phi}_m - \phi_m) = \frac{1}{2\pi}\int_{-\infty}^{+\infty} \exp(-i(\tilde{\phi}_m - \phi_m)\tau - \gamma|\tau|^\alpha)d\tau \tag{48}$$

According to [37], the FIM J for the parameters t_x and t_y is given by

$$J = \kappa_\alpha \sum_{m=1}^{M} \begin{bmatrix} \left(\frac{\partial \phi_m}{\partial t_x}\right)^2 & \frac{\partial \phi_m}{\partial t_x}\frac{\partial \phi_m}{\partial t_y} \\ \frac{\partial \phi_m}{\partial t_x}\frac{\partial \phi_m}{\partial t_y} & \left(\frac{\partial \phi_m}{\partial t_y}\right)^2 \end{bmatrix} \tag{49}$$

where κ_α is the Fisher information for the location of $f_\alpha(e)$,

$$\kappa_\alpha = \int_{-\infty}^{+\infty} \frac{(f'_\alpha(e))^2}{f_\alpha(e)} de \tag{50}$$

and the expressions of $\frac{\partial \phi_m}{\partial t_x}$ and $\frac{\partial \phi_m}{\partial t_x}$ are given by

$$\frac{\partial \phi_m}{\partial t_x} = -\frac{t_y - r_{y,m}}{\|t - r_m\|}, \quad \frac{\partial \phi_m}{\partial t_y} = \frac{t_x - r_{x,m}}{\|t - r_m\|} \tag{51}$$

Note that the value of κ_α depends on the parameter α. If $\alpha = 1$, $f_\alpha(e)$ follows a Cauchy PDF and κ_α equals to $\frac{3}{5}$ for $\gamma = 1$. In particular, when $\alpha = 2$, $f_\alpha(e)$ is Gaussian with variance 2γ and $\kappa_\alpha = \gamma$. Therefore, the CRLB of t in the case of $\alpha = 2$ is consistent with the CRLB result given in [8].

However, when $1 < \alpha < 2$, $f_\alpha(e)$ has no closed-form expressions. Hence, computing $f_\alpha(e)$ involves numerically evaluating the integrals in (48). We use the MATLAB function *STBLPDF* to compute the pdf of the $S\alpha S$ distribution. The derivative of $f_\alpha(e)$ is also calculated by using numerical differentiation methods.

6. Simulations

In this section, we present three simulation examples to illustrate the performance of the proposed IGED and GLM algorithms and compare them with the PLE and IRPLS algorithm, as well as the CRLB. The localization performance is characterized with the bias norm and the root-mean-square-error (RMSE). The RMSE for source localization is defined as

$$\text{RMSE} = \left(\frac{1}{N} \sum_{n=1}^{N} \|\hat{t}_n - t\|^2 \right)^{1/2} \tag{52}$$

where N represents the total number of Monte Carlo runs, and $\hat{t}_n$ denotes the source location estimate at the nth Monte Carlo run. The expression of the bias norm is given by

$$\text{BIAS} = \left\| \frac{1}{N} \sum_{n=1}^{N} (\hat{t}_n - t) \right\| \tag{53}$$

The RMSE given in (52) is bounded by the square root of the trace of the Cramér-Rao Lower Bound matrix

$$\text{RMSE} \geq \sqrt{tr(\text{CRLB}(t))} \tag{54}$$

where $tr(\cdot)$ denotes the trace of a matrix.

In the simulations, we examine the localization accuracy versus three kinds of parameters: (1) noise dispersion; (2) number of sensors; and (3) noise impulsiveness. All the sensors are uniformly placed in a 100×100 m^2 plane centered at $(50, 50)$ m. The source is located at $(100, 100)$ m. The IRPLS algorithm and the proposed IGED and GLM algorithms are iterative and they are all initialized by the PLE derived from (23) for ensuring a fair comparison. The termination parameters ϵ, ε and δ are set as $\epsilon = 10^{-5}$, $\varepsilon = 10^{-10}$ and $\delta = 10^{-5}$, respectively. For IRPLS, IGED and GLM, the maximum iteration is fixed at 200.

6.1. Various Levels of Noise Dispersion

In this example, we illustrate the RMSE and bias performance of PLE, IRPLS, IGED and GLM versus noise dispersion. For convenience, we use the generalized signal-to-noise ratio (GSNR) to characterize different noise levels. The GSNR is inversely proportional to the noise dispersion γ and is calculated as $1/\gamma$ after normalized. The range of $\gamma^{1/\alpha}$ is from $2\pi/180$ radian to $6\pi/180$ radian. The characteristic exponent α is set to 1.5 and the corresponding optimum value of p is 1.225.

Figure 4 plots the RMSE and bias curves of the PLE, IRPLS, IGED and GLM algorithms together with the CRLB. For RMSE curves shown in Figure 4a, the green line with Asterisk is the RMSE value for the PLE algorithm, the blue line with Point is for IRPLS, the black line with Cross is for IGED, the red line with Square is for GLM and the blue dash line is the CRLB. The PLE method can not give accurate target location estimates in the presence of impulsive noise. The large bias norm formed by the outlier data in turn leads to a poor performance for the PLE. The least-norm estimators, including IRPLS, IGED and GLM, yield much better localization performance. However, the IRPLS method still has a large estimation bias because it cannot overcome the bias attributed to the correlation between the system matrix A and the data vector b. On the other hand, the IGED and GLM algorithms utilizes the total Lp-norm optimization technique that can minimize the errors in A and b simultaneously. Therefore, the IGED and GLM algorithms are capable of outperforming the IRPLS in much lower RMSEs and biases.

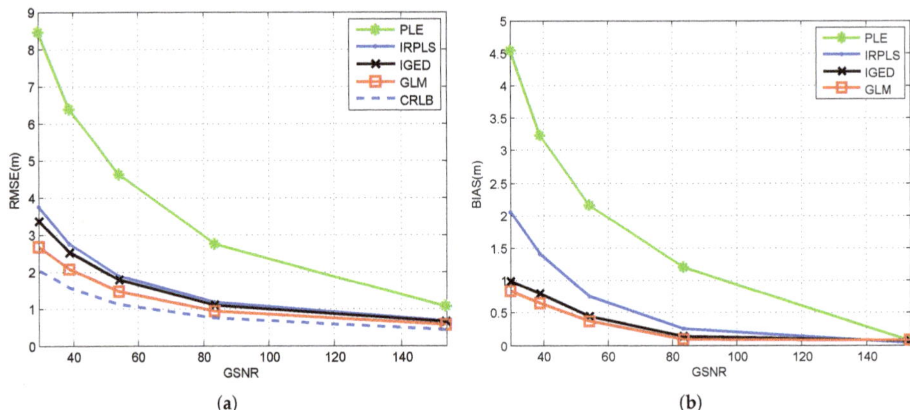

Figure 4. RMSE and bias performance comparison of PLE, IRPLS, IGED and GLM estimates with various GSNRs. (**a**) RMSE results. (**b**) Bias norm results.

Beyond the RMSE and bias performance, we need to explore the number of iterations and the computation time of IRPLS, IGED and GLM. The results are depicted in Figure 5. The algorithms run on a laptop with CPU i5-7200U @ 2.5 GHz and RAM 8 GB. The version of software is MATLAB 2017a. Collectively, the average number of iterations decreases when the GSNR increases. The GLM appears to have the least number of iterations. Its average number of iterations reduces from 1.05 to 1.01 when the GSNR ranges from 29.51 to 153.3. However, its computation time for single iteration is high (i.e., 0.1343 s), because the GLM algorithm has the steps of unconstrained optimization. On the contrary, the IGED algorithm has the least computing time since its computational complexity mainly lies in generalized eigenvalue decomposition. The IRPLS algorithm has high computing time due to the large number of iterations (over 30 iterations).

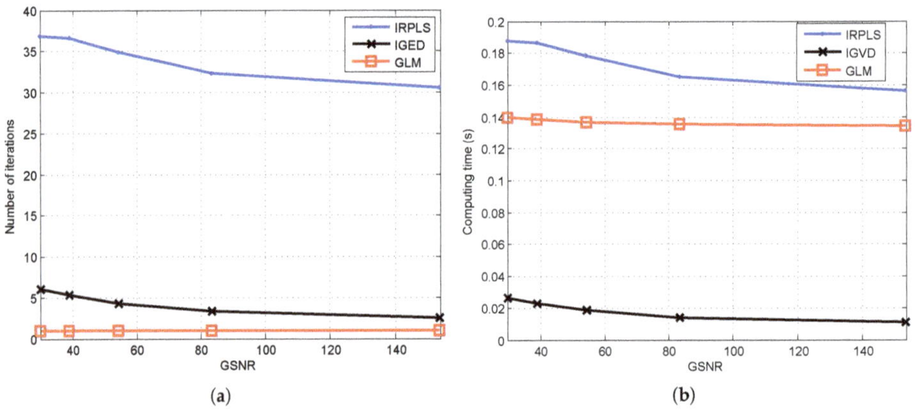

Figure 5. Number of iterations and computing time versus GSNR for the IRPLS, IGED and GLM algorithms. (**a**) Number of iterations. (**b**) Computing time.

6.2. Different Number of Sensors

In this subsection, we compare the RMSE and bias performance of the PLE, IRPLS, IGED and GLM algorithms over the number of sensors M ranging from 10 to 30, five at a time, when $\gamma^{1/\alpha}$ is kept at $4\pi/180$ radian. The other parameters remain the same. The simulation results are shown in Figures 6 and 7.

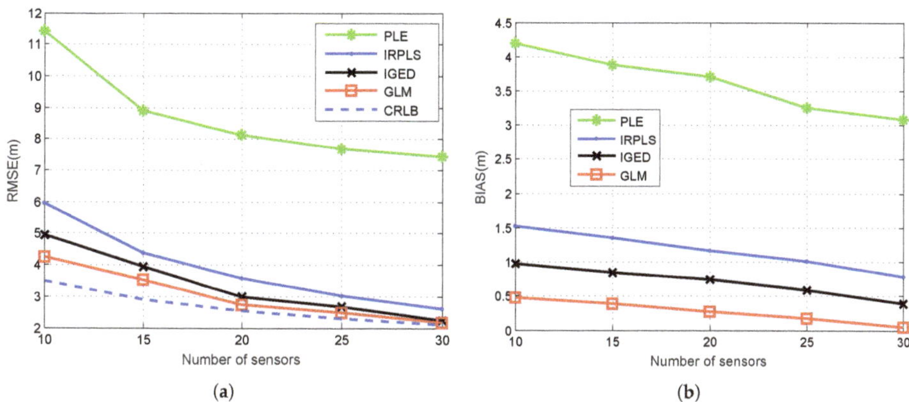

Figure 6. RMSE and bias performance comparison of PLE, IRPLS, IGED and GLM estimates with different number of sensors. (**a**) RMSE results. (**b**) Bias norm results.

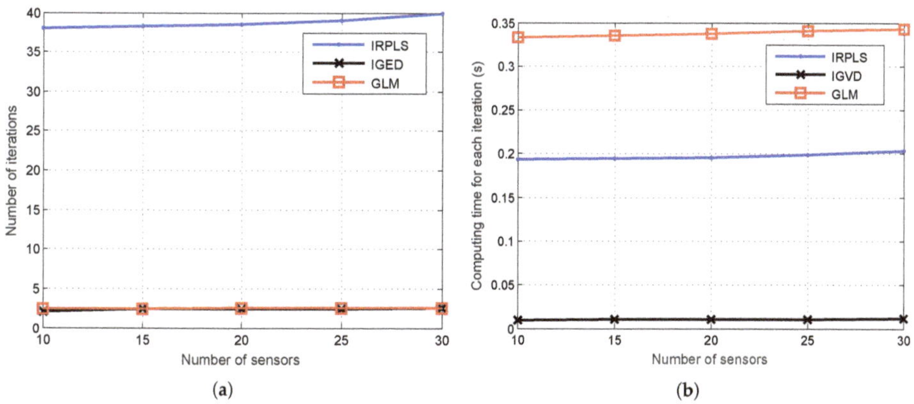

Figure 7. Number of iterations and computing time versus number of sensors for the IRPLS, IGED and GLM algorithms. (**a**) Number of iterations. (**b**) Computing time.

From Figure 6, we observe that the GLM algorithm has better RMSE and bias performance than that of PLE, IRPLS and IGED. When $M \geq 20$, the RMSE of GLM is close to the CRLB. The IGED algorithm slightly deviates from the CRLB, because the algorithm has not fully converged. Moreover, we also observe from Figure 7 that the average iterations is kept at 2.5 and is almost the same as that of the GLM algorithm. Meanwhile, the bias of IGED is higher than that of GLM. Setting more iterations would help the IGED algorithm achieve better performance. As expected, the PLE exhibits unreliable results since it is not robust to the impulsive noise. The bias of IRPLE does not vanish as the number of sensors increases. This phenomenon is also validated in (33). In terms of computational complexity, the total computing time for GLM is about 0.3378 s when the number of sensors is fixed at 20. The IGED has the least amounts of computation and it only requires 0.01032 s at $N = 20$.

6.3. Various Values of Noise Impulsiveness

To further verify the effectiveness of the proposed algorithms, we examine the performance of the algorithms for different levels of noise impulsiveness. In the following figures drawing the simulation results, the value of the noise impulsiveness level α deviates from

1.9 to 1.1 and the corresponding optimum values of p are set as 1.546, 1.430, 1.348, 1.282, 1.225, 1.174, 1.127, 1.083, 1.041 from (17). In this example, $\gamma^{1/\alpha}$ is fixed at $4\pi/180$ radian. The other parameters remain unchanged.

Figure 8 depicts the RMSE and bias performance with respect to various α. As shown in Figure 8, the RMSE and the bias of PLE is much higher than that of IRPLS, IGED and GLM, which is caused by the impulsive noise. As α decreases, the influence of the noise impulsiveness becomes more significant. The level of noise impulsiveness also affects the bias and RMSE performance of IRPLS. The GLM algorithm has relatively small bias and its RMSE is very close to the CRLB when $\alpha \geq 1.4$. The IGED has the comparable RMSE performance with the GLM. However, the bias of IGED slightly deviates from that of GLM as α decreases. From Figure 9, we observe that the number of iterations of IRPLS, GLM and IGED decrease as α increases, i.e., the number of average iterations reduces from 38.8 to 29.5 for IRPLS, from 4.4 to 1 for GLM and from 2.9 to 2.2 for IGED. The computation time also decreases when α increases. Again, the computation time of GLM is much higher than that of IGED.

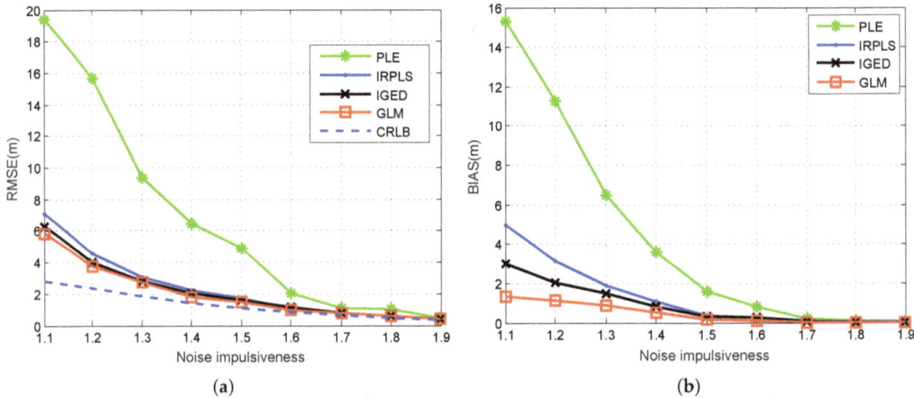

Figure 8. RMSE and bias performance comparison of PLE, IRPLS, IGED and GLM estimates with various noise impulsiveness. (**a**) RMSE results. (**b**) Bias norm results.

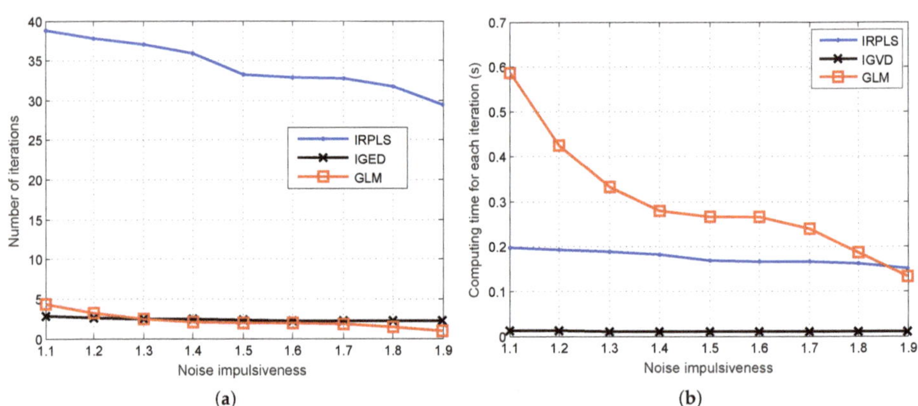

Figure 9. Number of iterations and computing time versus noise impulsiveness for the IRPLS, IGED and GLM algorithms. (**a**) Number of iterations. (**b**) Computing time.

6.4. Different Number of Iterations

In order to further verify the influence of the number of iterations on the proposed algorithm, we tested the performance of the algorithm for the increasing number of iterations. The simulation results are shown in Figure 10, where the number of iterations is set 50 for IRPLS and IGED, and 6 for GLM. The RMSE and bias are recorded every 10 iterations for IRPLS and IGED. The RMSE and bias performance of GLM is plotted for each iteration. In this example, $\gamma^{1/\alpha}$ is fixed at $4\pi/180$ radian. The number of sensors is kept at 20. The characteristic exponent α is set to 1.5 and the corresponding optimum value of p is 1.225.

Figure 10 describes the RMSE and bias performance of the IRPLS, IGED and GLM algorithms versus different iterations. As shown in Figure 10a, the RMSE result of IRPLS remains at 2.1 m after 30 iterations. We observe that the bias of the IRPLS algorithm decreases as the number of iterations increasing, as shown in Figure 10b. However, the bias does not vanish. It still has 0.41 m after 50 iterations. The RMSE value of IGED keeps at 1.8 m after 30 iterations and the bias of IGED reduces to 0.12 m using about 40 iterations. Differing from the IRPLS and IGED algorithms, the RMSE result of GLM reaches at 1.8 m using only four iterations. After 5 iteratios, the bias of GLM reduces to 0.12 m. From Figure 10, we can observe that the RMSE and bias performance of IGED is almost the same as that of GLM if IGED runs a sufficient number of iterations.

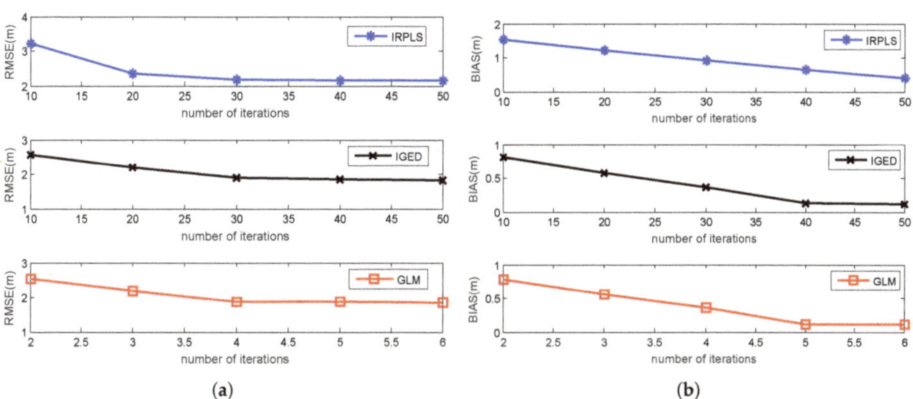

Figure 10. RMSE and bias performance versus the number of iterations for IRPLS, IGED and GLM. (**a**) RMSE results. (**b**) Bias norm results.

6.5. Scalability Evaluation

We analyze the scalability of the IRPLS, IGED and GLM algorithms in this subsection for increasing number of sensors, from 20 to 100, and decreasing levels of noise impulsiveness, from $\alpha = 1.1$ to $\alpha = 1.9$. To examine the scalability of the iterative algorithms, the IRPLS, IGED and GLM algorithms share the same stopping condition. Let $\hat{t}_i$ denote estimation result for the ith iteration. The criterion for stopping these iterative algorithms is given by $\|\hat{t}_i - \hat{t}_{i-1}\| < 10^{-5}$.

Figure 11 compares the number of iterations, computing time for each iteration, RMSE and bias performance of the IRPLS, IGED and GLM algorithms versus different numbers of sensors. The other parameters remain the same as Section 6.4. It can be observed from Figure 11 that the number of iterations of these algorithms does not reduce as the number of sensors increases. Among them, the number of iterations for IRPLS and IGED maintains at 35, while that of the GLM algorithm keeps at 10. The computing time for each iteration, however, slightly increases as the number of sensors increasing. The RMSE and bias performance decreases as the number of sensors increases due to the fact that more normal sensors can be utilized. This result is also demonstrated in Figure 6. We also observe that the RMSE and bias curves of IGED and GLM are consistently lower than those of IRPLS, which shows the performance advantage of the proposed algorithms.

Figure 12 shows the number of iterations, computing time for each iteration, RMSE and bias performance of the IRPLS, IGED and GLM algorithms versus different levels of noise impulsiveness. The number of sensors is set at 20. Other parameters remain unchanged. From Figure 12a, we observe that the number of iterations for IRPLS scales from 88 to 11, 43 to 7 for IGED. However, the iteration value does not change for GLM and the number keeps at 7. That is because GLM performs optimization at each iteration to reduce the impact of impulsive noise. Moreover, the RMSE and bias performance of these algorithms decreases as increasing the value of α. As expected, IGED and GLM exhibit almost the same performance. A small value of α results in a poorer IRPLE performance, as illustrated in Figure 12b. When α is greater than 1.6, the RMSE and bias curves of IRPLS reach those of IGED and GLM due to the fact that the level of noise impulsiveness has less impact on the bias performance.

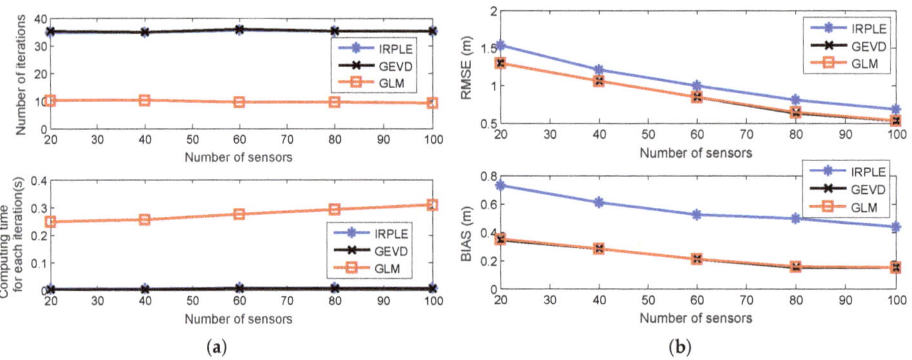

Figure 11. Scalability analysis for various number of sensors. (**a**) Number of iterations and computing time. (**b**) RMSE and bias performance.

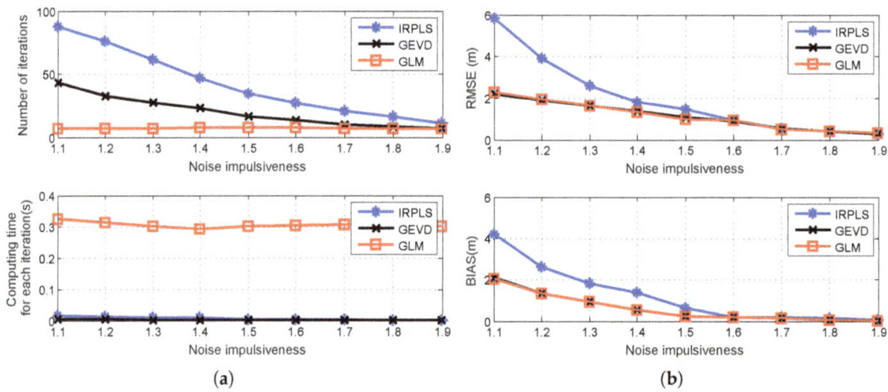

Figure 12. Scalability analysis for different levels of noise impulsiveness. (**a**) Number of iterations and computing time. (**b**) RMSE and bias performance.

7. Conclusions

In this paper, a total Lp-norm optimization method is presented to solve the BOSL problem and two algorithms, named IGED and GLM, are proposed to fulfill the optimization of the total least Lp-norm. By minimizing the errors in the system matrix and the data vector simultaneously for the least Lp-norm optimization, the proposed algorithms can overcome the bias arising from the correlation between the system matrix and the pseudolinear noise vector.

Simulation results show that the proposed IGED and GLM algorithms have much better RMSE and bias performance than the IRPLS algorithm. The number of iterations for GLM remains at a low level. With only a few iterations, the RMSEs of GLM can approach the CRLB. However, GLM expends more computation time than IGED under the same number of iterations, since GLM requires unconstrained optimization in each iteration.

Author Contributions: J.-A.L. studied the total Lp-norm optimization problem for bearing-only source localization and wrote the draft of the manuscript. C.-C.X. helped solve the total Lp-norm optimization problem and did the simulations, S.-T.H. and Y.-J.R. were involved in the discussion of the problem and revised the manuscript. All authors read and approved the final version of the manuscript.

Funding: This work was supported by the National Science Foundation for Young Scientists of China (No. 61703129) and the Science and Technology on Near-Surface Detection Laboratory Foundation, China (No. 6142414200203).

Institutional Review Board Statement: Not applicable.

Informed Consent Statement: Not applicable.

Data Availability Statement: Not applicable.

Conflicts of Interest: The authors declare that there is no conflict of interest regarding the publication of this article.

References

1. Li, X.; Luo, X.; Zhao, S. Globally Convergent Distributed Network Localization Using Locally Measured Bearings. *IEEE Trans. Control. Netw. Syst.* **2020**, *7*, 245–253. [CrossRef]
2. Kaplan, L.M.; Le, Q. On exploiting propagation delays for passive target localization using bearings-only measurements. *J. Frankl. Inst.* **2005**, *342*, 193–211. [CrossRef]
3. Volgyesi, P.; Balogh, G.; Nadas, A.; Nash, C.B.; Ledeczi, A. Shooter localization and weapon classification with soldier-wearable networked sensors. In Proceedings of the International Conference on Mobile Systems, Applications and Services, San Juan, Puerto Rico, 11–13 June 2007; pp. 113–126.
4. Ali, A.M.; Yao, K.; Collier, T.C.; Taylor, C.E.; Blumstein, D.T.; Girod, L. An empirical study of collaborative acoustic source localization. *J. Signal Process. Syst.* **2009**, *57*, 415–436. [CrossRef]
5. Wang, Y.; Wang, G.; Chen, S.; Ho, K.C.; Huang, L. An Investigation and Solution of Angle Based Rigid Body Localization. *IEEE Trans. Signal Process.* **2020**, *68*, 5457–5472. [CrossRef]
6. Doğançay, K. Bias compensation for the bearings-only pseudolinear target track estimator. *IEEE Trans. Signal Process.* **2006**, *54*, 59–68. [CrossRef]
7. Wang, Y.; Ho, K.C. An Asymptotically Efficient Estimator in Closed-Form for 3-D AOA Localization Using a Sensor Network. *IEEE Trans. Wirel. Commun.* **2015**, *14*, 6524–6535. [CrossRef]
8. Gavish, M.; Weiss, A.J. Performance analysis of bearing-only target location algorithms. *IEEE Trans. Aerosp. Electron. Syst.* **1992**, *28*, 817–828. [CrossRef]
9. Luo, J.A.; Shao, X.H.; Peng, D.L.; Zhang, X.P. A Novel Subspace Approach for Bearing-Only Target Localization. *IEEE Sens. J.* **2019**, *19*, 8174–8182. [CrossRef]
10. Swami, A.; Sadler, B.M. TDE, DOA and related parameter estimation problems in impulsive noise. In Proceedings of the IEEE Signal Processing Workshop on Higher-Order Statistics, Banff, AB, Canada; 21–23 July 1997; pp. 273–277.
11. Kozick R.J.; Sadler, B.M. Maximum-likelihood array processing in non-Gaussian noise with Gaussian mixtures. *IEEE Trans. Signal Process.* **2000**, *48*, 3520–3535.
12. Oh, H.; Seo, D.; Nam, H. Design of a Test for Detecting the Presence of Impulsive Noise. *Sensors* **2020**, *20*, 7135. [CrossRef]
13. Shao, M.; Nikias, C.L. Signal processing with fractional lower order moments: Stable processes and their applications. *Proc. IEEE* **1993**, *81*, 986–1010. [CrossRef]
14. Nguyen, N.H.; Doğançay, K.; Kuruoğlu, E.E. An Iteratively Reweighted Instrumental-Variable Estimator for Robust 3-D AOA Localization in Impulsive Noise. *IEEE Trans. Signal Process.* **2019**, *67*, 4795–4808. [CrossRef]
15. Zhong, X.; Premkumar, A.B.; Madhukumar, A.S. Particle Filtering for Acoustic Source Tracking in Impulsive Noise with Alpha-Stable Process. *IEEE Sens. J.* **2013**, *13*, 589–600. [CrossRef]
16. Tsakalides, P.; Nikias, C.L. Maximum likelihood localization of sources in noise modeled as a stable process. *IEEE Trans. Signal Process.* **1995**, *43*, 2700–2713. [CrossRef]

17. Luo, J.A.; Fang, F.; Shi, Y.F.; Shen-Tu, H.; Guo, Y.F. L1-Norm and Lp-Norm Optimization for Bearing-Only Positioning in Presence of Unreliable Measurements. In Proceedings of the Chinese Control And Decision Conference (CCDC), Hefei, China, 22–24 August 2020; pp. 1201–1205.
18. Maronna, B.R.A.; Martin, D.R.; Yohai,V.J. *Robust Statistics: Theory and Methods*; Wiley: New York, NY, USA, 2006.
19. Liu, Y.; Hu, Y.H.; Pan, Q. Distributed, robust acoustic source localization in a wireless sensor network. *IEEE Trans. Signal Process.* **2012**, *60*, 4350–4359. [CrossRef]
20. Jiang, Y.; Azimi-Sadjadi, M.R. A robust source localization algorithm applied to acoustic sensor network. In Proceedings of the IEEE International Conference on Acoustics, Speech and Signal Processing, Honolulu, HI, USA, 15–20 April 2007; pp. 1233–1236.
21. Panigrahi, T.; Panda, G.; Mulgrew, B.; Majhi, B. Robust incremental LMS over wireless sensor network in impulsive noise. In Proceedings of the International Conference on Computational Intelligence and Communication Networks, Bhopal, India, 26–28 November 2010; pp. 205–209.
22. Luo, J.A.; Xue, C.C.; Peng, D.L. Robust Bearing-Only Localization Using Total Least Absolute Residuals Optimization. *Complexity* **2020**, *2020*, 3456923. [CrossRef]
23. Satar, B.; Soysal, G.; Jiang, X.; Efe, M.; Kirubarajan, T. Robust Weighted $l_{1,2}$ Norm Filtering in Passive Radar Systems. *Sensors* **2020**, *20*, 3270. [CrossRef]
24. Chen, Y.; So, H.C.; Kuruoglu, E.E. Variance analysis of unbiased least lp-norm estimator in non-Gaussian noise. *Signal Process.* **2015**, *122*, 190–203. [CrossRef]
25. Jiang, X.; Chen, J.; So, H.C.; Liu, X. Large-Scale Robust Beamforming via ℓ_∞-Minimization. *IEEE Trans. Signal Process.* **2018**, *66*, 3824–3837. [CrossRef]
26. Wu, H.; Chen, S.X.; Zhang, Y.H.; Zhang, H.Y.; Ni, J. Robust structured total least squares algorithm for passive location. *J. Syst. Eng. Electron.* **2015**, *26*, 946–953. [CrossRef]
27. Picard, J.S.; Weiss, A.J. Bounds on the number of identifiable outliers in source localization by linear programming. *IEEE Trans. Signal Process.* **2010**, *58*, 2884–2895. [CrossRef]
28. Fragkos, G.; Apostolopoulos, P.A.; Tsiropoulou, E.E. ESCAPE: Evacuation strategy through clustering and autonomous operation in public safety systems. *Future Internet* **2019**, *11*, 20. [CrossRef]
29. Zhang, Y.; Guizani, M. *Game Theory for Wireless Communications and Networking*; CRC Press: Boca Raton, FL, USA, 2011.
30. Cambanis, S.; Miller, G. Linear Problems in Linear Problems in pth Order and Stable Processes. *SIAM J. Appl. Math.* **1981**, *41*, 43–69. [CrossRef]
31. Mallick, M. A Note on Bearing Measurement Model; ResearchGate. 2018. Available online: https://www.researchgate.net/publication/325214760_A_Note_on_Bearing_Measurement_Model (accessed on 16 September 2021).
32. Bishop, A.N.; Anderson, B.D.O.; Fidan, B.; Pathirana, P.N.; Mao, G. Bearing-Only Localization using Geometrically Constrained Optimization. *IEEE Trans. Aerosp. Electron. Syst.* **2009**, *45*, 308–320. [CrossRef]
33. Roger, F. *Practical Methods of Optimization*; John Wiley & Sons: New York, NY, USA, 1987.
34. Grinshpan, A.Z. Weighted inequalities and negative binomials. *Adv. Appl. Math.* **2010**, *45*, 564–606. [CrossRef]
35. Hestenes, M.R. Multiplier and gradient methods. *Optim. Theory Appl.* **1969**, *4*, 303–320. [CrossRef]
36. Kay, S.M. *Fundamentals of Statistical Signal Processing: Estimation Theory*; Prentice Hall PTR: Englewood Cliffs, NJ, USA, 1993.
37. Sadler, B.M.; Kozick, R.J.; Moore, T. Performance analysis for direction finding in non-Gaussian noise. In Proceedings of the IEEE International Conference on Acoustics, Speech, and Signal Processing, Phoenix, AZ, USA, 15–19 March 1999; pp. 2857–2860.

Article

Ground Target Tracking Using an Airborne Angle-Only Sensor with Terrain Uncertainty and Sensor Biases

Dipayan Mitra *, Aranee Balachandran and Ratnasingham Tharmarasa

Department of Electrical & Computer Engineering, McMaster University, Hamilton, ON L8S 4K1, Canada; a16@mcmaster.ca (A.B.); thamas@mcmaster.ca (R.T.)
* Correspondence: mitrad1@mcmaster.ca

Abstract: Airborne angle-only sensors can be used to track stationary or mobile ground targets. In order to make the problem observable in 3-dimensions (3-D), the height of the target (i.e., the height of the terrain) from the sea-level is needed to be known. In most of the existing works, the terrain height is assumed to be known accurately. However, the terrain height is usually obtained from Digital Terrain Elevation Data (DTED), which has different resolution levels. Ignoring the terrain height uncertainty in a tracking algorithm will lead to a bias in the estimated states. In addition to the terrain uncertainty, another common source of uncertainty in angle-only sensors is the sensor biases. Both these uncertainties must be handled properly to obtain better tracking accuracy. In this paper, we propose algorithms to estimate the sensor biases with the target(s) of opportunity and algorithms to track targets with terrain and sensor bias uncertainties. Sensor bias uncertainties can be reduced by estimating the biases using the measurements from the target(s) of opportunity with known horizontal positions. This step can be an optional step in an angle-only tracking problem. In this work, we have proposed algorithms to pick optimal targets of opportunity to obtain better bias estimation and algorithms to estimate the biases with the selected target(s) of opportunity. Finally, we provide a filtering framework to track the targets with terrain and bias uncertainties. The Posterior Cramer–Rao Lower Bound (PCRLB), which provides the lower bound on achievable estimation error, is derived for the single target filtering with an angle-only sensor with terrain uncertainty and measurement biases. The effectiveness of the proposed algorithms is verified by Monte Carlo simulations. The simulation results show that sensor biases can be estimated accurately using the target(s) of opportunity and the tracking accuracies of the targets can be improved significantly using the proposed algorithms when the terrain and bias uncertainties are present.

Keywords: angle-only sensor; terrain uncertainty; posterior Cramer–Rao lower bound; bias estimation; path planning

Citation: Mitra, D.; Balachandran, A.; Tharmarasa, R. Ground Target Tracking Using an Airborne Angle-Only Sensor with Terrain Uncertainty and Sensor Biases. *Sensors* **2022**, *22*, 509. https://doi.org/10.3390/s22020509

Academic Editor: Andrzej Stateczny

Received: 5 December 2021
Accepted: 5 January 2022
Published: 10 January 2022

Publisher's Note: MDPI stays neutral with regard to jurisdictional claims in published maps and institutional affiliations.

1. Introduction

Tracking a ground target using an airborne sensor platform is frequently used in various applications, such as surveillance, search, and rescue missions [1–3]. Airborne radar sensors are of particular interest in various surveillance missions, because of their 'day-and-night' operational capabilities [4]. Airborne synthetic aperture radars (SAR) are often used to acquire high-resolution images of ground targets [5,6]. In [7,8], authors presented the application of airborne sensors for magnetic anomaly detection (MAD). MAD is widely used in maritime surveillance, detection of shipwrecks, geophysical studies, etc. [7].

Estimating the state of a ground target using the measurements from an angle-only airborne sensor is one of the most practical applications. A number of works have extensively studied the angle-only tracking problem in the 2-D Cartesian Coordinate System (CCS) [9,10]. However, as the authors in [2] pointed out, the number of works on 3-D angle-only tracking problems is relatively low. Some of the earlier works involving tracking ground targets in the 3-D coordinate system using angle-only sensors are reported in [11,12]. In [11], tracking an air target using a ground-based angle-only sensor is considered. Tracking a ground target using an airborne angle-only sensor is a slightly different problem since additional information about the target height will be available. In this work, our focus is to track a ground target using an airborne angle-only sensor platform.

One of the major challenges in angle-only tracking is observability. Considering the additional information of the height of the target from the sea level, such observability issue can be addressed. Since the target is on the ground, the target's height is the same as the height of the terrain from the sea level, which can be obtained from pre-stored Digital Terrain Elevation Data (DTED) [13]. Most of the aforementioned works in 3-D angle-only tracking considered the height of the ground target from the sea level is known accurately [14–16]. However, in practical setups, such height information is associated with uncertainty due to the errors in the DTED data. Ignoring the height uncertainty (i.e., using a wrong height value) will lead to a bias in the estimated state. To the best of our knowledge, not many analysis is performed on tracking a ground target with terrain uncertainty. This is the motivation for this work to develop algorithms to handle the terrain uncertainty with angle-only sensors.

Apart from the terrain uncertainty, possible biases in the sensor measurements play a key role in determining the quality of the estimates. Possible sources of bias include sensor alignment bias, sensor altitude bias, location bias, etc. [17]. In this work, we consider only the measurement biases, i.e., the biases in the elevation and bearing angles. Sensor biases and terrain uncertainty should be handled jointly in order to obtain better tracking results.

In this work, we propose a filtering algorithm to track a target with bias and altitude uncertainties. With a larger bias uncertainty, a filter will take a longer time to reduce the target state bias. If we have an option to reduce the sensor bias uncertainty before we start tracking the target of interest, that will help to obtain a better estimate of the target faster. Usually, an airborne platform will fly from a base station to the Region Of Interest (ROI), where we have the target of interest. On the way to the ROI, a bias estimation could be performed by pointing the angle-only sensor toward one more multiple stationary ground object, called as targets of opportunity. In this work, this bias estimation is considered an optional step. Note that the bias uncertainty will not be completely removed even with this optional step. Hence, the filtering algorithm used for tracking the target should still consider the bias uncertainty.

In this work, we explore the possibility of improving the bias estimation using targets of opportunity by changing sensor trajectory. We also study the effect of increasing the number of targets of opportunity and changing their locations with respect to the sensor trajectory on bias estimation. Our proposed approach considers the bias estimation when the x and y coordinates of the target of opportunity are known as well as unknown. A number of bias estimation approaches are proposed in the literature [18–20]. However, the challenges in the bias estimation with the terrain uncertainty are not considered in any of the papers.

Predicting the performance of an estimator is essential to decide the optimal sensor trajectory or optimal targets of opportunity locations. The covariance of an unbiased estimator is bounded by the Posterior Cramer–Rao Lower Bound (PCRLB) [21,22]. When the estimator is biased, the estimated covariance can not be directly bounded by the PCRLB. In [23], the authors proposed a performance bound considering the gradients of the bias state. Such a performance bound on the total variance of the estimator is referred to as *biased PCRLB* [24]. The central idea behind using the gradient of the bias state is to have a dependency on the non-constant part of the bias. In other words, the bias can not be removed from the measurements by simple subtraction. In this work, PCRLB and a biased PCRLB are derived for angle-only tracking problems with bias and terrain uncertainties.

In this paper, we consider two possible scenarios: ground target with terrain uncertainty (1) remains stationary; (2) moves with a nearly constant velocity (CV) model [25]. We assume that the ground target (while moving in nearly CV) moves along the x–y plane, i.e., there is no velocity along the z-axis. Although the dynamic model of the target state is linear, the angle-only measurements are non-linear functions of the target and the sensor state [11]. For such an estimation problem, several non-linear filtering algorithms are proposed in the literature. Some of the examples include Extended Kalman Filter (EKF) [26], Cubature Kalman filter (CKF) [27], Unscented Kalman Filter (UKF) [28], and Feedback Particle Filter (FPF) [29].

The computational complexity of the UKF is of the same order as the EKF while providing improved estimation accuracy addressing the approximation issues of the EKF, as shown in [30,31]. As a result, we propose a filtering algorithm using UKF. However, UKF can easily be replaced by any other non-linear filter. One of the challenges in the ground target tracking problem is filter initialization with terrain uncertainty. In this work, we use the measurements obtained by the biased angle-only sensor to initialize the target state by incorporating the terrain uncertainty. The estimate errors of the target trajectory are compared with the PCRLB and a conventional approach to evaluate the accuracy and the benefit of the proposed algorithms. The simulation results show that the proposed approach provides better tracking results with all the given uncertainties.

The key contributions of our work can be stated as follows: (1) We derive the PCRLB for this problem to predict/evaluate the performance of the estimator and optimize bias estimation; (2) Bias estimation using a separate target(s) of opportunity is proposed with optimal platform trajectory and optimal target of opportunity location selection; (3) We propose a filtering approach to estimate a target with bias and terrain uncertainties.

This paper is organized as follows. We discuss the problem description in Section 2. System model detailing the coordinate system, measurement generation, and the system dynamics are introduced in Section 3. A discussion on performance bounds is presented in Section 4. Bias estimation approaches and related analysis are detailed in Section 5. Filter initialization, optional bias compensation, and ground target tracking are discussed in Section 6. Simulation results are shown in Section 7 and the paper ends with the concluding remarks in Section 8.

2. Problem Description

In this paper, our main objective is to track a ground target in 3-D using a biased airborne angle-only sensor. The ground target can either remain stationary or move at a nearly constant velocity. The height of the ground target from the sea level is obtained from DTED; hence it has uncertainty. The two major sources of uncertainty are measurement bias and terrain uncertainty.

To reduce the measurement bias uncertainty, the possible biases could be estimated using separate target(s) of opportunity on the way to the region of interest from the base station. Two possible cases that can happen with the target(s) of opportunity are (1) x and y coordinates of the target(s) are known accurately, but the z coordinate is obtained from the DTED data, which has error; (2) x and y coordinates of the target(s) are unknown, but the z coordinate is obtained from the DTED data as in the first case. Platform trajectory and the location of the target(s) of opportunity can be optimized to obtain a better bias estimate by minimizing the additional time required to reach the destination. That is, we prefer if we do not need to change the trajectory of the platform.

In this work, we make the following assumptions:

- Only a single target is considered. However, the proposed algorithm can be used for multiple well-separated targets without any modification;
- The height of the ground target from the sea level is fixed, but not known accurately;
- The ground target can either remain stationary or move with a nearly constant velocity;
- Bias affecting the angle-only measurements are unknown constant and additive. Time-varying bias is not considered in this work. However, the proposed approach can be easily extended for time-varying biases;
- Data association issues are not considered, i.e., false alarms are not considered.

In the next section, we introduce the coordinate system, the system dynamics and the measurement model.

3. System Model

3.1. Coordinate System

The states of the ground target and the own-ship at time step k are defined as $\mathbf{x}_k^t = [x_k^t, \dot{x}_k^t, y_k^t, \dot{y}_k^t, z_k^t]^T$ and $\mathbf{x}_k^o = [x_k^o, \dot{x}_k^o, y_k^o, \dot{y}_k^o, z_k^o]^T$, respectively, where $(.)^T$ denote the transpose operation. Note that the state of a stationary target is expressed as $\mathbf{x}^t = [x^t, y^t, z^t]^T$. Here, (x, y, z) represent the three axes of 3-D CCS and the superscripts 't' and 'o' are reserved to denote the target and the own-ship (airborne sensor platform), respectively. In this manuscript, the terms own-ship and airborne sensor platform have been used interchangeably. The velocity is denoted as $[\dot{x}, \dot{y}]$. Note that there is no velocity across z-axis for the target and the own-ship, i.e., $\dot{z}_k^t = 0$ and $\dot{z}_k^o = 0$. It is assumed that the ground is flat through out the region where the target moves, hence the target's z does not change over time, i.e., $z_k^t = z^t$.

In this work, we considered a locally flat earth model for all our calculations.

3.2. System Dynamics

In this work, a constant velocity (CV) model is used to model the system dynamics of the ground target. Given the discrete-time state of a moving target $\mathbf{x}_k^t$, the state evolution is expressed as,

$$\mathbf{x}_{k+1}^t = \mathbf{F}_k \mathbf{x}_k^t + \mathbf{G}_k \mathbf{v}_k \tag{1}$$

where $\mathbf{F}_k$ and $\mathbf{G}_k$ are the state transition and the gain matrices, respectively. The process noise $\mathbf{v}_k$ is a zero-mean Gaussian with covariance $\mathbf{Q}_k$, i.e., $\mathbf{v}_k \sim \mathcal{N}(\mathbf{v}_k; 0, \mathbf{Q}_k)$. For a stationary ground target, $\mathbf{F}_k$ is an identity matrix and the noise part is zero, hence $\mathbf{x}_{k+1}^t = \mathbf{x}_k^t$.

Motion legs with CV and Constant Turn (CT) are used to model the system dynamics of the own-ship trajectory. Transition matrices of CV and CT models are given in (A1) and (A2), respectively.

3.3. Measurement Model

Figure 1 shows the target-sensor geometry in 3-D CCS. The on-board angle-only sensor provides the bearing $\theta_k \in [-\pi, \pi]$ and elevation $\gamma_k \in [-\frac{\pi}{2}, \frac{\pi}{2}]$ measurements. As the own-ship trajectory is deterministic, z_k^o is known for all time-steps.

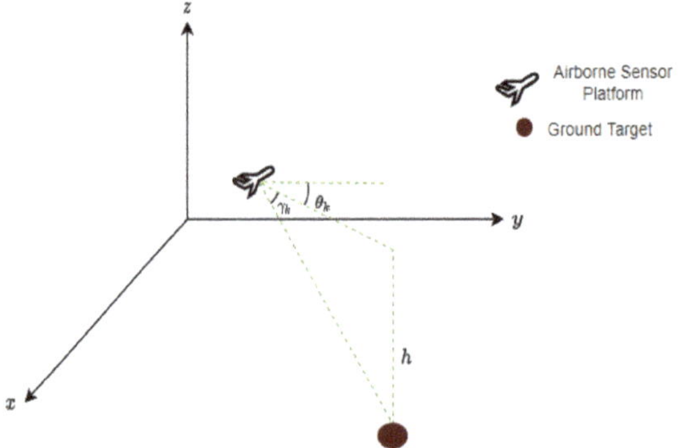

Figure 1. Target-sensor geometry.

Based on the assumptions from Section 2, a measurement is available for the target height from the sea level, z_m,

$$z_m = z^t + v^z \tag{2}$$

where v^z is a Gaussian noise with zero-mean and standard deviation σ_{z^t}, i.e., $v^z \sim \mathcal{N}(v^z; 0, \sigma_{z^t}^2)$.

The expressions for the true bearing and elevation angles are given by,

$$\theta_k^{\text{true}} \;=\; \tan^{-1}\left[(x_k^t - x_k^o),(y_k^t - y_k^o)\right] \tag{3}$$

$$\gamma_k^{\text{true}} \;=\; \tan^{-1}\left[(z_k^t - z_k^o),\sqrt{(x_k^t - x_k^o)^2 + (y_k^t - y_k^o)^2}\right] \tag{4}$$

Let us denote the bias vector of the sensor as $\mathbf{b}_k = [\theta_{b_k}, \gamma_{b_k}]^T$. In this work, we consider the measurement biases to be additive and unknown constants. As a result, a separate state transition matrix for the bias state evolution is not necessary. Measurement model for acquiring bearing and elevation from the angle-only sensor is expressed as,

$$\mathbf{z}_k = \mathbf{h}(x_k^t, x_k^o) + \mathbf{b}_k + \mathbf{w}_k \tag{5}$$

where $\mathbf{h}(x_k^t, x_k^o) = [\theta_k^{\text{true}}, \gamma_k^{\text{true}}]^T$ and measurement noise $\mathbf{w}_k \sim \mathcal{N}(\mathbf{w}_k; 0, \mathbf{R}_k)$ with covariance $\mathbf{R}_k = \text{diag}(\sigma_\theta^2, \sigma_\gamma^2)$. Here, σ_θ and σ_γ are the standard deviations for the bearing and the elevation measurements, respectively.

In the next section, we introduce the necessary bounds to evaluate the performance of an estimator.

4. Performance Bound

4.1. Posterior Cramer–Rao Lower Bound

Let us define $\mathbf{Z}_K$ as the collection of all the measurements for time $k = 1, \ldots, K$, i.e., $\mathbf{Z}_K = [\mathbf{z}_1, \mathbf{z}_2, \ldots, \mathbf{z}_K]$. The objective of an estimator is to find the conditional probability distribution $p(x_k^t | \mathbf{Z}_K)$. Associated covariance matrix $(\mathbf{C}_{\hat{x}^t})$ of an unbiased estimate of the target state x_k^t can be lower bounded by,

$$\mathbf{C}_{\hat{x}^t} = \mathbb{E}\left[(\hat{x}_k^t - x_k^t)(\hat{x}_k^t - x_k^t)^T\right] \geq \mathbf{J}_k^{-1} \tag{6}$$

where $\mathbf{J}_k$ is the Fisher Information Matrix (FIM) [21]. The above bound is called the PCRLB. In [32] Tichavsky et al. proposed a recursive formulation to obtain the FIM. In our problem, the target state evaluation is modeled by a discrete-time linear system. Knowing the state transition matrix $\mathbf{F}_k$, the process noise covariance $\mathbf{Q}_k$ and the measurement noise covariance $\mathbf{R}_k$, a simplified expression for the recursive formulation of the FIM is expressed as,

$$\mathbf{J}_{k+1} = \left(\mathbf{F}_k \mathbf{J}_k^{-1} \mathbf{F}_k^T + \mathbf{Q}_k\right)^{-1} + \mathbf{J}_z(k+1) \tag{7}$$

where $\mathbf{J}_z(k+1)$ is called the measurement contribution to the FIM, which is evaluated as

$$\mathbf{J}_z(k) = \mathbb{E}\left[q_k \mathbf{H}_k^T \mathbf{R}_k^{-1} \mathbf{H}_k\right] \tag{8}$$

Here $\mathbf{H}_k$ is the Hessian matrix of the measurement equation (see (A3) for details on the matrix construction) and q_k is called Information Reduction Factor (IRF). In this work, we do not consider the possibility of false alarms or miss detection. Hence, data association uncertainty is absent, i.e., IRF $q_k = 1$.

Finally, substituting $\mathbf{J}_z(k+1)$ from (8) into (7), the FIM is evaluated as,

$$\mathbf{J}_{k+1} = \left(\mathbf{F}_k \mathbf{J}_k^{-1} \mathbf{F}_k^T + \mathbf{Q}_k\right)^{-1} + \mathbf{H}_k^T \mathbf{R}_k^{-1} \mathbf{H}_k \tag{9}$$

Assuming the initial target state covariance to be $\mathbf{P}_1$, FIM is initialised as $\mathbf{J}_1 = \mathbf{P}_1^{-1}$.

4.2. Presence of Measurement Bias

It is well known that satisfying certain regulatory conditions, the performance of an unbiased estimator is bounded by the PCRLB. However, in this problem, we consider the additional uncertainty caused by the measurement bias. Hence, the estimates obtained by the biased angle-only measurements can not be bounded directly by the PCRLB without accounting for the additional bias uncertainty in the FIM (shown in (9)). In this section, we discuss the following modifications to the FIM for obtaining a suitable performance bound for this work.

Recalling the governing equation of $\mathbf{J}_z(k)$, it is clear that the measurement bias is going to impact the measurement noise covariance $\mathbf{R}_k$, whereas the Hessian matrix remains unchanged. First, we form the covariance matrix for the bias as $\mathbf{R}_{\text{bias}} = \text{diag}\left(\sigma_\theta^2, \sigma_\gamma^2\right)$. To provide bias compensation we obtain the bias compensated measurement covariance matrix as,

$$\mathbf{R}_k^b = \mathbf{R}_k + \mathbf{R}_{\text{bias}} \tag{10}$$

$$= \begin{bmatrix} \sigma_\theta^2 + \sigma_{\theta_b}^2 & 0 \\ 0 & \sigma_\gamma^2 + \sigma_{\gamma_b}^2 \end{bmatrix} \tag{11}$$

Once the bias compensated measurement covariance matrix is obtained, substituting $\mathbf{R}_k^b$ into $\mathbf{R}_k$ of (9) we obtain,

$$\mathbf{J}_{k+1} = \left(\mathbf{F}_k \mathbf{J}_k^{-1} \mathbf{F}_k^T + \mathbf{Q}_k\right)^{-1} + \mathbf{H}_k^T \left(\mathbf{R}_k^b\right)^{-1} \mathbf{H}_k \tag{12}$$

Now, we focus on modifying the FIM to account for the terrain uncertainty.

4.3. Presence of Terrain Uncertainty

The prior information about the terrain height (2) is used to initialize the FIM at $k = 1$. Note that this measurement can not be used more than once. In order to initialize the FIM at $k = 1$ using the terrain height information and the first target measurement, we consider a stacked measurement $[\theta_k, \gamma_k, z_m]$ and corresponding covariance matrix $\mathbf{R}_z = \text{diag}\left(\sigma_\theta^2, \sigma_\gamma^2, \sigma_{z^t}^2\right)$. The subscript 'z' indicates the initial time step where the terrain uncertainty is considered. We use this notation throughout this paper. Combining with $\mathbf{R}_{\text{bias}}$ the bias compensated covariance matrix is expressed as,

$$\mathbf{R}_z^b = \mathbf{R}_z + \mathbf{R}_{\text{bias}} \tag{13}$$

$$= \begin{bmatrix} \sigma_\theta^2 + \sigma_{\theta_b}^2 & 0 & 0 \\ 0 & \sigma_\gamma^2 + \sigma_{\gamma_b}^2 & 0 \\ 0 & 0 & \sigma_{z^t}^2 \end{bmatrix} \tag{14}$$

Once $\mathbf{R}_z^b$ is formed, corresponding Hessian matrix $\mathbf{H}_z$ is evaluated as shown in (A4). Using $\mathbf{R}_z^b$ and the newly evaluated $\mathbf{H}_z$, we can initialize FIM at $k = 1$ as

$$\mathbf{J}_1 = \left(\mathbf{F}_k \mathbf{J}_k^{-1} \mathbf{F}_k^T + \mathbf{Q}_k\right)^{-1} + (\mathbf{H}_z)^T \left(\mathbf{R}_z^b\right)^{-1} \mathbf{H}_z \tag{15}$$

4.4. Biased Posterior Cramer–Rao Lower Bound

A bound on the covariance of biased target state estimate $\hat{\mathbf{x}}_k^t$ is defined as biased PCRLB. In [21,23,24], the biased PCRLB is expressed as,

$$\mathbf{C}_{\hat{\mathbf{x}}^t} \geq \left(\mathbf{I} + \mathbf{D}_k\right)^T \mathbf{J}_k^{-1} \left(\mathbf{I} + \mathbf{D}_k\right) \tag{16}$$

where $\mathbf{D}_k$ is the bias gradient matrix. The FIM is evaluated following (9). The bias in the state is written as $\mathbf{b}_{\mathbf{x}_k^t} = \left(\mathbb{E}[\hat{\mathbf{x}}_k^t] - \mathbf{x}_k^t \right)$. Please note that the bias in the state is denoted as $\mathbf{b}_{\mathbf{x}_k^t}$, which is not to be confused with the measurement bias $\mathbf{b}$. Given the target state $\mathbf{x}_k^t$, bias gradient matrix is defined as,

$$\mathbf{D}_k = \frac{\partial \mathbf{b}_{\mathbf{x}_k^t}}{\partial \mathbf{x}_k^t} \tag{17}$$

In order to empirically obtain $\mathbf{D}_k$ from (17), first we have to form the bias state. For a stationary ground target, the bias in the state vector is denoted as $\mathbf{b}_{\mathbf{x}_k^t} = [b_{x_k^t}, b_{y_k^t}, b_{z_k^t}]$. Analytical form of the individual bias state coordinates can be obtained as,

$$b_{x_k^t} = \frac{z_k^t \sin(\theta_k^{\text{true}} + \theta_{b_k})}{\tan(\gamma_k^{\text{true}} + \gamma_{b_k})} - \frac{z_k^t \sin(\theta_k^{\text{true}})}{\tan(\gamma_k^{\text{true}})} \tag{18}$$

$$b_{y_k^t} = \frac{z_k^t \cos(\theta_k^{\text{true}} + \theta_{b_k})}{\tan(\gamma_k^{\text{true}} + \gamma_{b_k})} - \frac{z_k^t \cos(\theta_k^{\text{true}})}{\tan(\gamma_k^{\text{true}})} \tag{19}$$

The bias gradient $\mathbf{D}_k$ is evaluated as shown below,

$$\mathbf{D}_k = \begin{bmatrix} \frac{\partial b_{x_k^t}}{\partial x_k^t} & \frac{\partial b_{x_k^t}}{\partial y_k^t} & \frac{\partial b_{x_k^t}}{\partial z_k^t} \\ \frac{\partial b_{y_k^t}}{\partial x_k^t} & \frac{\partial b_{y_k^t}}{\partial y_k^t} & \frac{\partial b_{y_k^t}}{\partial z_k^t} \\ \frac{\partial b_{z_k^t}}{\partial x_k^t} & \frac{\partial b_{z_k^t}}{\partial y_k^t} & \frac{\partial b_{z_k^t}}{\partial z_k^t} \end{bmatrix} \tag{20}$$

The derivatives of (20) can be evaluated easily and not shown here in this paper. Additionally, modifying (20) for moving ground target is straightforward and not shown in this paper.

Intuitively the bias gradient represents the part of the bias which can not be removed by simple subtraction, i.e., the non-additive component of the bias. As a result, the biased PCRLB, shown in (16), depends on the gradient matrix $\mathbf{D}_k$. Performance of the ground target tracking, both stationary and moving, is validated against both PCRLB and biased PCRLB in Section 7.2.

In the next section, we use the derived PCRLB to optimize the platform trajectory to estimate the sensor biases using targets of opportunity.

5. Bias Estimation Using Targets of Opportunity

Recalling the discussions from Section 2, an optional two-step bias estimation using targets of opportunity is presented in this section. To bring clarity and avoid confusion with the original target, throughout this paper, we reserve the term 'target of opportunity' to indicate the target for which some prior information is known, or we are not interested in estimating that target's state. Note that this step is usually performed on the way to the target region from the base station. Hence, we need to consider only a reduced bias error when estimating the target state.

The two steps involved in our proposed bias estimation approach are as follows. The first step is to identify one or more stationary targets of opportunity on the sensor's path to the tracking region. The next step is to estimate the biases in the measurements using the identified targets of opportunity. Such a bias estimation approach has two following benefits:

- The uncertainty in the sensor bias is reduced before the original ground target appears in the sensor's field of view. Hence, the tracking can provide better estimates from the beginning. Otherwise, we may need to make more maneuvers to reduce the biases in the state estimate to obtain a reasonable tracking accuracy;

- The sensor can choose multiple targets of opportunity to improve the convergence of the bias estimates. When we use only the interested target to correct the bias, it will take a longer time to converge.

In this section, we analyze the impact on bias estimation caused by the change in sensor trajectory, the number of targets of opportunity used and targets' proximity to the sensor trajectory. We consider both scenarios where the locations of the targets of opportunity are known as well as unknown, but the terrain heights are not known accurately. PCRLB is used to quantify the estimation quality in order to find the optimal platform trajectory and the locations of the targets of opportunity. The terms 'target of opportunity' and 'target of opportunity with terrain uncertainty' are used interchangeably in this section.

5.1. Known Location with Terrain Uncertainty

In this section, the bias estimation is performed with a known location of the target of opportunity. To emphasize, the x and y coordinates of the target of opportunity are known, and the height (z) information is obtained from DTED. Hence, there is an uncertainty in the target's z value. The following factors impact the bias estimation: the number of targets of opportunity used, change in sensor trajectory and the proximity of the target of opportunity with the sensor trajectory. In this section, two possible scenarios are analyzed, and a conclusion on the optimal bias estimation is presented using the PCRLB. In the first scenario, the sensor bias is estimated using one target of opportunity with various sensor trajectories. In the second scenario, the sensor bias is estimated using two targets of opportunity with possibly different terrain heights.

Let us denote the state vector of the stationary target of opportunity as $\mathbf{x}_k^{to} = [x_k^{to}, y_k^{to}, z_k^{to}]^T$. Here, the superscript 'to' indicates the target of opportunity. The uncertainty in the height of the target of opportunity is modeled as a zero-mean Gaussian with a standard deviation of $\sigma_{z^{to}}$. Although the location of the target of opportunity is known, z_k^{to} is needed to be estimated because of the presence of the terrain uncertainty. Therefore, the state vector of the bias estimation problem at the time step k is expressed as $\mathbf{x}_k^{aug1} = [z_k^{to}, \theta_{b_k}, \gamma_{b_k}]^T$. Here, the post-fix 'aug1' refers to the augmented state for the first scenario, i.e., known location of the target of opportunity.

Our next step is to initialize the filter for bias estimation. Modeling the error associated with the terrain uncertainty by a Gaussian with zero-mean and $\sigma_{z^{to}}$ standard deviation, we can write $z_m^{to} \sim \mathcal{N}(z_m^{to}; z^{to}, \sigma_{z^{to}}^2)$. Hence, we can initialize the height of the target of opportunity with the DTED information z_m^{to}. The bias states are initialized as $[0,0]^T$. Therefore, the state vector is initialized as $\mathbf{x}_1^{aug1} = [z_m^{to}, 0, 0]^T$. Moreover, the initial state covariance $\mathbf{P}_1^{aug1}$ is calculated using the FIM evaluated at the initial time step as $\mathbf{P}_1^{aug1} = \left(\mathbf{J}_1^{aug1}\right)^{-1}$. After filter initialization, the non-linear filter (UKF in our work) is used to estimate the augmented state. Note that, for initialization $\mathbf{H}_z^{aug1}$ is expressed as,

$$\mathbf{H}_z^{aug1} = \begin{bmatrix} \frac{\partial z_k^{to}}{\partial z_k^{to}} & \frac{\partial z_k^{to}}{\partial \theta_{b_k}} & \frac{\partial z_k^{to}}{\partial \gamma_{b_k}} \\ \frac{\partial \theta_k}{\partial z_k^{to}} & \frac{\partial \theta_k}{\partial \theta_{b_k}} & \frac{\partial \theta_k}{\partial \gamma_{b_k}} \\ \frac{\partial \gamma_k}{\partial z_k^{to}} & \frac{\partial \gamma_k}{\partial \theta_{b_k}} & \frac{\partial \gamma_k}{\partial \gamma_{b_k}} \end{bmatrix} = \begin{bmatrix} 1 & 0 & 0 \\ 0 & 1 & 0 \\ \frac{\sqrt{x_k^{rto2} + y_k^{rto2}}}{x_k^{rto2} + y_k^{rto2} + z_k^{rto2}} & 0 & 1 \end{bmatrix} \tag{21}$$

The biased measurement covariance matrix, during initialization, of the augmented state is expressed as $\mathbf{R}_z^{aug1} = \mathrm{diag}\left(\sigma_\theta^2, \sigma_\gamma^2, \sigma_{z^{to}}^2\right)$. Once $\mathbf{H}_z^{aug1}$ and $\mathbf{R}_z^{aug1}$ are obtained for $k = 1$, the FIM is evaluated from (9) as,

$$\mathbf{J}_1^{aug1} = \left(\mathbf{F}_k \mathbf{J}_k^{-1} \mathbf{F}_k^T + \mathbf{Q}_k\right)^{-1} + \left(\mathbf{H}_z^{aug1}\right)^T \left(\mathbf{R}_z^{aug1}\right)^{-1} \mathbf{H}_z^{aug1} \tag{22}$$

After initialization, we obtain the PCRLB for the time steps $k > 1$. First, we obtain the Hessian matrix $\mathbf{H}_k^{\text{aug1}}$ of the bearing and elevation measurement model as,

$$
\mathbf{H}_k^{\text{aug1}} = \begin{bmatrix} \frac{\partial \theta_k}{\partial z_k^{to}} & \frac{\partial \theta_k}{\partial \theta_{b_k}} & \frac{\partial \theta_k}{\partial \gamma_{b_k}} \\ \frac{\partial \gamma_k}{\partial z_k^{to}} & \frac{\partial \gamma_k}{\partial \theta_{b_k}} & \frac{\partial \gamma_k}{\partial \gamma_{b_k}} \end{bmatrix} = \begin{bmatrix} 0 & 1 & 0 \\ \frac{\sqrt{x_k^{rto2} + y_k^{rto2}}}{x_k^{rto2} + y_k^{rto2} + z_k^{rto2}} & 0 & 1 \end{bmatrix}
\tag{23}
$$

where the relative state $\mathbf{x}_k^{rto} = \mathbf{x}_k^{to} - \mathbf{x}_k^{o}$. The measurement contribution $\mathbf{J}_z^{\text{aug1}}(k) = \left(\mathbf{H}_k^{\text{aug1}}\right) \left(\mathbf{R}_k^{\text{aug1}}\right)^{-1} \mathbf{H}_k^{\text{aug1}}$, where measurement noise covariance is $\mathbf{R}_k^{\text{aug1}} = \text{diag}\left(\sigma_\theta^2, \sigma_\gamma^2\right)$. With $\mathbf{J}_z^{\text{aug1}}(k)$, $\mathbf{H}_k^{\text{aug1}}$ and $\mathbf{R}_k^{\text{aug1}}$, PCRLB is evaluated using (9). As the sensor bias and the target height are constants, we consider $\mathbf{F}_k = \mathbf{I}$, i.e., identity matrix of appropriate dimension.

In order to improve the bias estimation further, we consider two different cases below.

5.1.1. Change in Sensor Trajectory

For a given target of opportunity, we restrict our observations to two types of sensor trajectories. The sensor can either follow the CV model and make a fly-by while estimating the bias state or follow a combination of CV and CT models to make a turn around the target to estimate the bias. In this work, the combination of CV and CT models is referred to as CV-CT model. The aforementioned two types of sample sensor trajectories are shown in Figure 2a,b, respectively.

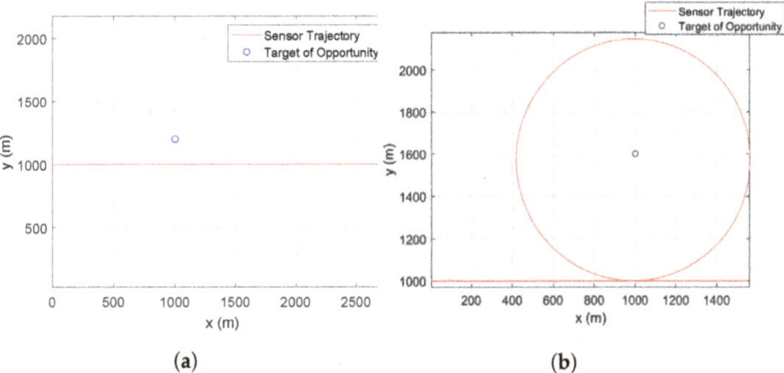

(**a**)   (**b**)

Figure 2. Sensor trajectories for bias estimation using a stationary target with known location and terrain uncertainty. Sensor trajectory follows: (**a**) the CV model, (**b**) the CV-CT model.

Let us assume that the target of opportunity remains in the sensor's field of view for K time steps. The x-axis indicates the sensor heading at the start time step. In other words, when the sensor follows the CV model, $\dot{y}_k^o = 0$.

Let us now consider the case when the sensor follows the CV-CT model. We denote K_{CV} as the total number of time steps the sensor follows the CV model in this CV-CT model. The platform switch to the CT model when the platform reaches the closest distance from the target of opportunity. Considering the sensor velocity to be $V\,\text{m/s}$ and the sampling rate to be T, we can write the total number of samples obtained while the sensor remains in the CV model to be $\left(\frac{1}{T} \frac{|x_1^{to} - x_1^o|}{V}\right)$. Once the sensor starts following the CT model, the total number of measurements obtained by completing one full cycle is denoted by $\left(\frac{1}{T} \frac{2\pi}{\omega}\right)$, where ω is the turn rate. We consider the total number of time steps needed to obtain $\left(\frac{1}{T} \frac{2\pi}{\omega}\right)$ samples to be K_{CT}.

Now we shift our focus into finding ω, for the sensor to follow the CT model. Let us denote the x–y coordinates of the sensor at $k = K_{CV}$ as $(x_{K_{CV}}^o, y_{K_{CV}}^o)$. From the discussions of the previous paragraph, the maximum change in the y-coordinate occurs at $k = \frac{K_{CT}}{2}$.

In this work, the idea behind finding ω is to ensure that the target of opportunity is inside the circle formed by the CT model,

$$\left| y^o_{K_{\frac{CT}{2}}} - y^o_{K_{CV}} \right| \geq \left| y^o_{K_{CV}} - y^{to}_1 \right| \tag{24}$$

Now, using the CT model (A1), we can show that

$$y^o_{K_{\frac{CT}{2}}} - y^o_{K_{CV}} = \frac{2}{\omega} \dot{x}^o_{K_{CV}} \tag{25}$$

From (25) and (24), we can obtain,

$$\frac{2}{\omega} \dot{x}^o_{K_{CV}} \geq \left| y^o_{K_{CV}} - y^{to}_1 \right|$$

$$\omega \leq \frac{2}{\left| y^o_{K_{CV}} - y^{to}_1 \right|} \dot{x}^o_{K_{CV}} \tag{26}$$

However, the platform has a constraint on maximum turn, ω_{max}, that it can make, hence we pick the ω as $\min\left(\omega_{max}, \frac{2}{\left| y^o_{K_{CV}} - y^{to}_1 \right|} \dot{x}^o_{K_{CV}} \right)$ to reduce the additional time required to reach the region of original target.

From (9), it is evident that the Hessian matrix $\mathbf{H}^{aug1}_k$ significantly affects the PCRLB. When considering (23), the elements corresponding to the differentiation involving θ_k in $\mathbf{H}^{aug1}_k$ are constant. Therefore, the measurement contribution from the bearing bias does not depend on the sensor trajectory. On the other hand, the terms corresponding to the differentiation involving γ_k in $\mathbf{H}^{aug1}_k$ depends on the location of both the sensor and the target of opportunity. As a sensor trajectory formed by the CV-CT model reduces the relative distance between the target of opportunity and the sensor, x^{rto}_k and y^{rto}_k reduces. As a result, reduction in $\left(\frac{\sqrt{x^{rto2}_k + y^{rto2}_k}}{x^{rto2}_k + y^{rto2}_k + z^{rto2}_k} \right)$ (from (23)) leads to the reduction in PCRLB. Thus, the sensor trajectory formed by the CV-CT model provides a better elevation bias estimation when compared to that of the sensor trajectory formed by the CV model. A comparative analysis between bias estimation performance while the sensor follows both the CV and the CV-CT model is shown in Section 7.3.1. Note that we need to spend more time on this bias estimation when we use the CV-CT model. With the CV model, no additional time is needed to scan the target of opportunity since there is no change in the platform trajectory.

We now expand our analysis to show the effect of using two targets of opportunity with known locations and additional terrain uncertainty.

5.1.2. Bias Estimation with Multiple Targets of Opportunity

In the previous section, we concluded that the sensor following a CV-CT trajectory improves the estimate $\hat{\gamma}_{b_k}$, when compared to the case where the sensor follows only a CV trajectory. In this section, we analyze the significance of adding a second stationary target of opportunity in the sensor's field of view. The goal here is to analyze whether the presence of the second target of opportunity coupled with the sensor following the CV model provides better $\hat{\gamma}_{b_k}$ so that we do not need to change the platform trajectory. Note that the second target of opportunity with a different terrain height could be located anywhere in the sensor's field of view. Figure 3 shows an example of two targets of opportunity located at a distance of d_1m and d_2m from the sensor trajectory.

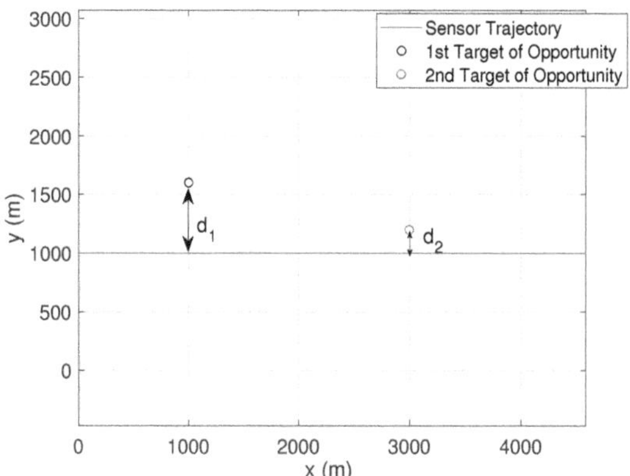

Figure 3. Sensor trajectory following CV model for bias estimation using two targets with known locations and terrain uncertainty.

Based on the assumption that only one target of opportunity is tracked at any given time, the second target of opportunity is picked far away from the first target so that the second target will be in the sensor's field of view for a reasonable time after completing the tracking of the first target. For instance, let us assume that the total number of time steps taken by the sensor for estimating the bias using one target of opportunity with the CV-CT model is K. Now, let us consider that the bias estimation is performed using two targets of opportunity. Denoting K_1 (where $K_1 < K$) as the total number of time steps for which the first target of opportunity is used, we obtain the total number of time steps used for the second target of opportunity as $K_2 = K - K_1$. Note that an equal number of time-steps are considered in both cases for a fair comparison. However, in practice, the total number of time steps depends on the sensor's field of view and the target locations.

A better elevation bias estimate can be obtained while the target of opportunity is located closer to the sensor trajectory. To explain such a result, we analyze the reduction in PCRLB.

To validate the above notion, we provide the following experimental analysis. Denoting the first target of opportunity as $t1$ and the second target target of opportunity as $t2$, from Figure 3, we can write the x-coordinates as $x^{t1} = 1000$ m and $x^{t2} = 3000$ m. Note that, in presence of multiple targets of opportunity, we denote first, second, and third target of opportunity with lower-case letter '$t1$', '$t2$', and '$t3$', respectively. Such notation is used to avoid conflict with the sampling time of sensors, which is denoted by the upper-case letter 'T'. The y-coordinates are changed (while keeping x^{t1} and x^{t2} unchanged) to locate both the '$t1$' and '$t2$' at various relative distances from the sensor trajectory. A comparison of the PCRLB of γ_b (in degree) at the end of the bias estimation is shown in Table 1. In this analysis, the performance of the CV model with two targets is compared with a CV-CT model with one target ($t1$), as described in the previous section.

Table 1. PCRLB (in degrees) of γ_b (in degrees) with different sensor trajectories and various locations of two targets of opportunity.

y-Coordinates		Sensor Trajectories	
y^{t2}	y^{t1}	CV Model	CV-CT Model
1300	1300	**0.0621**	0.0703
	1600	**0.0656**	0.0734
	1900	**0.0687**	0.0742
1600	1300	**0.0659**	0.0703
	1600	**0.0701**	0.0734
	1900	**0.0740**	0.0742
1900	1300	**0.0691**	0.0703
	1600	0.0740	**0.0734**
	1900	0.0786	**0.0742**

From Table 1, the following conclusions are drawn:

- The addition of the second target of opportunity with different terrain height (i.e., different error in the assumed height) provides additional information to the estimator. Therefore, in most of the above simulation scenarios, estimation of $\hat{\gamma}_b$ with the CV model for sensor trajectory and two targets of opportunity provides better performance than that of the CT model for sensor trajectory and one target of opportunity;
- Bias estimation accuracy diminishes with the distance of the targets from the sensor. If the second target of opportunity is further away from the sensor trajectory, the additional information contributed to the bias estimation is insignificant. In such a scenario, the location of the first target of opportunity plays a significant role in the performance of bias estimation. As shown on the {8 and 9}-th row of Table 1, when the targets are relatively far away from the sensor trajectory, sensor trajectory with CV-CT model and one target of opportunity outperforms bias estimation obtained by the CV model along with two targets of opportunity;
- The bias estimation also depends on the error of the assumed height of the targets of opportunity. For our analysis in Table 1, different height errors are used for different targets. As shown on the 3-rd and 7-th rows of Table 1, we obtain different PCRLB estimates even with same y values (($y^{t1} = 1300$, $y^{t2} = 1900$) and ($y^{t1} = 1900$, $y^{t2} = 1300$)).

Additional results involving the Root Mean Square Error (RMSE) plots along with the scenarios of positioning the second target of opportunity on two opposing sides of the sensor trajectory are shown in Section 7.3.

From this analysis, we can conclude that we can obtain a better estimate of the biases by using multiple targets of opportunity without wasting additional time that we discussed in the previous section for the CV-CT model with a single target.

Although the bias estimation discussed in this section only considers targets of opportunity having known locations, we may need to pick an unknown stationary object as a target of opportunity. In the next section, bias estimation with an unknown location of the target of opportunity is introduced.

5.2. Unknown Location with Terrain Uncertainty

Following the same notations from Section 5.1, we consider a stationary target of opportunity with unknown x^{to} and y^{to} to estimate the bias. Let us denote the unknown stationary target and bias state as $\mathbf{x}_k^{to} = [x_k^{to}, y_k^{to}, z_k^{to}]^T$ and $\mathbf{b}_k = [\theta_{b_k}, \gamma_{b_k}]^T$, respectively. In order to estimate both $\hat{\mathbf{x}}^{to}$ and $\hat{\mathbf{b}}_k$ simultaneously, we form the augmented state vector as $\mathbf{x}_k^{\text{aug2}} = [x_k^{to}, y_k^{to}, z_k^{to}, \theta_{b_k}, \gamma_{b_k}]^T$. Our goal here is to estimate the augmented state vector $\hat{\mathbf{x}}_k^{\text{aug2}} = [\hat{x}_k^{to}, \hat{y}_k^{to}, \hat{z}_k^{to}, \hat{\theta}_{b_k}, \hat{\gamma}_{b_k}]^T$, even though we are not interested in the target location.

We can write the initial augmented state vector $\mathbf{x}_1^{\text{aug2}} = [x_1^{to}, y_1^{to}, z_m^{to}, 0, 0]^T$, where x_1^{to} and y_1^{to} are converted position coordinates and z_m^{to} is the assumed target height (equations are provided later in (29) and ((30) of Section 6.2). The measurement covariance matrix $\mathbf{R}_z^{\text{aug2}} = \text{diag}\left(\sigma_\theta^2, \sigma_\gamma^2, \sigma_{z^{to}}^2\right)$ is used to find the initial covariance. Details about the construction of the Hessian matrix, i.e., $\mathbf{H}_z^{\text{aug2}}$, is shown in (A6). Once $\mathbf{H}_z^{\text{aug2}}$ and $\mathbf{R}_z^{\text{aug2}}$ are obtained, we obtain $\mathbf{J}_1^{\text{aug2}}$ from (22). Initial covariance is evaluated as $\mathbf{P}_1^{\text{aug2}} = \left(\mathbf{J}_1^{\text{aug2}}\right)^{-1}$. Non-linear filter is used to update the state $\hat{\mathbf{x}}_k^{\text{aug2}}$.

In order to evaluate the bias estimation performance, PCRLB is evaluated following the formulation of Section 4. First, we evaluate the Hessian matrix $\mathbf{H}_k^{\text{aug2}}$ (details about the matrix construction is shown in (A5)) and the measurement covariance matrix $\mathbf{R}_k^{\text{aug2}} = \text{diag}\left(\sigma_\theta^2, \sigma_\gamma^2\right)$. With $\mathbf{H}_k^{\text{aug2}}$ and $\mathbf{R}_k^{\text{aug2}}$, the measurement contribution of the PCRLB is evaluated as $\mathbf{J}_z^{\text{aug2}}(k) = \left(\mathbf{H}_k^{\text{aug2}}\right)\left(\mathbf{R}_k^{\text{aug2}}\right)^{-1}\mathbf{H}_k^{\text{aug2}}$. Substituting $\mathbf{H}_k^{\text{aug2}}$, $\mathbf{R}_k^{\text{aug2}}$ and $\mathbf{J}_z^{\text{aug2}}(k)$ into (9), PCRLB is evaluated.

In this section also, we study the possibility of improving bias estimation by,

- Changing sensor trajectory from the CV model to the CV-CT model;
- Choosing more than one target of opportunity in the sensor's field of view.

5.2.1. Change in Sensor Trajectory

For a known location of the target of opportunity, in Section 5.1.1, we concluded that a sensor trajectory comprised of the CV-CT model provides a relatively improved bias estimation when compared to that of the CV model. Now, considering the targets of opportunity with unknown locations are considered, we use PCRLB to explain the performance of bias estimation. Considering $\mathbf{H}_k^{\text{aug2}}$ from (A5) (Appendix A), the differentiation involving θ_k is dependent on x_k^{rto} and y_k^{rto}. As a result, a sensor trajectory formed by the CV-CT model provides a reduced estimation error of $\hat{\theta}_k$ by reducing the relative distance between the sensor and the target of opportunity. Note that this was not the case with the known target location. However, for the differentiation involving γ_k, the relative height z_k^{rto} is present in the numerator. As the terrain uncertainty is considered, a similar conclusion on a preferred trajectory can not be drawn for the estimation of $\hat{\gamma}_k$, as opposed to the estimation of $\hat{\theta}_k$.

Similar to Section 5.1, we now expand our analysis by introducing multiple targets of opportunity with unknown locations and terrain uncertainty.

5.2.2. Bias Estimation with Multiple Targets of Opportunity

In this section, we investigate the possibility of improving the bias estimation by introducing multiple targets of opportunity in the sensor's field of view. The sensor follows a trajectory formed by the CV model as opposed to the CV-CT model.

As discussed in Section 5.1, the bias estimation depends on the proximity of the targets of opportunity to the sensor trajectory. To analyze similar dependency for bias estimation with unknown locations, we perform a comparison of PCRLB. Let us consider the ground truth of x-coordinates of the first and second target of opportunity as $x^{t1} = 1000$ m and $x^{t2} = 3000$ m. For 3 different values of y^{t1} and y^{t2}, we obtain three different locations of the target of opportunity based on its proximity to the sensor trajectory. Note that by location, we refer to the ground truth needed to evaluate the PCRLB.

Let us analyze the PCRLB evaluations from Table 2. Firstly, we can draw the following conclusions for $\hat{\theta}_{b_k}$:

- When the sensor follows a trajectory formed by the CV-CT model, one target of opportunity is sufficient to estimate the bearing bias. See Section 5.1.2 for explanations.

Secondly, the following conclusions can be drawn for $\hat{\gamma}_{b_k}$:

- When both the targets of opportunity are relatively far away from the sensor trajectory, a better estimation of $\hat{\gamma}_{b_k}$ is obtained when the sensor follows a trajectory formed by

CV-CT model. The reason behind such result can be attributed to the reduction in relative distance x_k^{rto} and y_k^{rto}, which, in turn, reduces the PCRLB;

- When $t1$ is away from the sensor trajectory, we draw the same conclusions as the known locations of the targets of opportunity, discussed in Section 5.1.

Table 2. PCRLB (in degrees) for θ_{b_k} and γ_{b_k} estimation by changing sensor trajectories for various true locations of the 2 targets of opportunity.

		Sensor Trajectories			
y-Coordinates		for $\hat{\theta}_{b_k}$ (in Degrees)		for $\hat{\gamma}_{b_k}$ (in Degrees)	
y^{t1}	y^{t2}	CV Model	CV-CT Model	CV Model	CV-CT Model
	1800	0.1083	**0.0480**	0.0726	**0.0714**
1400	1600	0.0944	**0.0480**	**0.0694**	0.0714
	1400	0.0828	**0.0480**	**0.0661**	0.0714
	1800	0.1202	**0.0476**	0.0771	**0.0733**
1600	1600	0.1054	**0.0476**	**0.0733**	0.0733
	1400	0.0892	**0.0477**	**0.0695**	0.0733
	1800	0.1286	**0.0472**	0.0817	**0.0740**
1800	1600	0.1107	**0.0472**	0.0770	**0.0739**
	1400	0.0931	**0.0473**	0.0729	**0.0740**

Following the above discussions, following the CV-CT model with one target of opportunity is a better choice than following the CV model with two targets of opportunity. However, adding one more target of opportunity after completing the CV-CT model with the first target of opportunity on the way to the destination with the CV model will help to improve the elevation bias estimate.

6. Tracking Target with Terrain Uncertainty and Sensor Biases

In this section, we provide the algorithm for tracking stationary and moving targets with terrain uncertainty and sensor biases. Note that even if we perform the bias estimation using targets of opportunity, we do not find the exact bias to do the de-biasing before passing the measurements to the tracker. The variance of the bias at the start of the tracking will be larger if we do not perform the step proposed in Section 5. However, the tracking of the target of interest should incorporate possible biases.

Since false alarms are not considered in this paper, the tracking includes initialization and filtering.

6.1. Bias Compensation

For bias compensation, the bias state and the standard deviations are required to be known. In this work, we consider two following cases: bias compensation based on a bias prior and bias estimation using a target of opportunity.

- Let us assume that the bias state is known with a reasonable level of accuracy. In this text, the term reasonable level of accuracy does not refer to any formal definition of accuracy. We denote such a state with the superscript 'deduced'. In other words $\mathbf{b}^{\text{deduced}} = [\theta_b^{\text{deduced}}, \gamma_b^{\text{deduced}}]$ is known a-priori. Additionally, the corresponding bias standard deviations $\sigma_{\theta_b}^{\text{deduced}}$ and $\sigma_{\gamma_b}^{\text{deduced}}$ are known;
- However, in most applications, obtaining such prior information about the bias state and the standard deviation is not practical. Hence, estimating the bias state $\hat{\mathbf{b}} = [\hat{\theta}_b, \hat{\gamma}_b]$ and $\hat{\sigma}_{\theta_b}, \hat{\sigma}_{\gamma_b}$ using a target of opportunity is a more suitable alternative, which is discussed detail in Section 5.

Once the bias state and the corresponding standard deviations are obtained, by either of the two ways described above, we obtain the bias compensated measurements $\mathbf{z}_k^c$ as,

$$\mathbf{z}_k^c = \mathbf{z}_k - \mathbf{z}_k^{\text{prior}} \tag{27}$$

where $\mathbf{z}_k^{\text{prior}}$ is the measurement obtained by substituting $\mathbf{b}_k$ with $\mathbf{b}_k^{\text{deduced}}$ or $\hat{\mathbf{b}}_k$ in (5). Similarly, the modified measurement covariance matrix $\mathbf{R}_k^c$ is obtained as,

$$\mathbf{R}_k^c = \begin{bmatrix} \sigma_\theta^2 + \hat{\sigma}_{\theta_b}^2 & 0 \\ 0 & \sigma_\gamma^2 + \hat{\sigma}_{\gamma_b}^2 \end{bmatrix} \tag{28}$$

When the bias standard deviations are known a priori, $\hat{\sigma}_{\theta_b}$ and $\hat{\sigma}_{\gamma_b}$ are substituted with $\sigma_{\theta_b}^{\text{deduced}}$ and $\sigma_{\gamma_b}^{\text{deduced}}$, respectively.

6.2. Initialization

Recalling the discussions from Section 3, the height of the ground target is obtained from DTED and it can be used for track initialization. Note that we should not use this information again in the filtering steps to avoid double counting. From the target-sensor geometry shown in Figure 1, we can write the converted Cartesian coordinates of the target state as,

$$x_1^t = \frac{z_m^t}{\tan(\gamma_1)} \sin(\theta_1) \tag{29}$$

$$y_1^t = \frac{z_m^t}{\tan(\gamma_1)} \cos(\theta_1) \tag{30}$$

With the x and y coordinates from (29) and (30) along with z_m^t, the initial state of the ground target is written as $\mathbf{x}_1^t = [x_1^t, 0, y_1^t, 0, z_m^t]^T$. Note that the initial velocity along the x and y-axis, $\dot{x}_1^t$ and $\dot{y}_1^t$, are considered to be 0 in this one-point initialization [33]. Similarly, for the stationary target we can write $\mathbf{x}_1^t = [x_1^t, y_1^t, z_m^t]^T$.

In order to initialize the covariance $\mathbf{P}_1$, let us first introduce the covariance associated with measurements, including the terrain height, as $\mathbf{R}_z^c = \text{diag}\left(\sigma_\theta^2 + \hat{\sigma}_{\theta_b}^2, \sigma_\gamma^2 + \hat{\sigma}_{\gamma_b}^2, \sigma_{z^t}^2\right)$. Similar to the other sections of this paper, we reserve the use of subscript 'z' to denote the initial time step when the height information is considered to evaluate the measurement covariance and the Hessian matrices. Note that $\mathbf{R}_z^c$ is different from that of the noise covariance $\mathbf{R}_k^c$, as the terrain uncertainty is considered only for the filter initialization. Now following (9), we evaluate $\mathbf{J}_1$ to obtain $\mathbf{P}_1 = \mathbf{J}_1^{-1}$. First, the measurement contribution for initialization $\mathbf{J}_z(1)$ can be evaluated as $\mathbf{J}_z(1) = \mathbf{H}_z^T (\mathbf{R}_z^c)^{-1} \mathbf{H}_z$, where the initial Jacobian matrix $\mathbf{H}_z$ is,

$$\mathbf{H}_z = \begin{bmatrix} \frac{\partial \theta_1}{\partial x_1^t} & \frac{\partial \theta_1}{\partial \dot{x}_1^t} & \frac{\partial \theta_1}{\partial y_1^t} & \frac{\partial \theta_1}{\partial \dot{y}_1^t} & \frac{\partial \theta_1}{\partial z_1^t} \\ \frac{\partial \gamma_1}{\partial x_1^t} & \frac{\partial \gamma_1}{\partial \dot{x}_1^t} & \frac{\partial \gamma_1}{\partial y_1^t} & \frac{\partial \gamma_1}{\partial \dot{y}_1^t} & \frac{\partial \gamma_1}{\partial z_1^t} \\ \frac{\partial z_1^t}{\partial x_1^t} & \frac{\partial z_1^t}{\partial \dot{x}_1^t} & \frac{\partial z_1^t}{\partial y_1^t} & \frac{\partial z_1^t}{\partial \dot{y}_1} & \frac{\partial z_1^t}{\partial z_1^t} \end{bmatrix} \tag{31}$$

Evaluating the partial derivatives from (31), we can write the analytical form of $\mathbf{H}_z$ as,

$$\mathbf{H}_z = \begin{bmatrix} \frac{y_1^r}{x_1^{r2} + y_1^{r2}} & 0 & \frac{-x_1^r}{x_1^{r2} + y_1^{r2}} & 0 & 0 \\ \frac{-x_1^r z_1^t}{\sqrt{x_1^{r2} + y_1^{r2}}\left(x_1^{r2} + y_1^{r2} + z_1^{r2}\right)} & 0 & \frac{-y_1^r z_1^t}{\sqrt{x_1^{r2} + y_1^{r2}}\left(x_1^{r2} + y_1^{r2} + z_1^{r2}\right)} & 0 & \frac{\sqrt{x_1^{r2} + y_1^{r2}}}{x_1^{r2} + y_1^{r2} + z_1^{r2}} \\ 0 & 0 & 0 & 0 & 1 \end{bmatrix} \tag{32}$$

where x_1^r, y_1^r and z_1^r denote the coordinates of the relative state at time step $k = 1$ ($\mathbf{x}_1^{rt} = \mathbf{x}_1^t - \mathbf{x}_1^o$).

For the moving target, maximum possible target velocity, v_{max}, is used to evaluate $\mathbf{J}_p = \text{diag}\left(0, \frac{1}{v_{max}^2}, 0, \frac{1}{v_{max}^2}, 0\right)$. The initial covariance $\mathbf{P}_1$ is obtained by $\mathbf{P}_1 = (\mathbf{J}_z(1) + \mathbf{J}_p)^{-1}$. Evidently, for the stationary target $\mathbf{J}_1 = \mathbf{J}_z(1)$ and initial covariance $\mathbf{P}_1 = \mathbf{J}_z^{-1}$. Note that the Hessian matrix, $\mathbf{H}_z^S$, for the stationary target is expressed as,

$$
\mathbf{H}_z^S = \begin{bmatrix} \frac{y_1^r}{x_1^{r2}+y_1^{r2}} & \frac{-x_1^r}{x_1^{r2}+y_1^{r2}} & 0 \\ \frac{-x_1^r z_1^r}{\sqrt{x_1^{r2}+y_1^{r2}}\left(x_1^{r2}+y_1^{r2}+z_1^{r2}\right)} & \frac{-y_1^r z_1^r}{\sqrt{x_1^{r2}+y_1^{r2}}\left(x_1^{r2}+y_1^{r2}+z_1^{r2}\right)} & \frac{\sqrt{x_1^{r2}+y_1^{r2}}}{x_1^{r2}+y_1^{r2}+z_1^{r2}} \\ 0 & 0 & 1 \end{bmatrix} \tag{33}
$$

The superscript 'S' in $\mathbf{H}_z^S$, of (33), indicates the stationary target.

6.3. Filtering

Once the filter is initialized and the measurements are bias compensated, UKF is used to handle the non-linearity and obtain the estimated target state $\hat{\mathbf{x}}_k^t$ and the associated covariance $\hat{\mathbf{P}}_k$ [30]. The effect of terrain uncertainty is already considered in obtaining the initial state and the associated covariance. Hence, only the bearing and elevation measurements are used to update the state. Non-linear filtering uses (1) for the state prediction and (5) for the measurement update. Proposed filtering approach using UKF involves the following steps: calculation of the sigma points, measurement prediction and update. These processes are well known and not discussed in this paper.

Now, in the next section, we present the simulation results for tracking both the stationary and moving targets with terrain uncertainty using a biased angle-only sensor.

7. Simulations

7.1. Parameters

The target-sensor geometry is shown in Figure 4. We consider both mobile and stationary targets for this simulation as shown in Figure 4a,b, respectively. The own-ship follows a trajectory formed by the CV-CT model. Parameters used in this simulation are provided in Table 3. Note that in all initial states, the positions along x, y, and z-coordinates are in meter whereas the velocities $\dot{x}$, $\dot{y}$ and $\dot{z}$ are in m/s.

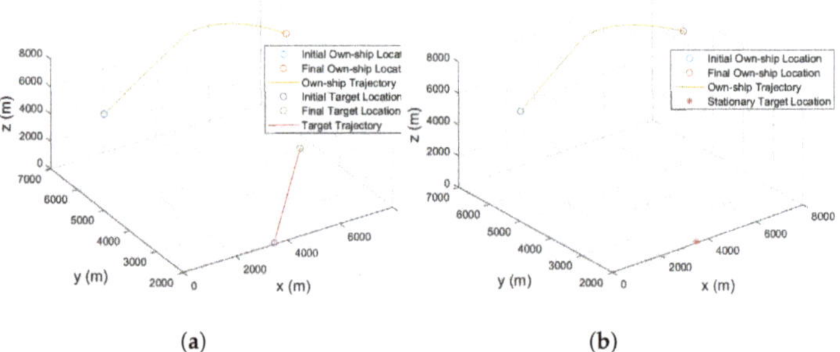

(a)   (b)

Figure 4. Geometry of the airborne sensor and the ground target with terrain uncertainty. (a) Target moves in nearly constant velocity. (b) Target remains stationary.

Table 3. Parameters.

Parameters	Value
Initial state of the own-ship $\left(\mathbf{x}_1^o = [x_1^o, \dot{x}_1^o, y_1^o, \dot{y}_1^o, z_1^o]\right)$	$[20, 40, 5000, 15, 7000]^T$
Initial state of the mobile target $\left(\mathbf{x}_1^t = [x_1^t, \dot{x}_1^t, y_1^t, \dot{y}_1^t, z_1^t]\right)$	$[3500, 20, 2000, 15, 100]^T$
Initial state of the stationary target $\left(\mathbf{x}_1^t = [x_1^t, y_1^t, z_1^t]\right)$	$[3500, 2000, 100]^T$
Standard deviation of terrain uncertainty (σ_{z^t})	10 m
Bearing measurement noise standard deviation (σ_θ)	$0.4°$
Elevation measurement noise standard deviation (σ_γ)	$0.2°$
Sampling time (T)	1 s
Maximum velocity of the ground target (v_{max})	35 m/s
Bearing bias (θ_{b_k})	$3°$
Elevation bias (γ_{b_k})	$1°$
Total simulation time	200 s

Considering $\sigma_{z^t} = 10$ m, we obtain the error associated with the terrain uncertainty as $v^z \sim \mathcal{N}(v^z; 0, 10^2)$. The moving target trajectory follows the CV model.

7.2. Performance Bound

Figure 5a,b show the position error plots for tracking the stationary and moving ground target, respectively. The position errors are obtained by performing square root on the sum of the diagonal elements of the covariance matrix (elements representing the x, y, and z coordinates). Position errors for both the biased PCRLB (see Section 4.4) and the PCRLB (see Section 4.2) are evaluated to compare with that of the proposed filtering approach. In order to evaluate the position error obtained by the estimated target states using the proposed filtering approach, we performed 500 Monte Carlo simulations.

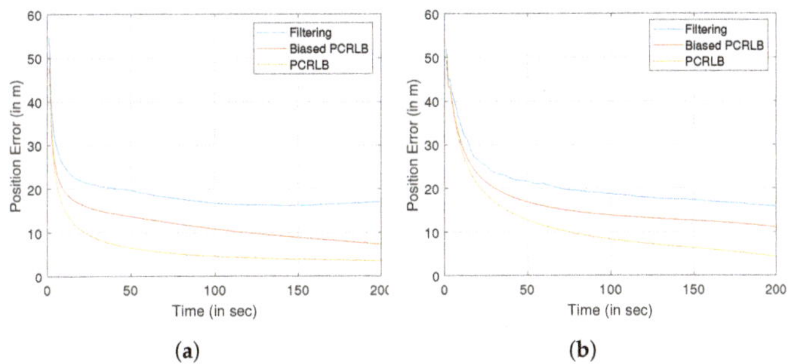

Figure 5. Comparison of the performance bounds. (**a**) Stationary ground target. (**b**) Mobile ground target.

From Figure 5, it is clear that the biased PCRLB would provide a tighter bound on position error when compared to the standard PCRLB.

7.3. Bias Estimation

In this section, we analyze the bias estimation for the true values of $\theta_{b_k} = 1°$ and $\gamma_{b_k} = 1°$. Non-linear filtering based on UKF is used to estimate the bias state using the measurements obtained from the stationary target of opportunity. Note that, all the targets of opportunity considered in this section have terrain uncertainty. We show the simulation results for the two following scenarios. In scenario 1, we consider the location of the target of opportunity is known with terrain uncertainty. In scenario 2, the location of the target of opportunity is unknown.

7.3.1. Scenario 1

Recalling the discussions from Section 5, one (or multiple) target of opportunity is chosen to estimate the bias. Figure 6 shows the RMSE values for the bias estimation using one stationary target of opportunity with known location. We consider the initial state of the target of opportunity as $\mathbf{x}_1^{to} = [x_1^{to}, y_1^{to}, z_1^{to}]^T = [1000, 1200, 100]^T$. The initial state of the own-ship is $\mathbf{x}_1^{o} = [x_1^{o}, \dot{x}_1^{o}, y_1^{o}, \dot{y}_1^{o}, z_1^{o}]^T = [0, 10, 1000, 0, 2500]^T$.

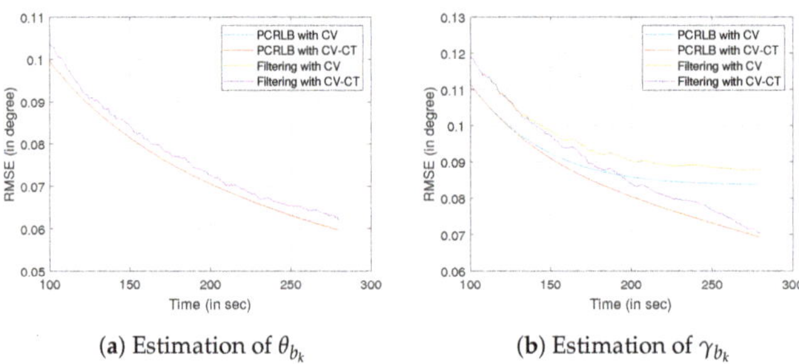

(a) Estimation of θ_{b_k} (b) Estimation of γ_{b_k}

Figure 6. RMSE evaluation for bias estimation using one target of opportunity with known location and terrain uncertainty.

To simulate the effect of change in sensor trajectory, discussed in Section 5.1.1, the CV model and the CV-CT model are used in this simulation. We obtain $K_{CV} = \frac{1000}{10} = 100$ sec as the time-step where the own-ship trajectory switches from the CV model to the CT model (when the CV-CT model is considered). As discussed in Section 5.1.1, the CV-CT model reduces the relative distances between the sensor and the target of opportunity. This leads to the reduction in x_k^{rto} and y_k^{rto}, causing the reduction in PCRLB. Hence, justifiably, the CV-CT model improves the estimate $\hat{\gamma}_{b_k}$ when compared to that of the sensor trajectory formed by the CV model. Recalling the discussions from Section 5.1.1, differentiation with respect to θ_k for obtaining the Hessian matrix is constant. As a result, for estimating $\hat{\theta}_{b_k}$ there is no change in the RMSE evaluation (or PCRLB) while the trajectories are changed. Note that the difference in RMSE, for both the CV and the CV-CT, is negligible before the time step $k = 100$ s. Hence, to produce a meaningful RMSE comparison for the said sensor trajectories, the last 181 time steps (starting from $k = 100$ s) are shown in Figure 6.

Now, we simulate the bias estimation results considering multiple targets of opportunity. Recalling the conclusions from Section 5.1.2, two targets of opportunity with known x and y coordinates, each having terrain uncertainty, are chosen. The sensor follows a trajectory formed by the CV model. The first target of opportunity is initialized as $\mathbf{x}_1^{t1} = [x_1^{t1}, y_1^{t1}, z_1^{t1}]^T = [1000, 1600, 100]^T$. Keeping $z_1^{t2} = z_1^{t1}$, the second target of opportunity is initialized as $\mathbf{x}_1^{t2} = [x_1^{t2}, y_1^{t2}, z_1^{t2}]^T = [3000, 1200, 100]^T$. For this simulation, we assume that $t2$ is introduced in the sensor's field of view at $K_1 = 2 \times K_{CV}$. Evidently, from Figure 7b, it is clear that the RMSE, as well as the PCRLB, for the estimate $\hat{\gamma}_{b_k}$ drops when $t2$ is introduced at K_1 time-step. However, the addition of $t2$ has no impact on the RMSE of $\hat{\theta}_{b_k}$ estimate.

The RMSE (as well as the PCRLB) obtained using one target of opportunity, while the sensor follows the CV-CT model, are also plotted. When compared with the PCRLB obtained using two targets of opportunity with the CV model, one target of opportunity with the CV-CT model provides inferior $\hat{\gamma}_{b_k}$ estimates. The reason behind such simulation results can be largely attributed to the information gain facilitated by the addition of the target of opportunity t_2. However, for the θ_{b_k} estimation no performance difference is observed. This result further confirms the conclusions obtained from Figure 6a, where we verified the change in the sensor trajectory does not impact on the θ_{b_k} estimation. Figure 7 also validates the conclusions drawn in Section 5.1.2. 100 Monte Carlo runs are performed to obtain these simulation results. Note that the differences in RMSE and PCRLB, for all the considered cases, are negligible before the time step $k = 100$ s. Hence, to produce a meaningful comparison between the bias estimation approaches, only the last 360 time steps (starting from $k = 100$ s) are considered in Figure 7.

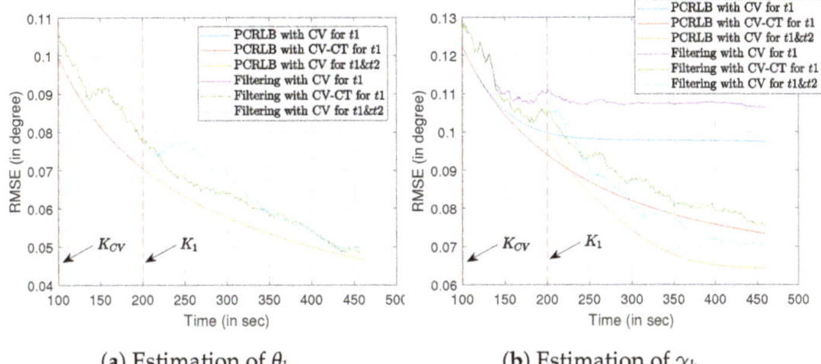

(a) Estimation of θ_{b_k} (b) Estimation of γ_{b_k}

Figure 7. RMSE evaluation for bias estimation using two targets of opportunity with a known location. Terrain uncertainty is present with the target height information.

Next, we analyze the significance of its close proximity to the sensor trajectory. We consider the second target of opportunity is introduced at $K_1 = 200$ s time step.

Table 1 (from Section 5.1.2) showed the RMSE comparison while the $t2$ is located either close or away from the sensor trajectory. Now we consider the possibility of $t2$ to be located on either side of the sensor trajectory. In Figure 8a, $t2$ is located at a distance of $d_2 = 300$ m from the sensor trajectory. The initial state vector of $t2$, is given as $\mathbf{x}_1^{t2} = [x_1^{t2}, y_1^{t2}, z_1^{t2}]^T = [3000, 1300, 100]^T$. Whereas, we consider $t3$ as the second target of opportunity being located $d_3 = 300$ m away from the sensor trajectory, with the initial state vector expressed as $\mathbf{x}_1^{t3} = [x_1^{t3}, y_1^{t3}, z_1^{t3}]^T = [3000, 700, 100]^T$. We consider $\sigma_{z^{t2}} = \sigma_{z^{t3}} = 10$ m, for the simulation. Once $\mathbf{x}_1^{t2}$ and $\mathbf{x}_1^{t3}$ are initialized, the proposed filtering approach is used to estimate the bias state $\hat{\mathbf{b}}_k$. Note that the first target of opportunity ($t1$) is located at a distance of $d_1 = 600$ m from the sensor trajectory.

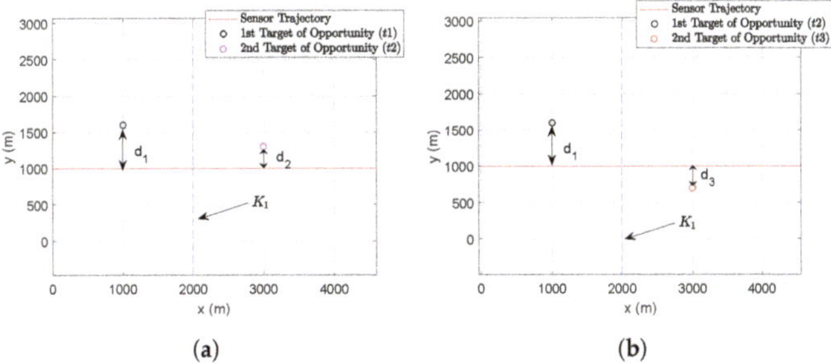

(a)   (b)

Figure 8. Sensor trajectory formed by the CV model to estimate bias using two known targets of opportunity with additional terrain uncertainty. (a) Second target of opportunity ($t2$) located d_2m away from sensor trajectory. (b) Second target of opportunity ($t3$) located d_3m away the sensor trajectory.

The PCRLB is a function of the relative distance between the sensor and the target of opportunity. Hence, the bias estimates are not impacted by which side the target of opportunity is located with respect to the sensor trajectory. RMSE evaluation results, presented in Figure 9b, validate such notion. Close proximity of the target of opportunity to the sensor trajectory reduces the relative distances, which, in turn, impacts the bias estimation as shown in Table 1.

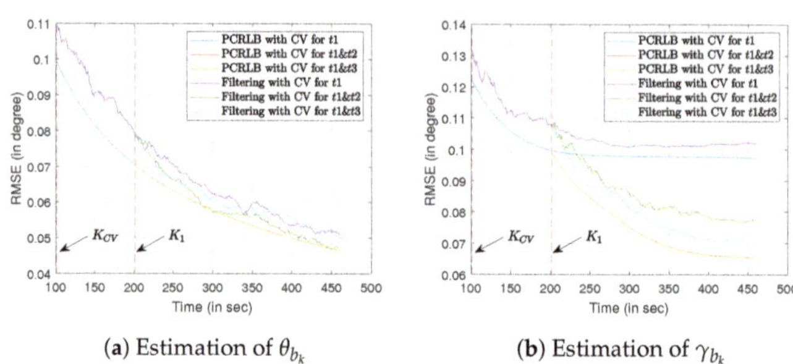

(**a**) Estimation of θ_{b_k}

(**b**) Estimation of γ_{b_k}

Figure 9. RMSE evaluation for bias estimation using two targets of opportunity with known location and terrain uncertainty. The second target of opportunity is located on either side of the sensor trajectory.

Similar to Figure 7, to produce a meaningful comparison between the bias estimation approaches, only the last 360 time steps (starting from $k = 100$ s) are considered in Figure 9. In total, 100 Monte Carlo simulations are performed to obtain all the Figures shown in Scenario 1.

7.3.2. Scenario 2

We consider two targets of opportunity along with two types of sensor trajectories that are formed with the CV and the CV-CT models. Figure 10 shows the RMSE and PCRLB values of the bias estimation when the x and y coordinates of the targets of opportunity are unknown.

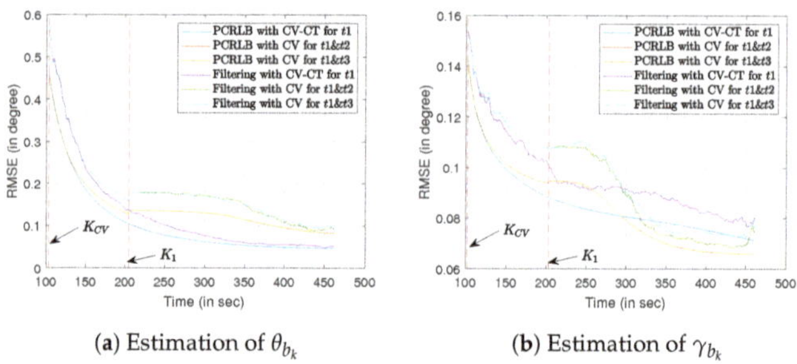

(**a**) Estimation of θ_{b_k}

(**b**) Estimation of γ_{b_k}

Figure 10. RMSE evaluation for bias estimation using two targets of opportunity with unknown location and terrain uncertainty. The second target of opportunity is located on either side of the sensor trajectory.

Ground truth for the first target of opportunity ($t1$) is $\mathbf{x}_1^{t1} = [x_1^{t1}, y_1^{t1}, z_1^{t1}]^T = [1000, 1400, 100]^T$. The ground truth for the second target of opportunity, which is $d_2 = 300$ m away from the sensor trajectory, $\mathbf{x}_1^{t2} = [x_1^{t2}, y_1^{t2}, z_1^{t2}]^T = [3000, 1400, 100]^T$. Similarly, the ground truth of the target of opportunity $t3$, located $d_3 = 300$ m away from the sensor trajectory on the opposite side, can be expressed as $\mathbf{x}_1^{t3} = [x_1^{t3}, y_1^{t3}, z_1^{t3}]^T = [3000, 600, 100]^T$. Note that only one of the two targets from $t2$ and $t3$ is used with target $t1$. The proposed filter is initialized with the converted measurements obtained at time step $k = 1$.

Similar to Scenario 1, the own-ship trajectory switches from the CV to the CT model at the time step $K_{CV} = 100$ s (shown in Figure 10a). The sensor starts tracking the second target of opportunity at $K_1 = 200$ s time step. Similar to Scenario 1, it is assumed that the second target of opportunity (either $t2$ or $t3$) is made available to the sensor's field of view at $K_1 = 2 \times K_{CV}$ time step. Unlike the estimation of θ_{b_k} in scenario 1, RMSE (as well as the PCRLB) obtained by using one target of opportunity while following the CV-CT model is lower than that of using two targets of opportunity and the CV model. As the x and y coordinates are unknown for both the targets of opportunity, adding $t2$ (or $t3$) in addition to that of $t1$ does not improve the estimation of θ_{b_k}. However, as the height information of both the targets of opportunity is known, the addition of $t2$ (or $t3$) improves the estimation of γ_{b_k}, as shown in Figure 10b. However, because of the presence of the terrain uncertainty for the first few time steps, the CV-CT model with one target of opportunity provides a superior estimation of γ_{b_k}. With time, as the filter converges, the improvement caused by the addition of the second target of opportunity while following the CV model becomes evident. Similar to that of Section 1, no improvement is observed in the PCRLB estimates while using $t2$ as opposed to $t3$.

Similar to Scenario 1, only the last 360 time steps (starting from $k = 100$ s) are considered in Figure 10 to produce a meaningful comparison between the bias estimation approaches. We performed 100 Monte Carlo simulations to obtain all the results of Scenario 2.

Note that the ground truth of the bearing and the elevation biases of $\theta_{b_k} = 1°$ and $\gamma_{b_k} = 1°$ are chosen to validate our proposed bias estimation approaches. This is not to be confused with the ground truth values of $\theta_{b_k} = 3°$ and $\gamma_{b_k} = 1°$, as introduced in Table 3, for tracking the original target. In the next section, for bias compensation, we choose the bias estimation approach involving two targets of opportunities with known x and y coordinates and the CV sensor trajectory.

7.4. Tracking Using Angle-Only Measurements

Let us recall the original tracking problem, where a ground target with terrain uncertainty is tracked using a biased airborne angle-only sensor. Parameters for both the sensor and the target initialization are presented in Table 3. The estimates of the biases from a sample run are $\hat{\theta}_{b_k} = 2.89°$ and $\hat{\gamma}_{b_k} = 1.03°$ with the associated variances $\sigma_{\hat{\theta}_{b_k}} = 0.0597°$ and $\sigma_{\hat{\gamma}_{b_k}} = 0.0834°$.

Once bias compensation is performed to reduce the bias uncertainty, proposed filtering approach is used for tracking both the moving or stationary target with terrain uncertainty. Figures 11a and 12a show the 3D CCS representation of the sensor trajectories and the estimated trajectories (or the estimates for the stationary target). The corresponding 2D CCS representations are shown in Figures 11b and 12b, respectively. The RMSE plots along with the corresponding PCRLBs are shown in Figure 13. To establish a baseline, we evaluate the RMSE and the PCRLB considering no measurement bias, i.e., $\theta_{b_k} = 0°$ and $\gamma_{b_k} = 0°$. Corresponding RMSE plots considering both moving and stationary ground targets, with terrain uncertainty, are shown in Figure 14a,b, respectively.

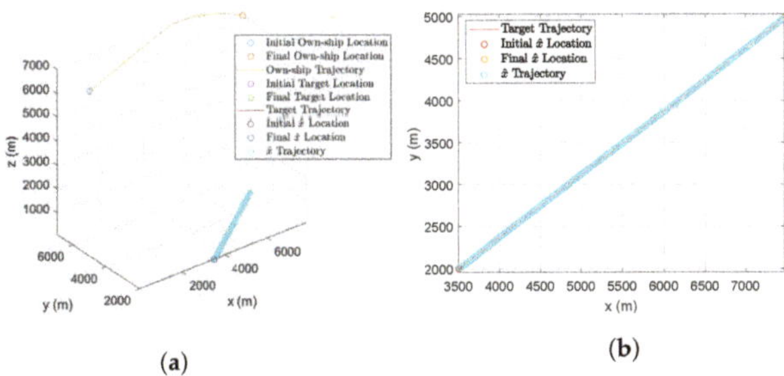

(a)   (b)

Figure 11. Tracking a mobile ground target with terrain uncertainty using a biased angle-only airborne sensor. (**a**) 3D CCS representation of the target-sensor geometry with the estimated trajectory. (**b**) 2D CCS representation of the target and the estimated trajectory.

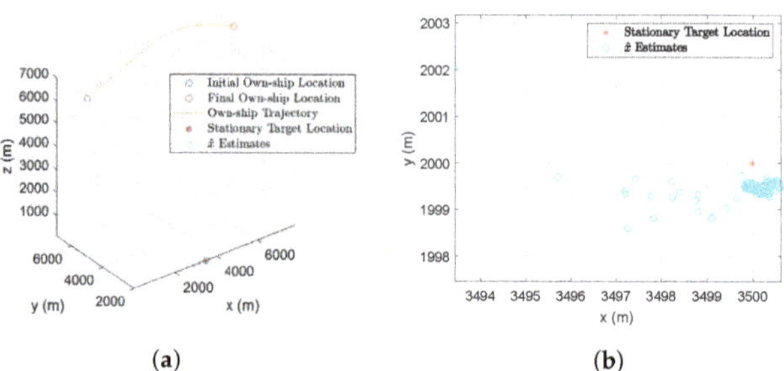

(a)   **(b)**

Figure 12. Tracking a stationary ground target with terrain uncertainty using an airborne angle-only sensor. (**a**) 3D CCS representation of the target own-ship geometry along with the estimates. (**b**) 2D CCS representation of the stationary target and the estimates.

The effect of bias uncertainty on the estimates is observed in Figure 12b. A handful of initial estimates are away from the original target. However, as the filter converges, the effect of bias uncertainty is minimized. Although a similar estimation error is caused by the bias uncertainty while considering the moving target tracking, it is not clearly observed in Figure 11b because of the scaling of the y-axis. The RMSE analysis, shown in Figure 13, validates our approach for both moving and stationary targets. All the simulation results in this section are obtained by performing an average of over 500 Monte Carlo runs.

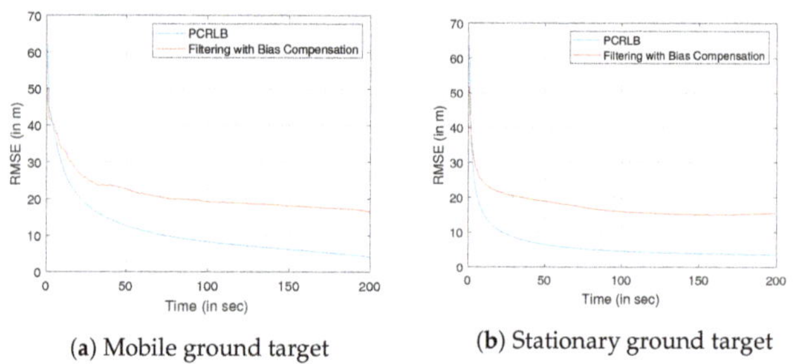

(a) Mobile ground target   **(b)** Stationary ground target

Figure 13. RMSE evaluation for tracking a ground target using airborne sensor platform with angle-only bias-compensated measurements.

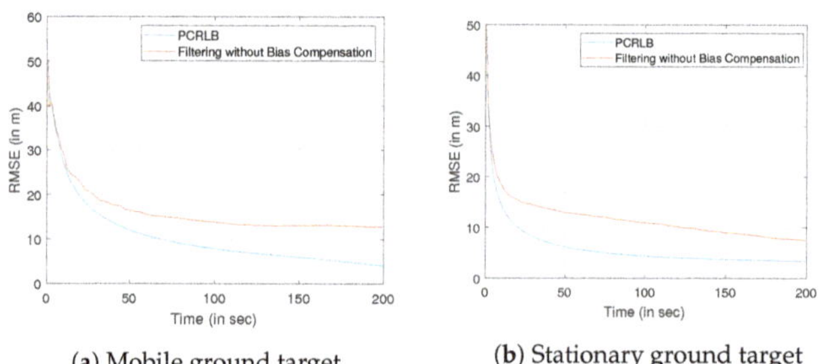

(a) Mobile ground target   **(b)** Stationary ground target

Figure 14. RMSE evaluation for tracking a ground target using an airborne sensor platform with unbiased angle-only measurements.

Now, we analyze the performance of the proposed approach with a conventional method, where the terrain uncertainty is ignored, for tracking both the stationary and moving ground targets. The RMSE comparison with $\sigma_{z^t} = 30$ m is shown in Figure 15. As expected, our proposed approach provides better RMSE for both the stationary and the moving ground targets. Although we ignore the terrain uncertainty, we use a wrong value for the height of the target. Hence, it results in a bias in the estimate and that bias will not even be reduced with more measurements. That is what we see in Figure 15 for both cases. For moving target, because of the target sensor geometry, significant change in the RMSE is noticed only after the first 300 s. Both the simulations were performed by taking an average of over 100 Monte Carlo runs.

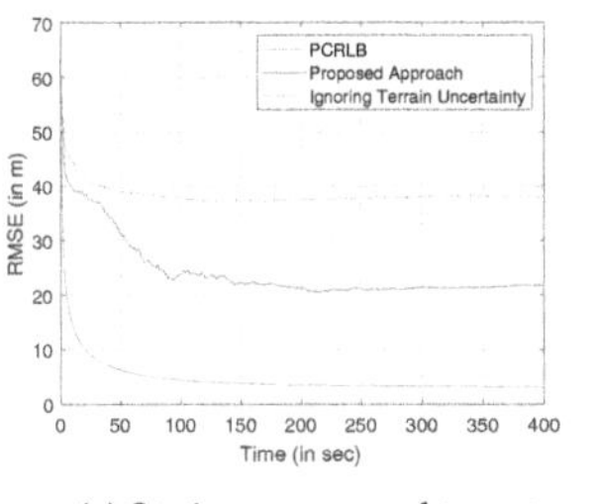

(**a**) Stationary ground target

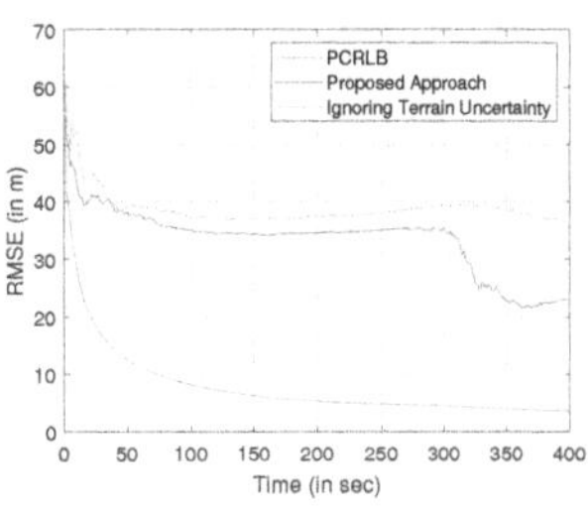

(**b**) Mobile ground target

Figure 15. Comparison of the RMSE of the proposed approach with an approach that ignores the terrain uncertainty, when the $\sigma_{z^t} = 30$ m.

8. Conclusions

In this paper, we have considered the 3-D tracking of a ground target with terrain uncertainty using a biased angle-only airborne sensor. We derived the PCRLB bound for the problem with sensor bias and terrain uncertainty. We provided a bias gradient-based PCRLB formulation to find a tighter bound under biases. We showed that the biased PCRLB provides a tighter lower bound when compared with the PCRLB while evaluating position error. Using the derived PCRLB, we proposed a method to pick an optimal target(s) of opportunity and optimal platform trajectory to estimate the bias. We demonstrated that tracking of a ground target in 3-D could be performed with biased angle-only measurements using UKF as a preferred non-linear filtering method.

Author Contributions: Conceptualization, R.T. and D.M.; methodology, D.M., A.B. and R.T.; software, D.M. and A.B.; validation, D.M., A.B. and R.T.; formal analysis, D.M. and A.B.; investigation, D.M. and A.B.; resources, R.T.; data curation, D.M. and A.B.; writing—original draft preparation, D.M.; writing—review and editing, D.M. and R.T.; visualization, D.M. and A.B.; supervision, R.T.; project administration, R.T.; funding acquisition, R.T. All authors have read and agreed to the published version of the manuscript.

Funding: This work was supported by the Natural Sciences and Engineering Research Council of Canada (NSERC).

Institutional Review Board Statement: Not applicable.

Informed Consent Statement: Not applicable.

Data Availability Statement: Not applicable.

Conflicts of Interest: The authors declare no conflict of interest.

Appendix A

Let us denote the state transition matrices for the CV and CT motion as $\mathbf{F}_{CV}$ and $\mathbf{F}_{CT}$. We also denote the state of the ground target and the own-ship as $\mathbf{x}_k^t = [x_k^t, \dot{x}_k^t, y_k^t, \dot{y}_k^t, z_k^t]^T$ and $\mathbf{x}_k^o = [x_k^o, \dot{x}_k^o, y_k^o, \dot{y}_k^o, z_k^o]^T$. Given the sampling time T and turn rate ω, we can write,

$$\mathbf{F}_{CV} = \begin{bmatrix} 1 & T & 0 & 0 & 0 \\ 0 & 1 & 0 & 0 & 0 \\ 0 & 0 & 1 & T & 0 \\ 0 & 0 & 0 & 1 & 0 \\ 0 & 0 & 0 & 0 & 1 \end{bmatrix} \quad \mathbf{F}_{CT} = \begin{bmatrix} 1 & \frac{\sin(\omega T)}{\omega} & 0 & \frac{\cos(\omega T)-1}{\omega} & 0 \\ 0 & \cos(\omega T) & 0 & -\sin(\omega T) & 0 \\ 0 & \frac{1-\cos(\omega T)}{\omega} & 1 & \frac{\sin(\omega T)}{\omega} & 0 \\ 0 & \sin(\omega T) & 0 & \cos(\omega T) & 0 \\ 0 & 0 & 0 & 0 & 1 \end{bmatrix} \tag{A1}$$

Note, $\mathbf{F}_{CV}$ is used to model the state transition of both the own-ship and the ground target (when the ground target moves with nearly CV). However, $\mathbf{F}_{CT}$ is used to model the state transition of the own-ship for the bias estimation when the CV-CT model is used.

In order to model the system dynamics of the target, gain matrix G is required to obtain the process noise. Gain matrix $\mathbf{G}$ is written as,

$$\mathbf{G} = \begin{bmatrix} \frac{T^2}{2} & 0 \\ T & 0 \\ 0 & \frac{T^2}{2} \\ 0 & T \\ 0 & 0 \end{bmatrix} \tag{A2}$$

We denote the relative state vector as $\mathbf{x}_k^r = \mathbf{x}_k^t - \mathbf{x}_k^o$, where $\mathbf{x}_k^t$ and $\mathbf{x}_k^o$ are the state vector of the target and the own-ship, respectively. Considering the angle-only measurements $\mathbf{z}_k = [\theta_k, \gamma_k]$, the Jacobian matrix for the measurements can be formed as,

$$\mathbf{H}_k = \begin{bmatrix} \frac{\partial \theta_k}{\partial x_k^t} & \frac{\partial \theta_k}{\partial \dot{x}_k^t} & \frac{\partial \theta_k}{\partial y_k^t} & \frac{\partial \theta_k}{\partial \dot{y}_k^t} & \frac{\partial \theta_k}{\partial z_k^t} \\ \frac{\partial \gamma_k}{\partial x_k^t} & \frac{\partial \gamma_k}{\partial \dot{x}_k^t} & \frac{\partial \gamma_k}{\partial y_k^t} & \frac{\partial \gamma_k}{\partial \dot{y}_k^t} & \frac{\partial \gamma_k}{\partial z_k^t} \end{bmatrix}$$

$$= \begin{bmatrix} \frac{y_k^r}{x_k^{r2}+y_k^{r2}} & 0 & \frac{-x_k^r}{x_k^{r2}+y_k^{r2}} & 0 & 0 \\ \frac{-x_k^r z_k^r}{\sqrt{x_k^{r2}+y_k^{r2}}\left(x_k^{r2}+y_k^{r2}+z_k^{r2}\right)} & 0 & \frac{-y_k^r z_k^r}{\sqrt{x_k^{r2}+y_k^{r2}}\left(x_k^{r2}+y_k^{r2}+z_k^{r2}\right)} & 0 & \frac{\sqrt{x_k^{r2}+y_k^{r2}}}{x_k^{r2}+y_k^{r2}+z_k^{r2}} \end{bmatrix} \tag{A3}$$

For track initialization at $k = 1$, the Jacobian matrix of the angle-only measurements and the terrain height measurement can be formed as,

$$\mathbf{H}_z = \begin{bmatrix} \frac{\partial \theta_1}{\partial x_1^t} & \frac{\partial \theta_1}{\partial \dot{x}_1^t} & \frac{\partial \theta_1}{\partial y_1^t} & \frac{\partial \theta_1}{\partial \dot{y}_1^t} & \frac{\partial \theta_1}{\partial z_1^t} \\ \frac{\partial \gamma_1}{\partial x_1^t} & \frac{\partial \gamma_1}{\partial \dot{x}_1^t} & \frac{\partial \gamma_1}{\partial y_1^t} & \frac{\partial \gamma_1}{\partial \dot{y}_1^t} & \frac{\partial \gamma_1}{\partial z_1^t} \\ \frac{\partial z_1}{\partial x_1^t} & \frac{\partial z_1}{\partial \dot{x}_1^t} & \frac{\partial z_1}{\partial y_1^t} & \frac{\partial z_1}{\partial \dot{y}_1^t} & \frac{\partial z_1}{\partial z_1^t} \end{bmatrix}$$

$$= \begin{bmatrix} \frac{y_1^r}{x_1^{r2}+y_1^{r2}} & 0 & \frac{-x_1^r}{x_1^{r2}+y_1^{r2}} & 0 & 0 \\ \frac{-x_1^r z_1^r}{\sqrt{x_1^{r2}+y_1^{r2}}\left(x_1^{r2}+y_1^{r2}+z_1^{r2}\right)} & 0 & \frac{-y_1^r z_1^r}{\sqrt{x_1^{r2}+y_1^{r2}}\left(x_1^{r2}+y_1^{r2}+z_1^{r2}\right)} & 0 & \frac{\sqrt{x_1^{r2}+y_1^{r2}}}{x_1^{r2}+y_1^{r2}+z_1^{r2}} \\ 0 & 0 & 0 & 0 & 1 \end{bmatrix} \tag{A4}$$

To estimate the bias with unknown targets, the state vector is formulated as $\mathbf{x}_k^{\text{aug2}} = [x_k^{to}, y_k^{to}, z_k^{to}, \theta_{b_k}, \gamma_{b_k}]^T$. Knowing the measurement vector $\mathbf{z}_k = [\theta_k, \gamma_k]$, partial derivatives with respect to the augmented state is shown below,

$$\mathbf{H}_k^{aug2} = \begin{bmatrix} \frac{\partial \theta_k}{\partial x_k^{to}} & \frac{\partial \theta_k}{\partial y_k^{to}} & \frac{\partial \theta_k}{\partial z_k^{to}} & \frac{\partial \theta_k}{\partial \theta_{b_k}} & \frac{\partial \theta_k}{\partial \gamma_{b_k}} \\ \frac{\partial \gamma_k}{\partial x_k^{to}} & \frac{\partial \gamma_k}{\partial y_k^{to}} & \frac{\partial \gamma_k}{\partial z_k^{to}} & \frac{\partial \gamma_k}{\partial \theta_{b_k}} & \frac{\partial \gamma_k}{\partial \gamma_{b_k}} \end{bmatrix}$$

$$= \begin{bmatrix} \frac{y_k^{rto}}{x_k^{rto2}+y_k^{rto2}} & \frac{-x_k^{rto}}{x_k^{rto2}+y_k^{rto2}} & 0 & 1 & 0 \\ \frac{-x_k^{rto}z_k^{rto}}{\sqrt{x_k^{rto2}+y_k^{rto2}}\left(x_k^{rto2}+y_k^{rto2}+z_k^{rto2}\right)} & \frac{-y_k^{rto}z_k^{rto}}{\sqrt{x_k^{rto2}+y_k^{rto2}}\left(x_k^{rto2}+y_k^{rto2}+z_k^{rto2}\right)} & \frac{\sqrt{x_k^{rto2}+y_k^{rto2}}}{x_k^{rto2}+y_k^{rto2}+z_k^{rto2}} & 0 & 1 \end{bmatrix} \tag{A5}$$

For initialization, the Hessian matrix $\mathbf{H}_z^{aug2}$ is expressed as,

$$\mathbf{H}_z^{aug2} = \begin{bmatrix} \frac{\partial \theta_1}{\partial x_1^{to}} & \frac{\partial \theta_1}{\partial y_1^{to}} & \frac{\partial \theta_1}{\partial z_1^{to}} & \frac{\partial \theta_1}{\partial \theta_{b_1}} & \frac{\partial \theta_1}{\partial \gamma_{b_1}} \\ \frac{\partial \gamma_1}{\partial x_1^{to}} & \frac{\partial \gamma_1}{\partial y_1^{to}} & \frac{\partial \gamma_1}{\partial z_1^{to}} & \frac{\partial \gamma_1}{\partial \theta_{b_1}} & \frac{\partial \gamma_1}{\partial \gamma_{b_1}} \\ \frac{\partial z_1^{to}}{\partial x_1^{to}} & \frac{\partial z_1^{to}}{\partial y_1^{to}} & \frac{\partial z_1^{to}}{\partial z_1^{to}} & \frac{\partial z_1^{to}}{\partial \theta_{b_1}} & \frac{\partial z_1^{to}}{\partial \gamma_{b_1}} \end{bmatrix}$$

$$= \begin{bmatrix} \frac{y_1^{rto}}{x_1^{rto2}+y_1^{rto2}} & \frac{-x_1^{rto}}{x_1^{rto2}+y_1^{rto2}} & 0 & 1 & 0 \\ \frac{-x_1^{rto}z_1^{rto}}{\sqrt{x_1^{rto2}+y_1^{rto2}}\left(x_1^{rto2}+y_1^{rto2}+z_1^{rto2}\right)} & \frac{-y_1^{rto}z_1^{rto}}{\sqrt{x_1^{rto2}+y_1^{rto2}}\left(x_1^{rto2}+y_1^{rto2}+z_1^{rto2}\right)} & \frac{\sqrt{x_1^{rto2}+y_1^{rto2}}}{x_1^{rto2}+y_1^{rto2}+z_1^{rto2}} & 0 & 1 \\ 0 & 0 & 1 & 0 & 0 \end{bmatrix} \tag{A6}$$

References

1. Oliveira, T.; Encarnaçao, P. Ground target tracking for unmanned aerial vehicles. In Proceedings of the AIAA Guidance, Navigation, and Control Conference, Toronto, ON, Canada, 2–5 August 2010; p. 8082.
2. Mallick, M.; Arulampalam, S.; Mihaylova, L.; Yan, Y. Angle-only filtering in 3D using modified spherical and log spherical coordinates. In Proceedings of the 14th International Conference on Information Fusion, Chicago, IL, USA, 5–8 July 2011; pp. 1–8.
3. Datta Gupta, S.; Yu, J.Y.; Mallick, M.; Coates, M.; Morelande, M. Comparison of angle-only filtering algorithms in 3D using EKF, UKF, PF, PFF, and ensemble KF. In Proceedings of the 18th International Conference on Information Fusion (Fusion), Washington, DC, USA, 6–9 July 2015; pp. 1649–1656.
4. Kirscht, M.; Mietzner, J.; Bickert, B.; Dallinger, A.; Hippler, J.; Meyer-Hilberg, J.; Zahn, R.; Boukamp, J. An Airborne Radar Sensor for Maritime and Ground Surveillance and Reconnaissance—Algorithmic Issues and Exemplary Results. *IEEE J. Sel. Top. Appl. Earth Obs. Remote Sens.* **2016**, *9*, 971–979. [CrossRef]
5. Budillon, A.; Gierull, C.H.; Pascazio, V.; Schirinzi, G. Along-Track Interferometric SAR Systems for Ground-Moving Target Indication: Achievements, Potentials, and Outlook. *IEEE Geosci. Remote Sens. Mag.* **2020**, *8*, 46–63. [CrossRef]
6. Suwa, K.; Yamamoto, K.; Tsuchida, M.; Nakamura, S.; Wakayama, T.; Hara, T. Image-Based Target Detection and Radial Velocity Estimation Methods for Multichannel SAR-GMTI. *IEEE Trans. Geosci. Remote Sens.* **2017**, *55*, 1325–1338. [CrossRef]
7. Sithiravel, R.; Balaji, B.; Nelson, B.; McDonald, M.K.; Tharmarasa, R.; Kirubarajan, T. Airborne Maritime Surveillance Using Magnetic Anomaly Detection Signature. *IEEE Trans. Aerosp. Electron. Syst.* **2020**, *56*, 3476–3490. [CrossRef]
8. Dames, P.M.; Schwager, M.; Rus, D.; Kumar, V. Active Magnetic Anomaly Detection Using Multiple Micro Aerial Vehicles. *IEEE Robot. Autom. Lett.* **2016**, *1*, 153–160. [CrossRef]
9. Liu, J.; Han, C.; Vadakkepat, P. Process noise identification based particle filter: An efficient method to track highly maneuvering target. In Proceedings of the 13th International Conference on Information Fusion, Edinburgh, UK, 26–29 July 2010; pp. 1–6.
10. Mallick, M.; Chang, K.C.; Arulampalam, S.; Yan, Y.; La Scala, B. Heterogeneous Track-to-Track Fusion in 2D Using Sonar and Radar Sensors. In Proceedings of the 22th International Conference on Information Fusion (FUSION), Ottawa, ON, Canada, 2–5 July 2019; pp. 1–8.
11. Mallick, M.; Sinha, A.; Liu, J. Enhancements to bearing-only filtering. In Proceedings of the 20th International Conference on Information Fusion (Fusion), Xi'an, China, 10–13 July 2017; pp. 1–8.
12. Schmitt, L.; Fichter, W. Globally Valid Posterior Cramér–Rao Bound for Three-Dimensional Bearings-Only Filtering. *IEEE Trans. Aerosp. Electron. Syst.* **2019**, *55*, 2036–2044. [CrossRef]
13. Fan, L.; Atkinson, P.M. Accuracy of Digital Elevation Models Derived From Terrestrial Laser Scanning Data. *IEEE Geosci. Remote Sens. Lett.* **2015**, *12*, 1923–1927. [CrossRef]
14. Li, Z.; Hovakimyan, N.; Dobrokhodov, V.; Kaminer, I. Vision-based target tracking and motion estimation using a small UAV. In Proceedings of the 49th IEEE Conference on Decision and Control (CDC), Atlanta, GA, USA, 15–17 December 2010; pp. 2505–2510.

15. Farmani, N.; Sun, L.; Pack, D.J. A Scalable Multitarget Tracking System for Cooperative Unmanned Aerial Vehicles. *IEEE Trans. Aerosp. Electron. Syst.* **2017**, *53*, 1947–1961. [CrossRef]
16. Tyagi, P.; Kumar, Y.; Sujit, P. NMPC-based UAV 3D Target Tracking In The Presence Of Obstacles and Visibility Constraints. In Proceedings of the 2021 International Conference on Unmanned Aircraft Systems (ICUAS), Athens, Greece, 15–18 June 2021; pp. 858–867.
17. Belfadel, D.; Bar-Shalom, Y.; Willett, P. Single Space Based Sensor Bias Estimation Using a Single Target of Opportunity. *IEEE Trans. Aerosp. Electron. Syst.* **2020**, *56*, 1676–1684. [CrossRef]
18. Reis, J.O.; Batista, P.T.M.; Oliveira, P.; Silvestre, C. Source Localization Based on Acoustic Single Direction Measurements. *IEEE Trans. Aerosp. Electron. Syst.* **2018**, *54*, 2837–2852. [CrossRef]
19. Zhou, L.; Liu, X.; Hu, Z.; Jin, Y.; Shi, W. Joint estimation of state and system biases in non-linear system. *IET Signal Process.* **2017**, *11*, 10–16. [CrossRef]
20. Tian, W.; Huang, G.; Peng, H.; Wang, X.; Lin, X. Sensor Bias Estimation Based on Ridge Least Trimmed Squares. *IEEE Trans. Aerosp. Electron. Syst.* **2020**, *56*, 1645–1651. [CrossRef]
21. Van Trees, H.L. *Detection, Estimation, and Modulation Theory, Part I*; Wiley: New York, NY, USA, 1968.
22. Bacharach, L.; Fritsche, C.; Orguner, U.; Chaumette, E. A Tighter Bayesian CramÉR-rao Bound. In Proceedings of the ICASSP 2019—2019 IEEE International Conference on Acoustics, Speech and Signal Processing (ICASSP), Brighton, UK, 12–17 May 2019; pp. 5277–5281.
23. Hero, A.; Fessler, J.; Usman, M. Exploring estimator bias-variance tradeoffs using the uniform CR bound. *IEEE Trans. Signal Process.* **1996**, *44*, 2026–2041. [CrossRef]
24. Eldar, Y. Minimum variance in biased estimation: Bounds and asymptotically optimal estimators. *IEEE Trans. Signal Process.* **2004**, *52*, 1915–1930. [CrossRef]
25. Duan, Z.; Li, X.R. Constrained target motion modeling–Part I: Straight line track. In Proceedings of the 16th International Conference on Information Fusion, Istanbul, Turkey, 9–12 July 2013; pp. 2153–2160.
26. Kulikov, G.Y.; Kulikova, M.V. The Accurate Continuous-Discrete Extended Kalman Filter for Radar Tracking. *IEEE Trans. Signal Process.* **2016**, *64*, 948–958. [CrossRef]
27. Arasaratnam, I.; Haykin, S.; Hurd, T.R. Cubature Kalman Filtering for Continuous-Discrete Systems: Theory and Simulations. *IEEE Trans. Signal Process.* **2010**, *58*, 4977–4993. [CrossRef]
28. Menegaz, H.M.T.; Ishihara, J.Y.; Borges, G.A.; Vargas, A.N. A Systematization of the Unscented Kalman Filter Theory. *IEEE Trans. Autom. Control* **2015**, *60*, 2583–2598. [CrossRef]
29. Yang, T.; Mehta, P.G.; Meyn, S.P. Feedback Particle Filter. *IEEE Trans. Autom. Control* **2013**, *58*, 2465–2480. [CrossRef]
30. Wan, E.; Van Der Merwe, R. The unscented Kalman filter for nonlinear estimation. In Proceedings of the IEEE Adaptive Systems for Signal Processing, Communications, and Control Symposium, Lake Louise, AB, Canada, 4 October 2000; pp. 153–158.
31. Kandepu, R.; Foss, B.; Imsland, L. Applying the unscented Kalman filter for nonlinear state estimation. *J. Process Control* **2008**, *18*, 753–768. [CrossRef]
32. Tichavsky, P.; Muravchik, C.; Nehorai, A. Posterior Cramer-Rao bounds for discrete-time nonlinear filtering. *IEEE Trans. Signal Process.* **1998**, *46*, 1386–1396. [CrossRef]
33. Musicki, D.; Song, T.L. Track Initialization: Prior Target Velocity and Acceleration Moments. *IEEE Trans. Aerosp. Electron. Syst.* **2013**, *49*, 665–670. [CrossRef]

sensors

Article

Angle-Only Filtering of a Maneuvering Target in 3D

Mahendra Mallick [1,*], Xiaoqing Tian [2], Yun Zhu [3] and Mark Morelande [4]

[1] Independent Researcher, Anacortes, WA 98221, USA
[2] School of Automation Science and Engineering, Xi'an Jiaotong University, Xi'an 710049, China; tianxiaoqing2017@stu.xjtu.edu.cn
[3] School of Computer Science, Shaanxi Normal University, Xi'an 710062, China; yunzhu@snnu.edu.cn
[4] National Australia Bank, Melbourne, VIC 3000, Australia; m.morelande@gmail.com
* Correspondence: mmallick.us@gmail.com

Abstract: We consider the state estimation of a maneuvering target in 3D using bearing and elevation measurements from a passive infrared search and track (IRST) sensor. Since the range is not observable, the sensor must perform a maneuver to observe the state of the target. The target moves with a nearly constant turn (NCT) in the XY-plane and nearly constant velocity (NCV) along the Z-axis. The natural choice for the NCT motion is to allow perturbations in speed and angular rate in the stochastic differential equation, as has been pointed out previously for a 2D scenario using range and bearing measurements. The NCT motion in the XY-plane cannot be discretized exactly, whereas the NCV motion along the Z-axis is discretized exactly. We discretize the continuous-time NCT model using the first and second-order Taylor approximations to obtain discrete-time NCT models, and we consider the polar velocity and Cartesian velocity-based states for the NCT model. The dynamic and measurement models are nonlinear in the target state. We use the cubature Kalman filter to estimate the target state. Accuracies of the first and second-order Taylor approximations are compared using the polar velocity-based and Cartesian velocity-based models using Monte Carlo simulations. Numerical results for realistic scenarios considered show that the second-order Taylor approximation provides the best accuracy using the polar velocity or Cartesian velocity-based models.

Keywords: angle-only filtering in 3D; infrared search and track (IRST) sensor; maneuvering target tracking; cubature Kalman filter (CKF); Itô stochastic differential equation

Citation: Mallick, M.; Tian, X.; Zhu, Y.; Morelande, M. Angle-Only Filtering of a Maneuvering Target in 3D. *Sensors* **2022**, *22*, 1422. https://doi.org/10.3390/s22041422

Academic Editor: Andrzej Stateczny

Received: 9 January 2022
Accepted: 9 February 2022
Published: 12 February 2022

Publisher's Note: MDPI stays neutral with regard to jurisdictional claims in published maps and institutional affiliations.

1. Introduction

Angle-only filtering in 2D and 3D finds many important applications in passive tracking [1–13]. The advantage of passive tracking over active tracking is that the presence of the passive sensor cannot be detected by the target. Passive tracking arises in submarine tracking using a passive sonar [1,11], passive ranging using an infrared search and track (IRST) sensor [2,4,5,8,12], passive radar tracking [4], satellite-to-satellite passive tracking [14], video tracking [15], etc. In this paper, we focus on tracking an aircraft using an IRST sensor on another aircraft. This problem is more difficult than the case where multiple sensors are used. The bearings-only filtering (BOF) problem in 2D has been extensively studied, and a vast number of publications exist in the research literature [1,10,16–19], Chapter 6 of [11,20]. However, research in the angle-only filtering (AOF) problem in 3D is limited compared to that in the bearings-only filtering problem.

Observability is a major issue for the BOF problem in 2D [21] and AOF problem in 3D. In the 2D problem, a four or five-dimensional state is estimated from bearings-only measurements for a non-maneuvering and maneuvering target, respectively. To observe the state of the target, the sensor must perform maneuvers with a motion of higher order than that of the target [18]. If a four-dimensional Cartesian state is used for the BOF problem in 2D for a non-maneuvering target, it has been observed that the extended Kalman filter (EKF) [22,23] diverges. Modified polar coordinates (MPC) [1,24,25] were formulated to overcome the divergence of the EKF. The components of the MPC are bearing, bearing-rate, range-rate divided by range, and the inverse of range [1]. The first three components of MPC are observable even before an ownship maneuver. MPC decouple the observable and unobservable components of the state vector and provide improved performance of the EKF. Use of MPC makes the dynamic model for the nearly constant velocity (NCV) model nonlinear and complex, but the measurement model becomes linear. In addition to the EKF, the unscented Kalman filter (UKF) [26], cubature Kalman filter (CKF) [27], Gaussian sum filter (GSF) [28], particle filter (PF) [11], uncorrelated conversion based filter (UCF) [20], etc. have also been applied to the BOF problem in 2D. In order to address the observability problem, the multiple model-based range-parametrized (RP) EKF (RP-EKF) was proposed by Peach [19] and Kronhamn [16]. In addition to the EKF, other filters can also be used in the RP framework.

Most of the existing work on the angle-only filtering problem in 3D is for a non-maneuvering target using the NCV model. The sensor must perform a maneuver to observe the target state. For a non-maneuvering target, the EKF using the Cartesian state for the AOF problem in 3D does not diverge [8], even though the filter diverges for the corresponding BOF in 2D [11]. In analogy with the MPC in 2D, Stallard proposed the modified spherical coordinates (MSC) in 3D [12,13]. The components of the MSC are elevation, elevation-rate, bearing, bearing-rate times cosine of elevation, the inverse of range, and range-rate divided by range. As in the case of MPC, the dynamic model for the NCV motion using MSC is nonlinear and complex. However, the measurement model is linear since bearing and elevation are components of MSC. The log spherical coordinates (LSC) [29] can also be used as an alternate to the MSC. The first five components of the LSC are the same as those of the MSC, but the inverse of range (sixth component) is replaced by the logarithm of the range. Many studies have shown that the EKF using the MSC (EKF-MSC) provides a better state estimation accuracy than the Cartesian EKF (CEKF) for the NCV motion [2,12,13].

Starting with the work of Stallard, the EKF-MSC was used in [2,12,13,30]. Karlsson and Gustafsson used the PF and compared the PF-based algorithms with the multiple model-based range-parametrized EKF (RP-EKF) using the Cartesian state and MSC in a number of tracking scenarios [6,7]. Their results showed the superiority of the PF-based algorithms over the RP-EKF-based algorithms.

In our previous work [29], we compared the EKF-MSC and EKF-LSC using the continuous-discrete filtering approach with the discrete-time CEKF. The results of this study show that the EKF-MSC and EKF-LSC have comparable accuracy and perform better than the discrete-time CEKF for low measurement accuracy. For high measurement accuracy, the discrete-time CEKF has higher state estimation accuracy than the EKF-MSC and EKF-LSC. Prior to our work in [8,31,32], the process noise using the MSC was modeled approximately. We proposed new algorithms using the MSC to model the process noise exactly. The AOF for the NCV motion can be solved in the following three possible ways [8,31,32]:

- Use the discrete-time NCV model with the Cartesian state vector (with linear dynamic model) and nonlinear measurement model;
- Use the exact discrete-time NCV model with the MSC (with nonlinear dynamic model) and linear measurement model;
- Use the MSC with approximate discretization of the continuous-time dynamic model (with nonlinear dynamic model) and linear measurement model.

In [8], we performed a comprehensive study of the AOF problem for a non-maneuvering target in 3D using the EKF, UKF, and PF with Cartesian state vector and MSC. In this study, new algorithms using the EKF, UKF, and PF with the MSC were formulated, and improved filter initialization algorithms for the Cartesian state and MSC were presented. Four versions of the PF were used in this work: the Cartesian bootstrap filter (CBF), bootstrap filter using MSC with exact dynamic model (BF-MSC(E)), bootstrap filter using MSC with an approximate dynamic model (BF-MSC(A)), and the optimal importance density-based PF using MSC with an approximate dynamic model (ODIPF-MSC(A)). The initial range between the target and the sensor in the scenarios in [8] is higher than that in [6,7]. Numerical results from this study indicate that the state estimation accuracy of the PF-based algorithms is inferior compared with that of the EKF and UKF-based algorithms. For the BOF problem in 2D Chapter 6 of [11], the measurement SD is of the order of a degree and the measurement time interval is about 30–60 s. In this scenario, a PF has one of the best state estimation accuracies Chapter 6 of [11,20]. Secondly, compared to the EKF, the PF-based algorithms have about two orders of magnitude higher computational cost. Thirdly, It is now well established that, when the measurement accuracy and data rate are high (which is true for the current problem), PFs do not offer any advantage over the EKF, UKF, and CKF [33–35]. Therefore, we did not consider the PF in this study.

A novel batch Bayesian weighted instrumental variable estimator for the 3D target motion analysis problem using bearing and elevation measurements is presented in [36]. Results of this study show that the proposed algorithm outperforms its non-Bayesian counterpart. The CEKF, Cartesian UKF (CUKF), Cartesian CKF (CCKF), and the Cartesian new sigma point Kalman filter (CNSKF) were used in [3] to analyze the AOF problem in 3D for a non-maneuvering target. Results of this study shows that these five filters have nearly the same accuracy in operational scenarios. The particle flow filter (PFF), ensemble Kalman filter (EnKF), EKF, UKF, and PF were compared for the AOF problem in 3D for a non-maneuvering target in [37]. It was observed that the EKF-MSC, UKF-MSC, deterministic EnKF-MSC, and Cartesian PFF had the best performance in operational conditions.

In [38], we studied the passive sonar tracking problem when the submarine and the ownship move in different planes using the EKF, UKF, RP-UKF, and PF. Our results showed that the depth of the non-maneuvering target can be estimated accurately, and the PF had the best performance in the scenarios studied. The 3D instrumental variable-based Kalman filter (3D-IVKF) is applied to an underwater passive sonar tracking scenario in [39] for a non-maneuvering target using bearing and elevation measurements. It is observed that at low measurement standard deviations (SDs) ($<6°$) the performance of the 3D-IVKF is comparable to that of the UKF and CKF. However, at higher measurement SDs, the 3D-IVKF outperforms the UKF and CKF with lower computational cost.

To compare the accuracies of the filters used in the AOF problem with the best achievable accuracy, we computed the posterior Cramér-Rao lower bound (PCRLB) [40] for a non-maneuvering target using the NCV model in [41]. Our results show that when the measurement accuracy is high, the root mean square (RMS) position and velocity errors are close to the corresponding PCRLBs. The difference between RMS position and velocity errors and corresponding PCRLBs increases with a decrease in the measurement accuracy. In [42], a globally valid posterior Cramér–Rao lower bound was derived for the AOF problem. The authors claim the von Mises–Fisher distribution to be superior to the conventional approach using additive Gaussian noise in measured angular coordinates.

A maneuvering target refers to an accelerating target [43]. Common accelerated motions considered in tracking are the nearly constant acceleration (NCA), nearly constant turn (NCT), and jerk [4,22,43]. The NCA and jerk models are linear, whereas the NCT model is nonlinear in the target state. The number of publications for a maneuvering target in the AOF problem is quite limited. In [5], the NCT model was used in the passive ranging problem using an IRST sensor in air-to-air tracking scenarios. The authors used the RP-UKF using the multiple model method. However, algorithm details are not presented in the paper. The NCT model in the $XY-$plane has been studied extensively where the angular rate is estimated [4,22,43–46]. This problem arises in the air-traffic control (ATC) scenario [4,22,27,43,47]. In most cases, the conventional discrete-time NCT model is approximate, since the state transition matrix and process noise covariance matrix cannot be derived from the continuous-time model using a consistent procedure.

We consider the tracking of a maneuvering aircraft in 3D and assume that the aircraft moves in the XY-plane with the NCT motion and has a NCV motion along the Z-axis. The speed and angular rate are constant for the constant turn motion (CT) in the XY-plane. Thus, it is natural to perturb the speed and angular rate in the NCT motion with the continuous-time white noise (Wiener processes) [22]. We follow this approach from [45] to obtain the nonlinear stochastic differential equation (SDE) [48,49] for the NCT motion. Since the SDE is nonlinear, it cannot be discretized exactly. We discretize the SDE using the first and second-order weak Taylor (TS2) approximations [50] to obtain approximate discrete-time dynamic models. The first-order stochastic Taylor series approximation is also known as the Euler approximation. Two types of states for the NCT motion in the XY-plane, namely the polar velocity and the Cartesian velocity-based states [43–46], are used. The NCV motion along the Z-axis is discretized exactly. The Cartesian velocity state in NCT comprises the 2D position, 2D velocity, and angular rate. In the polar velocity state, the speed and heading replace the 2D Cartesian velocity.

An IRST sensor on another maneuvering aircraft collects azimuth and elevation measurements. The accuracy of the angle measurements by an IRST sensor is usually high (1 mrad). The data rate of an IRST sensor is also high (1 Hz). As sensor technology improves, these factors are expected to improve. The AOF algorithm is required to process the sensor measurements sequentially in real time. Thus, a batch algorithm is ruled out for this tracking scenario. As discussed previously, a PF is not considered for this problem due to its lack of state estimation accuracy and high computational cost. It has been observed in [33–35] that when the measurement accuracy and data rate are high (which is true for the current problem), the UKF and CKF have nearly the same accuracy, and the accuracy of the EKF is somewhat lower. If the dimension of the state is n, then the UKF and CKF have $2n + 1$ and $2n$ sigma points and cubature points, respectively. As a result, the computational cost of the CKF is lower than that of the UKF. If $n > 3$, then the first weight in the UKF becomes negative and the rest of the $2n$ weights are positive. On the contrary, each of the $2n$ weights in the CKF is positive and equal to $1/2n$. This negative weight may cause a filter-calculated covariance matrix to be non-positive definite in some cases [27]. The CKF was also successfully used in our previous work on AOF in [9]. Hence, we chose the third-degree spherical–radial cubature rule-based CKF [27] to estimate the seven-dimensional state of the maneuvering target. The CKFs using the Euler and TS2 approximations are called CKF1 and CKF2, respectively. Thus, we consider four CKF filters; CKF1P, CKF1C, CKF2P, and CKF2C, where the last letter in the filter refers to polar and Cartesian velocity states.

Notation convention: For clarity, we use italics to denote scalar quantities and boldface for vectors and matrices. A lower or upper-case Roman letter represents a name (e.g. "s" for "sensor," "RMS" for "root mean square," etc.). We use ":=" to define a quantity and $\mathbf{A}'$ denotes the transpose of the vector or matrix $\mathbf{A}$. The $n-$dimensional identity matrix, $m-$dimensional null matrix, and $m \times n$ null matrix are denoted by $\mathbf{I}_n$, $\mathbf{0}_m$, and $\mathbf{0}_{m \times n}$, respectively.

The paper is organized as follows. Section 2 presents the dynamic models for the target. Section 3 explains the discretization of target NCT models, and Section 4 describes sensor dynamic and measurement models. A summary of the four CKF filters is given in Section 5. Numerical simulations and results are presented in Section 6. Finally, Section 7 summarizes our contributions in the paper.

2. Target Dynamic Models

We assume that the IRST sensor trajectory is deterministic and the states of the sensors are known exactly at measurement times. To improve the observability of the target state, the IRST sensor performs maneuvers with a sequence of CV and constant turn (CT) motions [5,8,31].

Two types of coordinates are commonly used for the NCT in the XY-plane: Cartesian velocity and polar velocity-based models [43–46]. In addition to the 2D position and velocity, the turn-rate or angular velocity ω is also estimated in the NCT model.

Let $\mathbf{z}(t)$ denote the Cartesian state along the Z-axis with position and velocity components

$$\mathbf{z}(t) := \begin{bmatrix} z(t) & \dot{z}(t) \end{bmatrix}'. \tag{1}$$

For the NCT model in the XY-plane, we use $\eta(t)$ and $\xi(t)$ for state vectors where the angular velocity ω is estimated. The velocity in $\eta(t)$ and $\xi(t)$ has Cartesian and polar coordinates, respectively. Let $s(t)$ and $\theta(t)$ denote the speed and heading of the target in the XY-plane. In this paper, the heading is defined as the angle of the velocity in the XY-plane, measured from the X-axis in the counter-clockwise direction, as shown in Figure 1.

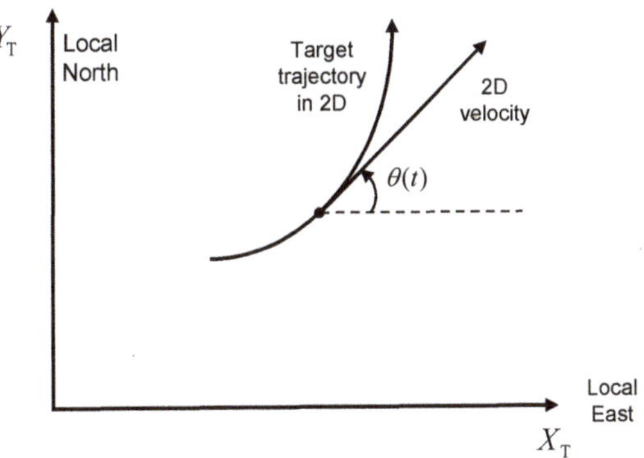

Figure 1. Definition of heading $\theta(t)$ in the tracker coordinate frame, $\theta(t) \in [0, 2\pi)$.

Then $\eta(t)$ and $\xi(t)$ are defined, respectively, by

$$\eta(t) := \begin{bmatrix} x(t) & y(t) & \dot{x}(t) & \dot{y}(t) & \omega(t) \end{bmatrix}', \tag{2}$$

$$\xi(t) := \begin{bmatrix} x(t) & y(t) & s(t) & \theta(t) & \omega(t) \end{bmatrix}'. \tag{3}$$

Three-dimensional state vectors where angular velocity is estimated are defined, respectively, by

$$\mathbf{x}_c(t) := \begin{bmatrix} \eta(t)' & \mathbf{z}(t)' \end{bmatrix}', \tag{4}$$

$$\mathbf{x}_p(t) := \begin{bmatrix} \xi(t)' & \mathbf{z}(t)' \end{bmatrix}'. \tag{5}$$

We assume that the measurement time interval is constant; i.e., $t_k - t_{k-1} = T$ for all k. In this paper, we use the discretized continuous-time models [22].

The discrete-time dynamic model for the NCV motion along the Z-axis is given by

$$\mathbf{z}_k = \mathbf{F}_1 \mathbf{z}_{k-1} + \mathbf{w}_{z,k-1}, \tag{6}$$

where $\mathbf{F}_1$ is the state transition matrix and $\mathbf{w}_{z,k-1}$ is a zero-mean white Gaussian process noise with covariance $\mathbf{Q}_z$,

$$\mathbf{F}_1 = \begin{bmatrix} 1 & T \\ 0 & 1 \end{bmatrix}, \tag{7}$$

$$\mathbf{Q}_z = q_z \mathbf{B}, \tag{8}$$

$$\mathbf{B} = \begin{bmatrix} T^3/3 & T^2/2 \\ T^2/2 & T \end{bmatrix}, \tag{9}$$

where q_z is the power spectral density (PSD) of the continuous-time acceleration process noise along the Z-axis [22].

The time derivative of $\xi(t)$ is given by

$$\dot{\xi}(t) = \begin{bmatrix} \dot{x}(t) & \dot{y}(t) & \dot{s}(t) & \dot{\theta}(t) & \dot{\omega}(t) \end{bmatrix}'. \tag{10}$$

We have

$$\dot{x}(t) = s(t)\cos\theta(t), \quad \dot{y}(t) = s(t)\sin\theta(t). \tag{11}$$

Since $\dot{\theta}(t) = \omega(t)$, (10) can be written as

$$\dot{\xi}(t) = \begin{bmatrix} s(t)\cos\theta(t) & s(t)\sin\theta(t) & \dot{s}(t) & \omega(t) & \dot{\omega}(t) \end{bmatrix}'. \tag{12}$$

The time derivative of $\eta(t)$ is

$$\dot{\eta}(t) = \begin{bmatrix} \dot{x}(t) & \dot{y}(t) & \ddot{x}(t) & \ddot{y}(t) & \dot{\omega}(t) \end{bmatrix}'. \tag{13}$$

In the constant turn (CT) model, the speed and turn rate are constant. The speed and turn rate can be modeled as nearly constant in the NCT motion. Examining (12) and (13), we find that for the NCT model, (12) is more suitable than (13), based on symmetry considerations. Using conventional models in the engineering literature [22], for the NCT model, we may write

$$\dot{s}(t) = w_s(t), \quad \dot{\omega}(t) = w_\omega(t), \tag{14}$$

where $w_s(t)$ and $w_\omega(t)$ are continuous-time zero-mean white acceleration and angular acceleration process noises with power spectral densities q_s and q_ω, respectively, [22]

$$E\{w_s(t)\} = 0, \quad E\{w_s(t)w_s(\tau)\} = q_s\delta(t - \tau), \tag{15}$$

$$E\{w_\omega(t)\} = 0, \quad E\{w_\omega(t)w_\omega(\tau)\} = q_\omega\delta(t - \tau), \tag{16}$$

where δ is the Dirac delta function [51]. We can write (14)–(16) mathematically rigorously by defining

$$ds(t) = \sqrt{q_s}d\beta_s(t), \quad d\omega(t) = \sqrt{q_\omega}d\beta_\omega(t), \tag{17}$$

where $d\beta_s(t)$ and $d\beta_\omega(t)$ are standard independent Wiener processes [45,48]

$$E\{d\beta_s(t)d\beta_s(t)\} = dt, \quad E\{d\beta_\omega(t)d\beta_\omega(t)\} = dt, \tag{18}$$

$$E\{d\beta_s(t)d\beta_\omega(t)\} = 0. \tag{19}$$

Define

$$\mathbf{f}_\mathrm{p}(\boldsymbol{\xi}(t)) := \begin{bmatrix} s(t)\cos\theta(t) & s(t)\sin\theta(t) & 0 & \omega(t) & 0 \end{bmatrix}', \tag{20}$$

$$\mathbf{w}_\mathrm{p}(t) := \begin{bmatrix} 0 & 0 & w_s(t) & 0 & w_\omega(t) \end{bmatrix}', \tag{21}$$

$$\mathbf{G}_\mathrm{p} := \begin{bmatrix} 0 & 0 & \sqrt{q_s} & 0 & 0 \\ 0 & 0 & 0 & 0 & \sqrt{q_\omega} \end{bmatrix}', \tag{22}$$

$$d\boldsymbol{\beta}(t) := \begin{bmatrix} d\beta_s(t) & d\beta_\omega(t) \end{bmatrix}'. \tag{23}$$

Then, conventionally, we can write $\dot{\boldsymbol{\xi}}(t)$ as [23,48]

$$\dot{\boldsymbol{\xi}}(t) = \mathbf{f}_\mathrm{p}(\boldsymbol{\xi}(t)) + \mathbf{w}_\mathrm{p}(t). \tag{24}$$

We can write the time derivative of the polar state vector mathematically rigorously using the Itô stochastic differential equation (SDE) [45,48,49]

$$d\boldsymbol{\xi}(t) = \mathbf{f}_\mathrm{p}(\boldsymbol{\xi}(t))dt + \mathbf{G}_\mathrm{p}d\boldsymbol{\beta}(t), \tag{25}$$

where

$$E\{d\boldsymbol{\beta}(t)d\boldsymbol{\beta}'(t)\} = \mathbf{I}_2 dt. \tag{26}$$

We assume that the prior distribution of $\boldsymbol{\xi}$ is Gaussian,

$$\boldsymbol{\xi}_0 = \boldsymbol{\xi}(t_0) \sim \mathcal{N}(\boldsymbol{\xi}_0; \bar{\boldsymbol{\xi}}_0, \mathbf{P}_0^{\boldsymbol{\xi}}). \tag{27}$$

The time derivative of η contains Cartesian accelerations $\ddot{x}$ and $\ddot{y}$ in (13). It is necessary to transform them to derivatives of speed and angular velocity. The 2D Cartesian velocity is given by

$$\mathbf{v}(t) = \begin{bmatrix} \dot{x}(t) & \dot{y}(t) \end{bmatrix}' \tag{28}$$

and the Cartesian acceleration is $\dot{\mathbf{v}}(t)$. Using (11), the Cartesian acceleration is expressed by

$$\dot{\mathbf{v}}(t) = \begin{bmatrix} \dot{s}(t)\cos\theta(t) - \omega(t)\dot{y}(t) & \dot{s}(t)\sin\theta(t) + \omega(t)\dot{x}(t) \end{bmatrix}'. \tag{29}$$

Using a similar approach, the Itô SDE [45,48,49] for the Cartesian state is

$$d\eta(t) = \mathbf{f}_c(\eta(t))dt + \mathbf{G}_c(\eta(t))d\beta(t), \tag{30}$$

where

$$\mathbf{f}_c(\eta(t)) := \begin{bmatrix} \dot{x}(t) & \dot{y}(t) & -\omega(t)\dot{y}(t) & \omega(t)\dot{x}(t) & 0 \end{bmatrix}', \tag{31}$$

$$\mathbf{G}_c(\eta(t)) := \begin{bmatrix} 0 & 0 & \sqrt{q_s}\dot{x}(t)/s(t) & \sqrt{q_s}\dot{y}(t)/s(t) & 0 \\ 0 & 0 & 0 & 0 & \sqrt{q_\omega} \end{bmatrix}'. \tag{32}$$

3. Discretization of Target NCT Models

3.1. The Euler Approximation

The Euler approximation is obtained by applying the Itô lemma [48] to the integral form of the SDE and retaining only single integral terms. Applying the Euler approximation [50] to the 2D polar velocity dynamic model, we obtain the stochastic difference equation [45]:

$$\xi_k = \xi_{k-1} + T\mathbf{f}_p(\xi_{k-1}) + \sqrt{T}\mathbf{G}_p\mathbf{w}_1, \tag{33}$$

where $\mathbf{f}_p(\xi)$ is defined in (20) and

$$\mathbf{w}_1 \sim \mathcal{N}(\mathbf{w}_1; \mathbf{0}_{2\times1}, \mathbf{I}_2). \tag{34}$$

The covariance of the polar velocity process noise $\mathbf{w}_{k-1}^\xi = \sqrt{T}\mathbf{G}_p\mathbf{w}_1$ is described by

$$\mathbf{w}_{k-1}^\xi \sim \mathcal{N}(\mathbf{w}_{k-1}^\xi; \mathbf{0}_{5\times1}, \mathbf{Q}^\xi), \tag{35}$$

$$\mathbf{Q}^\xi = \mathrm{diag}(0, 0, q_s T, 0, q_\omega T). \tag{36}$$

From (36), we see that the polar velocity process noise is independent of the state.

Similarly, applying the Euler approximation to the Cartesian velocity dynamic model, we obtain the stochastic difference equation [45]

$$\eta_k = \eta_{k-1} + T\mathbf{f}_c(\eta_{k-1}) + \sqrt{T}\mathbf{G}_c(\eta_{k-1})\mathbf{w}_1, \tag{37}$$

where $\mathbf{f}_c(\eta)$ is defined in (31). The Cartesian velocity process noise $\mathbf{w}_{k-1}^\eta = \sqrt{T}\mathbf{G}_c(\eta_{k-1})\mathbf{w}_1$ is described by

$$\mathbf{w}_{k-1}^\eta(\eta_{k-1}) \sim \mathcal{N}(\mathbf{w}_{k-1}^\eta; \mathbf{0}_{5\times1}, \mathbf{Q}_{k-1}^\eta), \tag{38}$$

$$\mathbf{Q}_{k-1}^\eta = TE\{\mathbf{G}_c(\eta_{k-1})\mathbf{G}_c'(\eta_{k-1})\}. \tag{39}$$

We make the following approximation in calculating $E\{\mathbf{G}_c(\eta_{k-1})\mathbf{G}_c'(\eta_{k-1})\}$,

$$E\{\mathbf{G}_c(\eta_{k-1})\mathbf{G}_c'(\eta_{k-1})\} \approx \mathbf{G}_c(\hat{\eta}_{k|k-1})\mathbf{G}_c'(\hat{\eta}_{k|k-1}), \tag{40}$$

where $\hat{\eta}_{k|k-1}$ is the predicted Cartesian velocity state estimate at time k. Then,

$$\mathbf{Q}_{k-1}^\eta \approx T\mathbf{G}_c(\hat{\eta}_{k|k-1})\mathbf{G}_c'(\hat{\eta}_{k|k-1}). \tag{41}$$

Simplification of (41) gives

$$\mathbf{Q}^{\eta}_{k-1} \approx \begin{bmatrix} \mathbf{0}_2 & \mathbf{0}_2 & \mathbf{0}_{2\times 1} \\ \mathbf{0}_2 & q_s T \mathbf{A}_{k-1}(\hat{\boldsymbol{\eta}}_{k|k-1}) & \mathbf{0}_{2\times 1} \\ \mathbf{0}_{1\times 2} & \mathbf{0}_{1\times 2} & q_\omega T \end{bmatrix}', \tag{42}$$

where

$$\mathbf{A}_{k-1}(\hat{\boldsymbol{\eta}}_{k|k-1}) = \frac{1}{\hat{s}^2_{k|k-1}} \begin{bmatrix} \hat{x}^2_{k|k-1} & \hat{x}_{k|k-1}\hat{y}_{k|k-1} \\ \hat{x}_{k|k-1}\hat{y}_{k|k-1} & \hat{y}^2_{k|k-1} \end{bmatrix}'. \tag{43}$$

From (42) and (43), we see that the Cartesian velocity-based process noise covariance is state-dependent.

From (4) and (5), we get the polar and Cartesian velocity-based states as

$$\mathbf{x}_{\mathrm{p},k} = \begin{bmatrix} \boldsymbol{\xi}'_k & \mathbf{z}'_k \end{bmatrix}', \tag{44}$$

$$\mathbf{x}_{\mathrm{c},k} = \begin{bmatrix} \boldsymbol{\eta}'_k & \mathbf{z}'_k \end{bmatrix}'. \tag{45}$$

The 3D discrete-time dynamic model for the polar velocity-based model is given by

$$\mathbf{x}_{\mathrm{p},k} = \mathbf{x}_{\mathrm{p},k-1} + T\tilde{\mathbf{f}}_{\mathrm{p}}(\mathbf{x}_{\mathrm{p},k-1}) + \mathbf{w}_{\mathrm{p},k-1}, \tag{46}$$

where $\tilde{\mathbf{f}}_{\mathrm{p}}(\mathbf{x}_{\mathrm{p}})$ is defined by

$$\tilde{\mathbf{f}}_{\mathrm{p}}(\mathbf{x}_{\mathrm{p}}) = \begin{bmatrix} s\cos\theta & s\sin\theta & 0 & \omega & 0 & \dot{z} & 0 \end{bmatrix}', \tag{47}$$

$$\mathbf{w}_{\mathrm{p},k-1} := \begin{bmatrix} (\mathbf{w}^{\xi}_{k-1})' & \mathbf{w}'_{z,k-1} \end{bmatrix}', \tag{48}$$

$$\mathbf{w}_{\mathrm{p},k-1} \sim \mathcal{N}(\mathbf{w}_{\mathrm{p},k-1}; \mathbf{0}_{7\times 1}, \mathbf{Q}_{\mathrm{p}}), \tag{49}$$

$$\mathbf{Q}_{\mathrm{p}} = \begin{bmatrix} \mathbf{Q}^{\xi} & \mathbf{0}_{5\times 2} \\ \mathbf{0}_{2\times 5} & \mathbf{Q}_z \end{bmatrix}. \tag{50}$$

Similarly, the 3D discrete-time dynamic model for the Cartesian velocity-based model is given by

$$\mathbf{x}_{\mathrm{c},k} = \mathbf{x}_{\mathrm{c},k-1} + T\tilde{\mathbf{f}}_{\mathrm{c}}(\mathbf{x}_{\mathrm{c},k-1}) + \mathbf{w}_{\mathrm{c},k-1}, \tag{51}$$

where $\tilde{\mathbf{f}}_{\mathrm{c}}(\mathbf{x}_{\mathrm{c}})$ is defined by

$$\tilde{\mathbf{f}}_{\mathrm{c}}(\mathbf{x}_{\mathrm{c}}) = \begin{bmatrix} \dot{x} & \dot{y} & -\omega\dot{y} & \omega\dot{x} & 0 & \dot{z} & 0 \end{bmatrix}', \tag{52}$$

$$\mathbf{w}_{\mathrm{c},k-1} := \begin{bmatrix} (\mathbf{w}^{\eta}_{k-1})' & \mathbf{w}'_{z,k-1} \end{bmatrix}', \tag{53}$$

$$\mathbf{w}_{\mathrm{c},k-1} \sim \mathcal{N}(\mathbf{w}_{\mathrm{c},k-1}; \mathbf{0}_{7\times 1}, \mathbf{Q}_{\mathrm{c},k-1}), \tag{54}$$

$$\mathbf{Q}_{\mathrm{c},k-1} = \begin{bmatrix} \mathbf{Q}^{\eta}_{k-1} & \mathbf{0}_{5\times 2} \\ \mathbf{0}_{2\times 5} & \mathbf{Q}_z \end{bmatrix}. \tag{55}$$

3.2. Order 2 Weak Taylor Approximation

Using the order 2 weak Taylor approximation [50] to the SDE, we obtain the discretized dynamic model for the polar velocity-based model as [45]

$$\xi_k = \xi_{k-1} + T\mathbf{f}_{p,2}(\xi_{k-1}) + \mathbf{G}_{p,2}(\xi_{k-1})\mathbf{w}_2, \tag{56}$$

where

$$\mathbf{f}_{p,2}(\xi_{k-1}) = \begin{bmatrix} s_{k-1}\cos(\theta_{k-1}) - Ts_{k-1}\omega_{k-1}\sin(\theta_{k-1})/2 \\ s_{k-1}\sin(\theta_{k-1}) + Ts_{k-1}\omega_{k-1}\cos(\theta_{k-1})/2 \\ 0 \\ \omega_{k-1} \\ 0 \end{bmatrix}, \tag{57}$$

$$\mathbf{G}_{p,2}(\xi_{k-1}) = \mathbf{E}_p(\xi_{k-1})\mathbf{V}(T), \tag{58}$$

$$\mathbf{E}_p(\xi_{k-1}) = \begin{bmatrix} \sqrt{q_s}\cos(\theta_{k-1}) & 0 & 0 & 0 \\ \sqrt{q_s}\sin(\theta_{k-1}) & 0 & 0 & 0 \\ 0 & 0 & \sqrt{q_s} & 0 \\ 0 & \sqrt{q_\omega} & 0 & 0 \\ 0 & 0 & 0 & \sqrt{q_\omega} \end{bmatrix}, \tag{59}$$

$$\mathbf{V}(T) = \begin{bmatrix} \sqrt{T^3/3} & 0 \\ \sqrt{3T}/2 & \sqrt{T}/2 \end{bmatrix} \otimes \mathbf{I}_2, \tag{60}$$

$$\mathbf{w}_2 \sim \mathcal{N}(\mathbf{w}_2; \mathbf{0}_{4\times1}, \mathbf{I}_4). \tag{61}$$

In (60), $\otimes$ refers to the Kronecker product [52].

The process noise $\mathbf{w}_{p,2,k-1} = \mathbf{G}_{p,2}(\xi_{k-1})\mathbf{w}_2$ and associated covariance $\mathbf{Q}_{p,2,k-1}$ for the second-order polar velocity-based model are described, respectively, by

$$\mathbf{w}_{p,2,k-1} \sim \mathcal{N}(\mathbf{w}_{p,2,k-1}; \mathbf{0}_{5\times1}, \mathbf{Q}_{p,2,k-1}), \tag{62}$$

$$\mathbf{Q}_{p,2,k-1} = E\{\mathbf{G}_{p,2}(\xi_{k-1})\mathbf{G}'_{p,2}(\xi_{k-1})\}. \tag{63}$$

Using a similar approximation as before, we obtain

$$\mathbf{Q}_{p,2,k-1} \approx \mathbf{G}_{p,2}(\hat{\xi}_{k|k-1})\mathbf{G}'_{p,2}(\hat{\xi}_{k|k-1}), \tag{64}$$

where $\hat{\xi}_{k|k-1}$ is the predicted polar velocity state estimate at time k.

Simplification of $\mathbf{G}_{p,2}(\xi_{k-1})\mathbf{G}'_{p,2}(\xi_{k-1})$ gives

$$\mathbf{G}_{p,2}(\xi_{k-1})\mathbf{G}'_{p,2}(\xi_{k-1}) = \begin{bmatrix} \frac{T^3\cos^2(\theta_{k-1})}{3}q_s & \frac{T^3\sin(2\theta_{k-1})}{6}q_s & \frac{T^2\cos(\theta_{k-1})}{2}q_s & 0 & 0 \\ \frac{T^3\sin(2\theta_{k-1})}{6}q_s & \frac{T^3\sin^2(\theta_{k-1})}{3}q_s & \frac{T^2\sin(\theta_{k-1})}{2}q_s & 0 & 0 \\ \frac{T^2\cos(\theta_{k-1})}{2}q_s & \frac{T^2\sin(\theta_{k-1})}{2}q_s & Tq_s & 0 & 0 \\ 0 & 0 & 0 & \frac{T^3}{3}q_\omega & \frac{T^2}{2}q_\omega \\ 0 & 0 & 0 & \frac{T^2}{2}q_\omega & Tq_\omega \end{bmatrix}. \tag{65}$$

The discretized dynamic model for the Cartesian velocity-based model using the TS2 approximation to the SDE is given by [45]

$$\eta_k = \eta_{k-1} + T\mathbf{f}_{c,2}(\eta_{k-1}) + \mathbf{G}_{c,2}(\eta_{k-1})\mathbf{w}_2, \tag{66}$$

where

$$\mathbf{f}_{c,2}(\eta_{k-1}) = \begin{bmatrix} \dot{x}_{k-1} - T\omega_{k-1}\dot{y}_{k-1}/2 \\ \dot{y}_{k-1} + T\omega_{k-1}\dot{x}_{k-1}/2 \\ -\omega_{k-1}\dot{y}_{k-1} - T\omega_{k-1}^2\dot{x}_{k-1}/2 \\ \omega_{k-1}\dot{x}_{k-1} - T\omega_{k-1}^2\dot{y}_{k-1}/2 \\ 0 \end{bmatrix}, \tag{67}$$

$$\mathbf{G}_{c,2}(\eta_{k-1}) = \mathbf{E}_c(\eta_{k-1})\mathbf{V}(T), \tag{68}$$

$$\mathbf{E}_c(\boldsymbol{\eta}_{k-1}) = \begin{bmatrix} \sqrt{q_s}\dot{x}_{k-1}/s_{k-1} & 0 & 0 & 0 \\ \sqrt{q_s}\dot{y}_{k-1}/s_{k-1} & 0 & 0 & 0 \\ 0 & -\sqrt{q_\omega}\dot{y}_{k-1} & \sqrt{q_s}(\dot{x}_{k-1}-T\omega_{k-1}\dot{y}_{k-1})/s_{k-1} & 0 \\ 0 & \sqrt{q_\omega}\dot{x}_{k-1} & \sqrt{q_s}(\dot{y}_{k-1}+T\omega_{k-1}\dot{x}_{k-1})/s_{k-1} & 0 \\ 0 & 0 & 0 & \sqrt{q_\omega} \end{bmatrix}. \tag{69}$$

The process noise $\mathbf{w}_{c,2,k-1} = \mathbf{G}_{c,2}(\boldsymbol{\eta}_{k-1})\mathbf{w}_2$, and corresponding covariance $\mathbf{Q}_{c,2,k-1}$ for the second-order Cartesian velocity-based model are given, respectively, by

$$\mathbf{w}_{c,2,k-1} \sim \mathcal{N}(\mathbf{w}_{c,2,k-1}; \mathbf{0}_{5\times 1}, \mathbf{Q}_{c,2,k-1}), \tag{70}$$

$$\mathbf{Q}_{c,2,k-1} = E\{\mathbf{G}_{c,2}(\boldsymbol{\eta}_{k-1})\mathbf{G}'_{c,2}(\boldsymbol{\eta}_{k-1})\}. \tag{71}$$

The approximate expression for the process noise is given by

$$\mathbf{Q}_{c,2,k-1} \approx \mathbf{G}_{c,2}(\hat{\boldsymbol{\eta}}_{k|k-1})\mathbf{G}'_{c,2}(\hat{\boldsymbol{\eta}}_{k|k-1}), \tag{72}$$

where $\hat{\boldsymbol{\eta}}_{k|k-1}$ is the predicted state estimate at time k. Simplification of $\mathbf{G}_{c,2}(\boldsymbol{\eta}_{k-1})\mathbf{G}'_{c,2}(\boldsymbol{\eta}_{k-1})$ gives

$$\mathbf{G}_{c,2}(\boldsymbol{\eta}_{k-1})\mathbf{G}'_{c,2}(\boldsymbol{\eta}_{k-1}) =$$

$$\begin{bmatrix} \frac{T^3}{3}a_1^2 q_s & \frac{T^3}{3}a_1 a_2 q_s & \frac{T^2}{2}a_1 a_3 q_s & \frac{T^2}{2}a_1 a_4 q_s & 0 \\ \frac{T^3}{3}a_1 a_2 q_s & \frac{T^3}{3}a_2^2 q_s & \frac{T^2}{2}a_2 a_3 q_s & \frac{T^2}{2}a_2 a_4 q_s & 0 \\ \frac{T^2}{2}a_1 a_3 q_s & \frac{T^2}{2}a_2 a_3 q_s & Ta_3^2 q_s + \frac{T^3}{3}\dot{y}_{k-1}^2 q_\omega & Ta_3 a_4 q_s - \frac{T^3}{3}\dot{x}_{k-1}\dot{y}_{k-1}q_\omega & -\frac{T^2}{2}\dot{y}_{k-1}q_\omega \\ \frac{T^2}{2}a_1 a_4 q_s & \frac{T^2}{2}a_2 a_4 q_s & Ta_3 a_4 q_s - \frac{T^3}{3}\dot{x}_{k-1}\dot{y}_{k-1}q_\omega & Ta_4^2 q_s + \frac{T^3}{3}\dot{x}_{k-1}^2 q_\omega & \frac{T^2}{2}\dot{x}_{k-1}q_\omega \\ 0 & 0 & -\frac{T^2}{2}\dot{y}_{k-1}q_\omega & \frac{T^2}{2}\dot{x}_{k-1}q_\omega & Tq_\omega \end{bmatrix}, \tag{73}$$

where

$$a_1 = \frac{\dot{x}_{k-1}}{s_{k-1}}, \quad a_2 = \frac{\dot{y}_{k-1}}{s_{k-1}}, \tag{74}$$

$$a_3 = \frac{\dot{x}_{k-1}-T\omega_{k-1}\dot{y}_{k-1}}{s_{k-1}}, \quad a_4 = \frac{\dot{y}_{k-1}+T\omega_{k-1}\dot{x}_{k-1}}{s_{k-1}}. \tag{75}$$

3.3. Comparison with Conventional NCT Model

We consider the NCT model using the Cartesian velocity-based state, when the angular rate is estimated. The NCT model using the direct discrete approach is described in Chapter 11 of [22]. The discretized continuous-time NCT model [27] is described by

$$\boldsymbol{\eta}_k = \mathbf{F}_{\text{NCT}}^{\text{C}}(\omega)\boldsymbol{\eta}_{k-1} + \mathbf{w}_{k-1}^{\text{C}}, \tag{76}$$

$$\mathbf{F}_{\text{NCT}}^{\text{C}}(\omega) = \begin{bmatrix} 1 & 0 & \frac{\sin(\omega T)}{\omega} & -\frac{1-\cos(\omega T)}{\omega} & 0 \\ 0 & 1 & \frac{1-\cos(\omega T)}{\omega} & \frac{\sin(\omega T)}{\omega} & 0 \\ 0 & 0 & \cos(\omega T) & -\sin(\omega T) & 0 \\ 0 & 0 & \sin(\omega T) & \cos(\omega T) & 0 \\ 0 & 0 & 0 & 0 & 1 \end{bmatrix}, \tag{77}$$

$$\mathbf{w}_{k-1}^{\text{C}} \sim \mathcal{N}(\mathbf{w}_{k-1}^{\text{C}}; \mathbf{0}, \mathbf{Q}_{k-1}^{\text{C}}), \tag{78}$$

$$\mathbf{Q}_{k-1}^{\text{C}} = \begin{bmatrix} qT^3/3 & 0 & qT^2/2 & 0 & 0 \\ 0 & qT^3/3 & 0 & qT^2/2 & 0 \\ qT^2/2 & 0 & qT & 0 & 0 \\ 0 & qT^2/2 & 0 & qT & 0 \\ 0 & 0 & 0 & 0 & q_\omega T \end{bmatrix}, \tag{79}$$

where q is the PSD of the acceleration process noise along the X or Y direction.

Remark 1. *The upper left 4×4 block of the state transition matrix in (77) is the state transition matrix for the NCV model using Cartesian state [22]. Similarly, the upper left 4×4 block of the process noise covariance matrix in (79) is the process noise covariance matrix for the NCV model using Cartesian state [22]. They cannot be derived from a continuous-time model of the NCT motion.*

The second-order model with the TS2 approximation and Cartesian velocity-based state was used in [53], and a superior RMSE was reported compared with the conventional model described above.

4. Sensor Dynamic and Measurement Models

4.1. Sensor Dynamic Models

We assume that the motion of the sensor is deterministic and the state of the sensor at each measurement time is exactly known. The sensor follows two types of motion: constant velocity (CV) in 3D and the second type of motion with known angular velocity Ω. For both types of motion, the Cartesian state vector of the sensor is appropriate and is defined by

$$\mathbf{x}^s(t) := \begin{bmatrix} x^s(t) & y^s(t) & \dot{x}^s(t) & \dot{y}^s(t) & z^s(t) & \dot{z}^s(t) \end{bmatrix}'. \tag{80}$$

The dynamic models of the sensor for the CV and CT are described, respectively, by [8,31]

$$\mathbf{x}_k^s = \mathbf{F}^{CV}(T)\mathbf{x}_{k-1}^s, \tag{81}$$

$$\mathbf{F}^{CV}(T) = \begin{bmatrix} \mathbf{I}_2 & \mathbf{I}_2 T & \mathbf{0}_2 \\ \mathbf{0}_2 & \mathbf{I}_2 & \mathbf{0}_2 \\ \mathbf{0}_2 & \mathbf{0}_2 & \mathbf{F}_1 \end{bmatrix}. \tag{82}$$

$$\mathbf{x}_k^s = \mathbf{F}^{CT}(T, \Omega_{k-1})\mathbf{x}_{k-1}^s, \tag{83}$$

where the state transition matrix $\mathbf{F}_1$ for CV is defined in (7), Ω_{k-1} is the angular velocity of the sensor during $[t_{k-1}, t_k)$ and the state transition matrix for CT is given by

$$\mathbf{F}^{CT}(T, \Omega) = \begin{bmatrix} 1 & 0 & \sin(\Omega T)/\Omega & -[1-\cos(\Omega T)]/\Omega & 0 & 0 \\ 0 & 1 & [1-\cos(\Omega T)]/\Omega & \sin(\Omega T)/\Omega & 0 & 0 \\ 0 & 0 & \cos(\Omega T) & -\sin(\Omega T) & 0 & 0 \\ 0 & 0 & \sin(\Omega T) & \cos(\Omega T) & 0 & 0 \\ 0 & 0 & 0 & 0 & 1 & T \\ 0 & 0 & 0 & 0 & 0 & 1 \end{bmatrix}. \tag{84}$$

In passive IRST tracking, the sensor moves with a sequence of CV and CT motions [8,31].

4.2. Measurement Model

Let $\mathbf{p}_k$ and $\mathbf{p}_k^s$ denote the target and sensor position vectors, respectively, at time t_k,

$$\mathbf{p}_k := [x_k \quad y_k \quad z_k]', \tag{85}$$

$$\mathbf{p}_k^s := [x_k^s \quad y_k^s \quad z_k^s]'. \tag{86}$$

An IRST sensor measures the bearing and elevation angles of a target [5,8], as shown in Figure 2. We note that the bearing (ϕ_k) and elevation (ϵ_k) angles depend on the relative position $\mathbf{p}_k - \mathbf{p}_k^s$ in Cartesian and polar velocity-based models. Hence, for both type of state vectors, the measurement model for the bearing and elevation angles is described by

$$\mathbf{y}_k = \mathbf{h}(\mathbf{p}_k, \mathbf{p}_k^s) + \mathbf{n}_k, \tag{87}$$

$$\mathbf{h}(\mathbf{p}_k, \mathbf{p}_k^s) := \begin{bmatrix} \phi_k \\ \epsilon_k \end{bmatrix} = \begin{bmatrix} \tan^{-1}(x_k - x_k^s, y_k - y_k^s) \\ \arctan((z_k - z_k^s)/\rho_k) \end{bmatrix}, \tag{88}$$

where ϕ_k and ϵ_k lie in $[0, 2\pi)$ and $(-\pi/2, \pi/2)$, respectively and the ground range ρ_k is defined by

$$\rho_k := \sqrt{(x_k - x_k^s)^2 + (y_k - y_k^s)^2}, \quad \rho_k > 0. \tag{89}$$

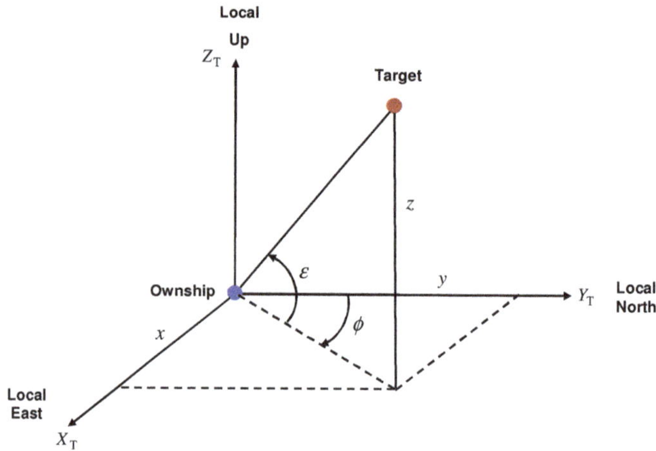

Figure 2. Definition of the tracker coordinate frame (T frame), bearing $\phi \in [0, 2\pi)$ and elevation $\epsilon \in (-\pi/2, \pi/2)$.

We assume that the measurement noise is zero-mean Gaussian with covariance $\mathbf{R}$

$$\mathbf{n}_k \sim \mathcal{N}(\mathbf{n}_k; \mathbf{0}, \mathbf{R}), \tag{90}$$

$$\mathbf{R} := \mathrm{diag}(\sigma_\phi^2, \sigma_\epsilon^2), \tag{91}$$

where σ_ϕ and σ_ϵ are the bearing and elevation measurement standard deviations (SDs), respectively.

5. Filtering Algorithms

We compare the performances of four CKF-based algorithms using the Euler and TS2 approximations with the polar and Cartesian velocity-based states. These four algorithms are called CKF1P, CKF1C, CKF2P, and CKF1P. The discrete-time dynamic and measurement models in these algorithms are nonlinear. The features of these algorithms are summarized in Table 1. In [54], the authors have considered the maneuvering target tracking problem using a CKF-based CDF filter with range, azimuth, and elevation measurements. They claim that this is a very challenging problem. They use the prior distribution to initialize the filter. The problem considered in our study is relatively harder since only azimuth, and elevation measurements are available.

Table 1. Features of CKF based algorithms.

Filter	2D State in NCT	Approximation	Process Noise
CKF1P	Polar velocity	Euler	State-independent
CKF1C	Cartesian velocity	Euler	State-dependent
CKF2P	Polar velocity	TS2	State-dependent
CKF2C	Cartesian velocity	TS2	State-dependent

6. Numerical Simulation and Results

The IRST sensor trajectory and parameters in the simulation are similar to those used in [8,31]. The target moves with an NCT motion in a plane parallel to the XY-plane and moves with an NCV motion along the Z-axis. Table 2 presents prior mean polar velocity-based state parameters of the target. The NCT motion has a centripetal acceleration $s_1\omega_1$ of 3 g, where $g = 9.8 \text{ m}^2\text{s}^{-2}$. This scenario was used in [5]. We use the same filter initialization with that in [54] in the current study. The prior variance of the 3D polar velocity-based state is chosen as

$$\mathbf{P}_{0,p,1} = \text{diag}(1000^2 \text{ m}^2,\ 1000^2 \text{ m}^2,\ 30^2 \text{ m}^2\text{s}^{-2},\ 0.0873 \text{ rad}^2,\ (4.95 \times 10^{-3})^2 \text{rad}^2 \text{ s}^{-2},$$
$$100^2 \text{ m}^2, 5^2 \text{ m}^2\text{s}^{-2}). \tag{92}$$

Using the Euler approximation, the process noise covariance in the polar velocity-based model for the NCT motion can be calculated exactly. Hence, we use the Euler approximation for the polar velocity-based model to generate true target trajectories for the NCT motion in the XY-plane using 100 sub-sampling intervals for the measurement time interval (T) of 1 s. The Z-component of the NCV trajectory is generated exactly. The process noise parameters used in the simulation are $q_s = 0.2 \text{ m}^2\text{s}^{-3}$, $q_\omega = 5e - 07 \text{ rad}^2\text{s}^{-3}$, and $q_z = 0.001 \text{ m}^2\text{s}^{-3}$. Figure 3 presents the true NCT trajectory of the target in the XY-plane from the first Monte Carlo run.

Table 2. Prior polar mean velocity-based 3D state parameters of target.

Variable	Value
$\bar{x}_0$ (m)	97,580.7358
$\bar{y}_0$ (m)	97,580.7358
$\bar{s}_0$ (m/s)	297.0
$\bar{\theta}_0$ (deg)	215.0
$\bar{\omega}_0$ (deg/s)	5.672
$\bar{z}_0$ (m)	9000.0
$\bar{\dot{z}}_0$ (m/s)	0.0

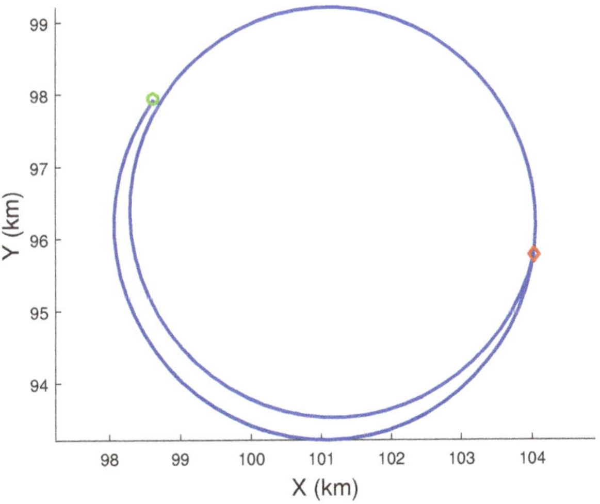

Figure 3. Target true trajectory in the XY-plane from the first Monte Carlo run. The green circle and the red diamond represent the start point and end point, respectively.

We assume that the motion of the sensor is deterministic. The sensor moves in a plane parallel to the XY-plane at a fixed height of 10 km and follows a sequence of CV and CT motions. The initial position and velocity vectors of the sensor are (0, 0, 10,000) m and (0, 264, 0) m/s, respectively. Table 3 presents the motion profile of the sensor. In Table 3, Δt represents the duration of the segment, $\Delta\phi_s$ is the total angular change during the segment, and Ω is the angular velocity of the sensor during the segment. The measurement time interval of the IRST sensor is 1 s and there are 101 measurements. The measurement error SDs for bearing and elevation have the same value. We use two angle SDs of 1 mrad and 2 mrad in this simulation. The sensor trajectory in the XY-plane is shown in Figure 4.

Table 3. Motion profile of the sensor.

Interval (s)	Δt (s)	$\Delta\phi_s$ (rad)	Motion Type	Ω (rad/s)
$[0, 15]$	15	0	CV	0
$[15, 31]$	16	$-\pi/4$	CT	$-\pi/64$
$[31, 43]$	12	0	CV	0
$[43, 75]$	32	$\pi/2$	CT	$\pi/64$
$[75, 86]$	11	0	CV	0
$[86, 102]$	16	$-\pi/4$	CT	$-\pi/64$

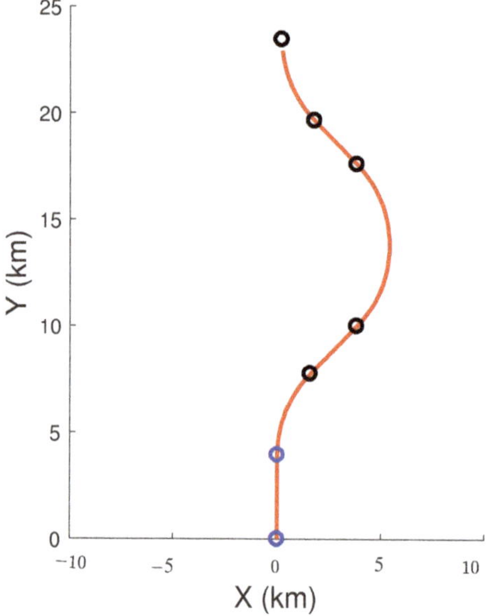

Figure 4. Sensor trajectory in the XY-plane.

6.1. Comparison of Filtering Algorithms

We used 500 Monte Carlo runs to compute the root mean square (RMS) position, velocity, and angular rate errors of the CKF1P, CKF1C, CKF2P, and CKF2C. Each filter is initialized using the prior mean and covariance. The RMS errors for these four filters for angle SDs of 1 mrad and 2 mrad are presented in Figures 5–7. Results in Figure 5 show that RMS position errors of the CKF1P, CKF2P, and CKF2C are close and they are nearly the same towards the end. On the contrary, the RMS position error of the CKF1C is significantly higher during some measurement intervals and also significantly lower during a time interval. We see in Figure 6 that the CKF2P and CKF2C have the best results and nearly the same RMS velocity errors. The RMS velocity error of the CKF1P is slightly higher than that CKF2P and CKF2C. The RMS velocity error of the CKF1C is significantly higher than that of the other three filters. It appears that the CKF1C diverges for this maneuvering target tracking scenario. A similar pattern is observed in the results of the angular rate errors in Figure 7.

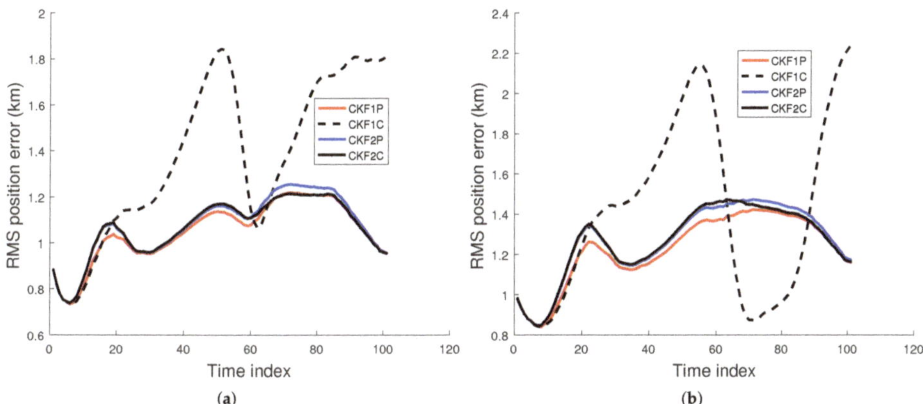

Figure 5. RMS position error using the prior variance $\mathbf{P}_{0,p,1}$ from 500 Monte Carlo runs. (**a**) Angle SD of 1 mrad, (**b**) Angle SD of 2 mrad.

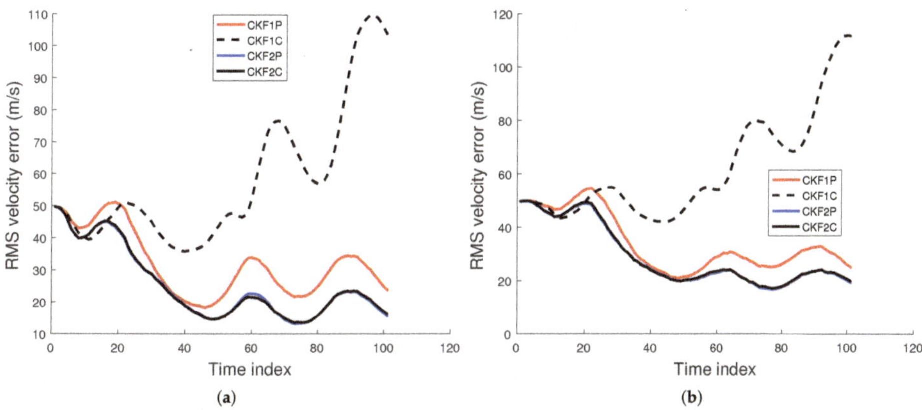

Figure 6. RMS velocity error using the prior variance $\mathbf{P}_{0,p,1}$ from 500 Monte Carlo runs. (**a**) Angle SD of 1 mrad, (**b**) Angle SD of 2 mrad.

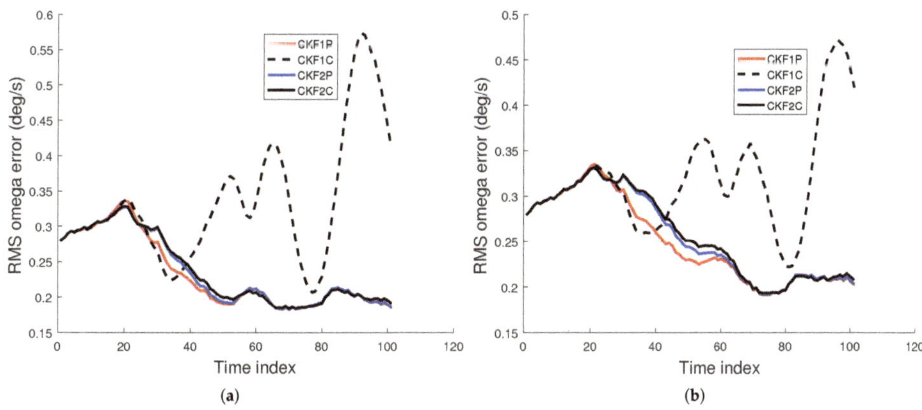

Figure 7. RMS angular rate error using the prior variance $\mathbf{P}_{0,p,1}$ from 500 Monte Carlo runs. (**a**) Angle SD of 1 mrad, (**b**) Angle SD of 2 mrad.

To evaluate the overall performance of a filter, we use the root time-averaged mean square (RTAMS) error [11] for position, velocity, and angular rate. Let $\mathbf{p}^t_{j,i}$ and $\hat{\mathbf{p}}^t_{j,i}$ denote the true and estimated position of the target, respectively, at time index j in the ith Monte Carlo run. The RTAMS position error [11] is defined by

$$\text{RTAMS}_{\text{pos}} = \sqrt{\frac{1}{N_f M} \sum_{j \in S_f} \sum_{i=1}^{M} \|\hat{\mathbf{p}}^t_{j,i} - \mathbf{p}^t_{k,j}\|^2}, \tag{93}$$

where S_f is a set of time indices with N_f indices, and M is the number of Monte Carlo runs. We have chosen time indices from 51 to 101 to define S_f. The RTAMS error [11] for velocity and angular rate are similarly defined. Table 4 presents the RTAMS error metric for position, velocity, and angular rate for measurement error SDs of 1 mrad and 2 mrad. Results in Table 4 show that the CKF2P and CKF2C have the best RTAMS errors for position, velocity, and angular rate, which are nearly the same.

Table 4. RTAMS position, velocity, and angular rate errors for CKF1P, CKF1C, CKF2P, and CKF2C using the prior variance $\mathbf{P}_{0,p,1}$.

Metric	Filter	1 mrad	2 mrad
	CKF1P	1.137	1.355
Position error (km)	CKF1C	1.596	1.582
	CKF2P	1.165	1.400
	CKF2C	1.146	1.389
	CKF1P	28.628	28.178
Velocity error (m/s)	CKF1C	75.853	77.691
	CKF2P	18.959	21.132
	CKF2C	18.867	21.334
	CKF1P	0.197	0.211
Angular rate error (deg/s)	CKF1C	0.394	0.347
	CKF2P	0.197	0.214
	CKF2C	0.197	0.216

Table 5 presents CPU times for each Monte Carlo run and CPU times relative to the CKF1P. Results in Table 5 show that the CKF1P has the fastest CPU time, being slightly faster than the CKF2P.

Table 5. CPU times (s) for each Monte Carlo run and CPU times relative to CKF1P for angle SD of 1 mrad.

Filter	CPU Time (s)	CPU Relative to CKF1P
CKF1P	0.0377	1.0000
CKF1C	0.0386	1.0129
CKF2P	0.0391	1.0356
CKF2C	0.0789	2.0910

Let $x_{k,i}$ and $\hat{x}_{k,i}$ be the true and filtered estimated X-coordinates at time k, respectively. Similar definitions apply for other position coordinates. Then, the sample position bias is given by [22,33]

$$b_{\text{pos},k} = \frac{1}{M} \sum_{i=1}^{M} [(x_{k,i} - \hat{x}_{k,i}) + (y_{k,i} - \hat{y}_{k,i}) + (z_{k,i} - \hat{z}_{k,i})]. \tag{94}$$

For simplicity, we use "bias" to represent sample bias. Similarly, the biases for velocity and angular rate can be defined. The bias at time k can be positive or negative. It is desirable to have a small bias in the state estimate. The sample bias for position, velocity, and angular rate are shown in Figures 8 and 9. Results in Figure 8 show that the position biases of the CKF1P, CKF2P, and CKF2C are small, and the velocity biases of CKF2P and CKF2C are nearly zero. The angular rate biases of CKF1P, CKF2P, and CKF2C become smaller with time and approach zero. The CKF1C has large position, velocity, and angular rate biases.

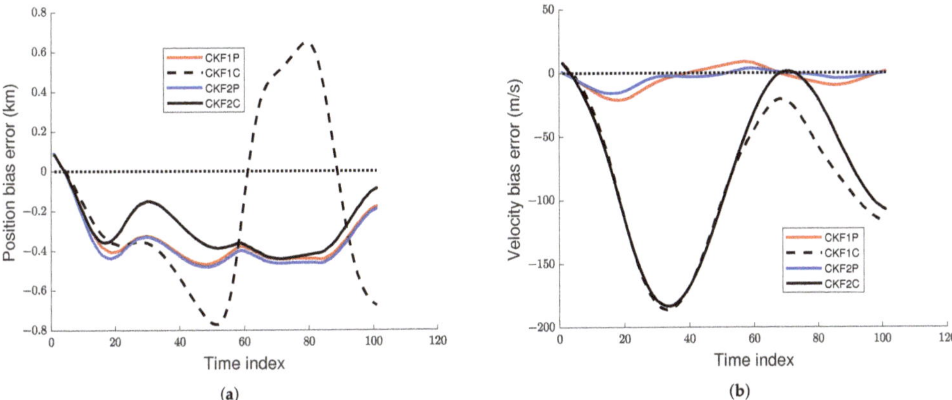

Figure 8. Position and velocity bias errors for the 1 mrad case. (**a**) Position bias, (**b**) Velocity bias.

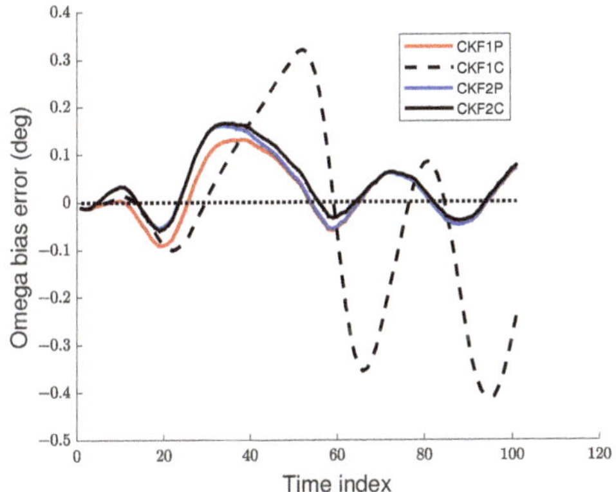

Figure 9. Angular rate bias error for the 1 mrad case.

6.2. Dependence of Filtering Accuracy on the Prior Distribution

In order to analyze the dependence of filtering accuracy on the prior distribution, we have chosen a larger prior variance for the 3D polar velocity-based state relative to that used in (92)

$$\mathbf{P}_{0,p,2} = \mathrm{diag}(5000^2\ \mathrm{m}^2,\ 5000^2\ \mathrm{m}^2,\ 90^2\ \mathrm{m}^2\mathrm{s}^{-2},\ (3*0.0873)\ \mathrm{rad}^2,\ (3*4.95\times10^{-3})^2\ \mathrm{rad}^2\mathrm{s}^{-2},$$
$$500^2\ \mathrm{m}^2, 15^2\ \mathrm{m}^2\mathrm{s}^{-2}). \tag{95}$$

The prior variance of Cartesian position is increased by 25 times, and the other components have been increased by 9 times. The prior mean is unchanged. The RMSE plots of position, velocity, and angular rate are presented in Figures 10 and 11 for the 1 mrad scenario.

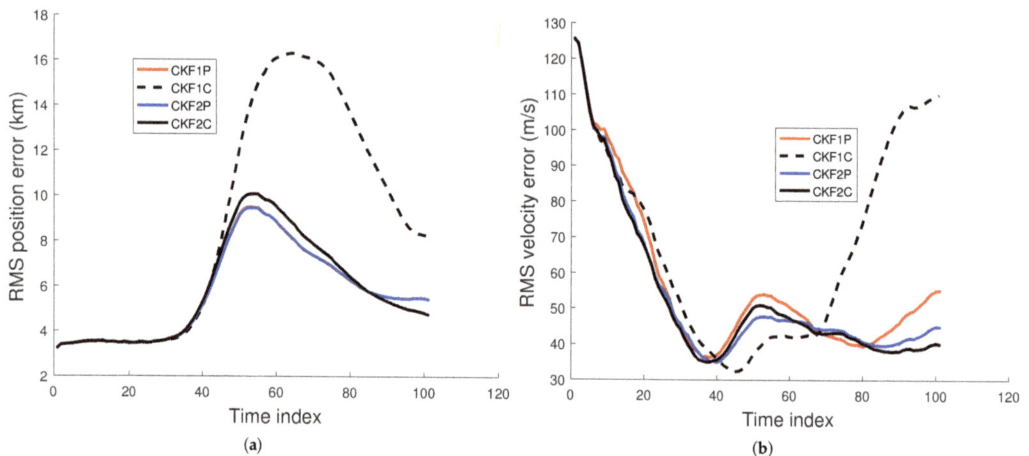

Figure 10. RMS position and velocity errors using the prior variance $\mathbf{P}_{0,p,2}$ with angle SD of 1 mrad. (**a**) RMS position error, (**b**) RMS velocity error.

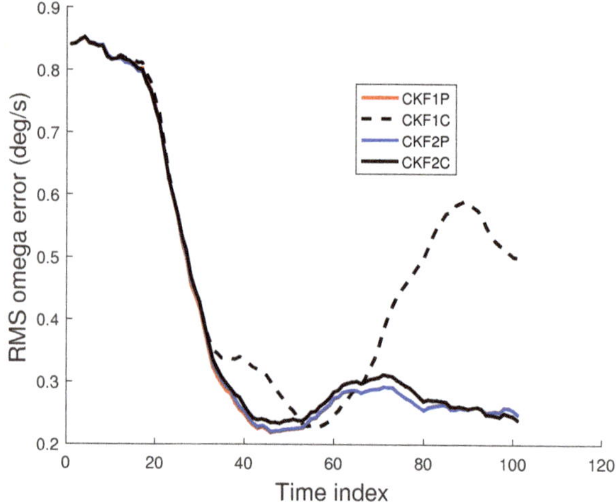

Figure 11. RMS angular rate error using the prior variance $\mathbf{P}_{0,p,2}$ with angle SD of 1 mrad.

Table 6 presents the RTAMS error for position, velocity, and angular rate for measurement error SDs of 1 mrad and the second prior distribution. Results in Table 6 show that the CKF2P and CKF2C have the best RTAMS errors for position, velocity, and angular rate, which are nearly the same.

6.3. Summary of Key Results

Based on RMS and RTAMS errors, the key results of our study are as follows:

- The CKF1P has the best position estimation accuracy. The position estimation accuracies of the CKF2P and CKF2C are close to that of the CKF1P;
- The CKF2P and CKF2C have the best velocity estimation accuracy;
- The state estimation accuracies of the CKF2P and CKF2C are comparable. However, the computational cost of the CKF2C is about twice that of the CKF2P;
- The CKF1C does not perform well for this problem and has high estimation errors.

Table 6. Comparison of RTAMS position, velocity, and angular rate errors for CKF1P, CKF1C, CKF2P, CKF2C using prior variances $\mathbf{P}_{0,p,1}$ and $\mathbf{P}_{0,p,2}$ with angle SD of 1 mrad.

Metric	Filter	$\mathbf{P}_{0,p,1}$	$\mathbf{P}_{0,p,2}$
Position error (km)	CKF1P	1.137	7.175
	CKF1C	1.596	13.559
	CKF2P	1.165	7.178
	CKF2C	1.146	7.454
Velocity error (m/s)	CKF1P	28.628	47.204
	CKF1C	75.853	75.192
	CKF2P	18.959	43.836
	CKF2C	18.867	42.973
Angular rate error (deg/s)	CKF1P	0.197	0.265
	CKF1C	0.394	0.440
	CKF2P	0.197	0.265
	CKF2C	0.197	0.274

7. Conclusions

We considered the challenging filtering problem of a maneuvering target in 3D using the bearing and elevation measurements from a maneuvering passive IRST sensor. Research on this problem is rather limited. The target moves with the NCT motion in the XY-plane and has an NCV motion along the Z-axis. We discretized the continuous-time stochastic differential equation for the NCT model using the first (Euler) and second-order Taylor approximations to obtain discrete-time NCT models. Discrete-time dynamic and measurement models are nonlinear. For each approximation, we used the polar and Cartesian velocity-based states for the NCT model. The CKF was used in each case giving rise to four filters: CKF1P, CKF1C, CKF2P, and CKF2C. Numerical results based on Monte Carlo simulations suggest that the second-order Taylor approximation-based filters CKF2P and CKF2C have the best state estimation accuracy for this scenario. Secondly, the accuracies of these two filters are nearly the same.

Our future work will develop filter initialization algorithms that can be used with real data. We shall also focus on calculating the PCRLB for the filtering problem to assess the best achievable accuracy.

Author Contributions: Conceptualization, M.M. (Mahendra Mallick), X.T., Y.Z. and M.M. (Mark Morelande); methodology, M.M. (Mahendra Mallick) and M.M. (Mark Morelande); software, M.M. (Mahendra Mallick), X.T. and M.M. (Mark Morelande); validation, M.M. (Mahendra Mallick), X.T. and Y.Z.; formal analysis, M.M. (Mahendra Mallick); writing—original draft preparation, M.M. (Mahendra Mallick); writing—review and editing, M.M. (Mahendra Mallick), X.T., Y.Z. and M.M. (Mark Morelande); supervision, M.M. (Mahendra Mallick). All authors have read and agreed to the published version of the manuscript.

Funding: This research received no external funding.

Institutional Review Board Statement: Not applicable.

Informed Consent Statement: Not applicable.

Data Availability Statement: Not applicable.

Conflicts of Interest: The authors declare no conflict of interest.

Abbreviations

The following abbreviations are used in this manuscript:

3D-IVKF	3D instrumental variable based Kalman filter
AOF	Angle-only filtering
ATC	Air-traffic control
BOF	Bearings-only filtering
CKF	Cubature Kalman filter
CKF1C	CKF using Euler approximation with Cartesian velocity
CKF1P	CKF using Euler approximation with polar velocity
CKF2C	CKF using order 2 weak Taylor approximation with Cartesian velocity
CKF2P	CKF using order 2 weak Taylor approximation with polar velocity
CCKF	Cartesian CKF
CEKF	Cartesian EKF
CNSKF	Cartesian new sigma point Kalman filter
CUKF	Cartesian UKF
CT	Constant turn
CV	Constant velocity
EKF	Extended Kalman filter
EKF-MSC	EKF using the MSC
EnKF	Ensemble Kalman filter
IRST	Infrared search and track
LSC	Log spherical coordinates
MPC	Modified polar coordinates
MSC	Modified spherical coordinates
MMSE	Minimum mean square error
NCT	Nearly constant turn
NCV	Nearly constant velocity
PFF	Particle flow filter
PCRLB	Posterior Cramér–Rao lower bound
PSD	Power spectral density
RP-EKF	Range-parametrized EKF
RP-UKF	Range-parametrized UKF
RMS	Root mean square
SD	Standard deviation
SDE	Stochastic differential equation
TS2	Order 2 weak Taylor
UKF	Unscented Kalman filter
UKF-MSC	UKF using the MSC

References

1. Aidala, V.J. Kalman filter behaviour in Bearings-Only Tracking Applications. *IEEE Trans. Aerosp. Electron. Syst.* **1979**, *15*, 29–39. [CrossRef]
2. Allen, R.R.; Blackman, S.S. Angle-only tracking with a MSC filter. In Proceedings of the Digital Avionics Systems Conference, Los Angeles, CA, USA, 14–17 October 1991; pp. 561–566.
3. Asfia, U.; Radhakrishnan, R.; Sharma, S.N. Three-Dimensional Bearings-Only Target Tracking: Comparison of Few Sigma Point Kalman Filters. In *Communication and Control for Robotic Systems*; Gu, J., Dey, R., Adhikary, N., Eds.; Springer: Singapore, 2021; pp. 273–289.
4. Blackman, S.; Popoli, R. *Design and Analysis of Modern Tracking Systems*; Artech House: Boston, MA, USA; London, UK, 1999.
5. Blackman, S.S.; Dempster, R.J.; Blyth, B.; Durand, C. Integration of Passive Ranging with Multiple Hypothesis Tracking (MHT) for Application with Angle-Only Measurements. In Proceedings of the SPIE 7698, Signal and Data Processing of Small Targets 2010, Orlando, FL, USA, 15 April 2010; pp. 769815-1–769815-11.
6. Karlsson, R.; Gustafsson, F. Range estimation using angle-only target tracking with particle filters. In Proceedings of the American Control Conference, Arlington, VA, USA, 25–27 June 2001; pp. 3743–3748.
7. Karlsson, R.; Gustafsson, F. Recursive Bayesian estimation: Bearings-only applications. *IEE Proc.-Radar Sonar Navig.* **2005**, *152*, 305–313. [CrossRef]

8. Mallick, M.; Morelande, M.; Mihaylova, L.; Arulampalam, S.; Yan, Y. Angle-only Filtering in Three Dimensions. Chapter 1. In *Integrated Tracking, Classification, and Sensor Management: Theory and Applications*; Mallick, M., Krishnamurthy, V., Vo, B.-N., Eds.; John Wiley & Sons: Hoboken, NJ, USA, 2012.

9. Mallick, M.; Chang, K.; Arulampalam, S.; Yan, Y. Heterogeneous Track-to-Track Fusion in 3D Using IRST Sensor and Air MTI Radar. *IEEE Trans. Aerosp. Electron. Syst.* **2019**, *55*, 3062–3079. [CrossRef]

10. Radhakrishnan, R.; Bhaumik, S.; Tomar, N.K. Gaussian sum shifted Rayleigh filter for underwater bearings-only target tracking problems. *IEEE J. Ocean. Eng.* **2018**, *44*, 492–501. [CrossRef]

11. Ristic, B.; Arulampalam, S.; Gordon, N. *Beyond the Kalman Filter*; Artech House: Boston, MA, USA, 2004.

12. Stallard, D.V. An angle-only tracking filter in modified spherical coordinates. In Proceedings of the AIAA Guidance, Navigation and Control Conference, Monterey, CA, USA, 17–19 August 1987; pp. 542–550.

13. Stallard, D.V. Angle-only tracking filter in modified spherical coordinates. *J. Guid. Control. Dyn.* **1991**, *14*, 694–696. [CrossRef]

14. Li, Q.; Shi, L.; Wang, H.; Guo, F. Utilization of Modified Spherical Coordinates for Satellite to Satellite Bearings-Only Tracking. *Chin. J. Space Sci.* **2009**, *29*, 627–634.

15. Maggio, E.; Cavallaro, A. *Video Tracking: Theory and Practice*, 1st ed.; John Wiley & Sons: Chichester, UK; West Sussex, UK, 2011.

16. Kronhamn, T.R. Bearings-only Target Motion Analysis Based on a Multihypothesis Kalman Filter and Adaptive Ownship Motion Control. *IEE Proc.-Radar Sonar Navig.* **1998**, *145*, 247–252. [CrossRef]

17. Kronhamn, T.R. Angle-only tracking of manoeuvring targets using adaptive-IMM multiple range models. In Proceedings of the Radar 2002, Edinburgh, UK, 15–17 October 2002; pp. 310–314.

18. Nardone, S.; Lindgren, A.; Gong, K. Fundamental properties and performance of conventional bearings-only target motion analysis. *IEEE Trans. Autom. Control* **1984**, *29*, 775–787. [CrossRef]

19. Peach, N. Bearings-only Tracking using a Set of Range-parameterised Extended Kalman Filters. *IEE Proc.-Control Theory Appl.* **1995**, *142*, 73–80. [CrossRef]

20. Zhang, Y.; Lan, J.; Mallick, M.; Li, X.R. Bearings-Only Filtering Using Uncorrelated Conversion Based Filters. *IEEE Trans. Aerosp. Electron. Syst.* **2021**, *57*, 882–896. [CrossRef]

21. Jauffret, C.; Pillon. D. Observability in passive target motion analysis. *IEEE Trans. Aerosp. Electron. Syst.* **1996**, *32*, 1290–1300. [CrossRef]

22. Bar-Shalom, Y.; Li, X.R.; Kirubarajan, T. *Estimation with Applications to Tracking and Navigation*; Wiley: New York, NY, USA, 2001.

23. Gelb, A. *Applied Optimal Estimation*; MIT Press: Cambridge, MA, USA, 1974.

24. Hoelzer, H.D.; Johnson, G.W.; Cohen, A.O. Modified Polar Coordinates—The Key to Well Behaved Bearings Only Ranging. In *IR&D Report 78-M19-OOOlA*; IBM Federal Systems Division, Shipboard and Defense Systems: Manassas, VA, USA, 1978.

25. Johnson, G.W.; Hoelzer, H.D.; Cohen, A.O.; Harrold, E.F. Improved coordinates for target tracking from time delay information. *J. Acoust. Soc. Am.* **1979**, *854*, M1–M32.

26. Julier, S.; Uhlmann, J.; Durrant-Whyte, H.F. A new method for the nonlinear transformation of means and covariances in filters and estimators. *IEEE Trans. Autom. Control* **2000**, *45*, 477–482. [CrossRef]

27. Arasaratnam, I.; Haykin, S. Cubature Kalman filters. *IEEE Trans. Autom. Control* **2009**, *54*, 1254–1269. [CrossRef]

28. Alspach, D.; Sorenson, H. Nonlinear Bayesian estimation using Gaussian sum approximations. *IEEE Trans. Autom. Control* **1972**, *17*, 439–448. [CrossRef]

29. Mallick, M.; Mihaylova, L.; Arulampalam, S.; Yan, Y. Angle-only Filtering in 3D Using Modified Spherical and Log Spherical Coordinates. In Proceedings of the 14th International Conference on Information Fusion, Chicago, IL, USA, 5–8 July 2011; pp. 1–8.

30. Robinson, P.N.; Yin, M.R. Modified spherical coordinates for radar. In Proceedings of the AIAA Guidance, Navigation and Control Conference, Scottsdale, AZ, USA, 1–3 August 1994; pp. 55–64.

31. Mallick, M.; Morelande, M.; Mihaylova, L.; Arulampalam, S.; Yan, Y. Comparison of Angle-only Filtering Algorithms in 3D using Cartesian and Modified Spherical Coordinates. In Proceedings of the 15th International Conference on Information Fusion, Singapore, 9–12 July 2012; pp. 1392–1399.

32. Mallick, M.; Morelande, M.; Mihaylova, L. Continuous-Discrete Filtering using EKF, UKF, and PF. In Proceedings of the 15th International Conference on Information Fusion, Singapore, 9–12 July 2012.

33. Mallick, M.; Arulampalam, S. Comparison of Nonlinear Filtering Algorithms in Ground Moving Target Indicator (GMTI) Target Tracking. In Proceedings of the SPIE 5204, Signal and Data Processing of Small Targets 2003, San Diego, CA, USA, 5 January 2004; pp. 630–647.

34. Mallick, M.; Tian, X.; Liu, J. Evaluation of Measurement Converted KF, EKF, UKF, CKF, and PF in GMTI Filtering. In Proceedings of the 10th International Conference on Control, Automation and Information Sciences (ICCAIS 2018), Xi'an, China, 14–17 October 2021.

35. Radhika, M.N.; Mallick, M.; Tian, X.; Liu, J. Performance Evaluation of IMM-based Nonlinear Filters for a Highly Maneuvering Aircraft. In Proceedings of the 10th International Conference on Control, Automation and Information Sciences (ICCAIS 2021), Xi'an, China, 14–17 October 2021.

36. Badriasl, L.; Arulampalam, S.; Finn, A. A novel batch Bayesian WIV estimator for three-dimensional TMA using bearing and elevation measurements. *IEEE Trans. Signal Process.* **2018**, *66*, 1023–1036. [CrossRef]

37. Gupta, S.D.; Yu, J.Y.; Mallick, M.; Coates, M.; Morelande, M. Comparison of Angle-only Filtering Algorithms in 3D Using EKF, UKF, PF, PFF, and Ensemble KF. In Proceedings of the 18th International Conference on Information Fusion, Washington, DC, USA, 6–9 July 2015; pp. 1649–1656.
38. Mallick, M.; Sinha, A.; Liu, J. Enhancements to Bearing-only Filtering. In Proceedings of the 20th International Conference on Information Fusion, Xi'an, China, 10–13 July 2017; pp. 1422–1449.
39. Nguyen, N.H.; Doğançay, K. Instrumental variable based Kalman filter algorithm for three-dimensional AOA target tracking. *IEEE Signal Process. Lett.* **2018**, *25*, 1605–1609. [CrossRef]
40. Tichavsky, P.; Muravchik, C.H.; Nehorai, A. Posterior Cramér-Rao bounds for discrete-time nonlinear filtering. *IEEE Trans. Signal Process.* **1998**, *46*, 1386–1396. [CrossRef]
41. Mallick, M.; Arulampalam, S.; Yan, Y. Posterior Cramér-Rao Lower Bound for Angle-only Filtering in 3D. In Proceedings of the 7th International Conference on Control, Automation and Information Sciences (ICCAIS 2021), Hangzhou, China, 24–27 October 2018; pp. 349–354.
42. Schmitt, L.; Fichter, W. Globally valid posterior Cramér–Rao bound for three-dimensional bearings-only filtering. *IEEE Trans. Aerosp. Electron. Syst.* **2019**, *55*, 2036–2044. [CrossRef]
43. Li, X.R.; Jilkov, V.P. Survey of maneuvering target tracking, Part I: Dynamic Models. *IEEE Trans. Aerosp. Electron. Syst.* **2003**, *39*, 1333–1364.
44. Gustafsson, F.; Isaksson, A.J. Best choice of state variables for tracking coordinated turns. In Proceedings of the 35th IEEE Conference on Decision and Control, Kobe, Japan, 13 December 1996; pp. 3145–3150.
45. Morelande, M.R.; Gordon, N.J. Target tracking through a coordinated turn. In Proceedings of the IEEE International Conference on Acoustics, Speech, and Signal Processing (ICASSP'05), Philadelphia, PA, USA, 23–23 March 2005; pp. iv/21–iv/24.
46. Roth, M.; Hendeby, G.; Gustafsson, F. EKF/UKF maneuvering target tracking using coordinated turn models with polar/Cartesian velocity Target tracking through a coordinated turn. In Proceedings of the 17th International Conference on Information Fusion (FUSION), Salamanca, Spain, 7–10 July 2014; pp. 1–8.
47. Li, X.R.; Bar-Shalom, Y. Design of an interacting multiple model algorithm for air traffic control tracking. *IEEE Trans. Autom. Control* **1993**, *1*, 186–194. [CrossRef]
48. Jazwinski, A.H. *Stochastic Processes and Filtering Theory*, 1st ed.; Academic Press: New York, NY, USA, 1970.
49. Särkkä, S.; Solin, A. *Applied Stochastic Differential Equations*; Cambridge University Press: Cambridge, MA, USA, 2019.
50. Kloeden, P.E.; Platen, E. *Numerical Solution of Stochastic Differential Equations*; Springer: Berlin/Heidelberg, Germany, 1992.
51. Arfken, G.B.; Weber.; H.J.; Harris, F.E. *Mathematical Methods for Physicists: A Comprehensive Guide*, 7th ed.; Elsevier: New York, NY, USA, 2013.
52. Horn, R.A.; Johnson, C.R. *Topics in Matrix Analysis*; Cambridge University Press: Cambridge, MA, USA, 1991.
53. Yuan, T.; Bar-Shalom, Y.; Tian, X. Heterogeneous Track-to-Track Fusion. *J. Adv. Inf. Fusion* **2011**, *6*, 131–149.
54. Arasaratnam, I.; Haykin, S.; Hurd, T.R. Cubature Kalman filtering for continuous-discrete systems: Theory and simulations. *IEEE Trans. Signal Process.* **2010**, *58*, 4977–4993. [CrossRef]

sensors

Article

2D and 3D Angles-Only Target Tracking Based on Maximum Correntropy Kalman Filters

Asfia Urooj [1], Aastha Dak [1], Branko Ristic [2,*] and Rahul Radhakrishnan [1]

[1] Department of Electrical Engineering, Sardar Vallabhbhai National Institute of Technology, Surat 395007, India; ds19el006@eed.svnit.ac.in (A.U.); p20ic001@eed.svnit.ac.in (A.D.); r.rahul@eed.svnit.ac.in (R.R.)
[2] School of Engineering, RMIT University, Melbourne, VIC 3000, Australia
*   Correspondence: branko.ristic@rmit.edu.au

Abstract: In this paper, angles-only target tracking (AoT) problem is investigated in the non Gaussian environment. Since the conventional minimum mean square error criterion based estimators tend to give poor accuracy in the presence of large outliers or impulsive noises in measurement, a maximum correntropy criterion (MCC) based framework is presented. Accordingly, three new estimation algorithms are developed for AoT problems based on the conventional sigma point filters, termed as MC-UKF-CK, MC-NSKF-GK and MC-NSKF-CK. Here MC-NSKF-GK represents the maximum correntropy new sigma point Kalman filter realized using Gaussian kernel and MC-NSKF-CK represents realization using Cauchy kernel. Similarly, based on the unscented Kalman filter, MC-UKF-CK has been developed. The performance of all these estimators is evaluated in terms of root-mean-square error (RMSE) in position and % track loss. The simulations were carried out for 2D as well as 3D AoT scenarios and it was inferred that, the developed algorithms performed with improved estimation accuracy than the conventional ones, in the presence of non Gaussian measurement noise.

Keywords: nonlinear filtering; non Gaussian noise; maximum correntropy criterion; Gaussian kernel; Cauchy kernel

Citation: Urooj, A.; Dak, A.; Ristic, B.; Radhakrishnan, R. 2D and 3D Angles-Only Target Tracking Based on Maximum Correntropy Kalman Filters. *Sensors* **2022**, *22*, 5625. https://doi.org/10.3390/s22155625

Academic Editor: Andrzej Stateczny

Received: 15 June 2022
Accepted: 22 July 2022
Published: 27 July 2022

Publisher's Note: MDPI stays neutral with regard to jurisdictional claims in published maps and institutional affiliations.

1. Introduction

In state estimation, Kalman filter (KF) is a recursive solution used in various applications, such as information fusion, system control, integrated navigation, target tracking, and GPS solutions [1–4]. Kalman filter gives optimal estimates provided the dynamical system is linear and the noises assumed are Gaussian. However, it is extended to nonlinear systems through suitable approximation of the nonlinear functions. Using the Taylor series to linearize the nonlinear functions, the popular extended Kalman filter (EKF) [5] was derived. Also various sigma point filters have been proposed in the literature such as unscented Kalman filter (UKF) [6], cubature Kalman filter (CKF) [7], new sigma point Kalman filter (NSKF) [8], to obtain improved estimation accuracy than the EKF.

Since these filters are based on minimum mean square error criterion, their performance is likely to get deteriorated in the presence of non Gaussian noises such as heavy tailed and impulsive noises [9]. This makes state estimation a very challenging problem in the presence of nonlinear models and non Gaussian noise. Other possible solutions that can provide robust state estimates are Gaussian sum filter (GSF) [10,11], particle filter (PF) [12], Huber's KF (HKF, also known as M-estimation) [13], H_∞ filter [14] etc.

In order to improve the robustness of nonlinear state estimators in the presence of non Gaussian noise, a local similarity measure called correntropy [15,16], based filter called correntropy filter (C-Filter) was first proposed in [17]. Since it was developed by replacing the minimum mean square error (MMSE) criterion with maximum correntropy criterion (MCC), it proved beneficial for non Gaussian systems, but only for linear systems [18]. This algorithm made use of least squares and fixed point iteration, but failed to incorporate covariance estimation. In order to avoid this, a maximum correntropy Kalman filter (MCKF) involving fixed point iteration and covariance propagation was proposed in [19]. Similar issue was also addressed in [20], which used a cost function consisting of weighted least square (WLS) and Gaussian kernel function, and hence was named as maximum correntropy criterion-Kalman filter (MCC-KF).

To deal with nonlinear systems, extensions to the existing conventional algorithms based on MCC criterion were also developed and were named as maximum correntropy EKF (MC-EKF) [21], maximum correntropy UKF (MC-UKF) [22] and maximum correntropy sparse grid Gauss-Hermite quadrature filter (MC-SGHQF) [23]. But in the presence of large outliers in measurements, these filters incurred analytical problems in calculating inverse of matrices. Thus, new algorithms involving new cost function, WLS and statistical linearisation were proposed in [24], which were called as MC-UKF-constant and MC-UKF-adaptive [25]. In developing the above mentioned estimators, Gaussian kernel played an important role in suppressing the non Gaussian measurement noise. In target tracking applications, we may receive measurements which have larger outliers. This could prove to be a challenging task in successful estimation of states using Gaussian kernel as it may be difficult to select a proper kernel bandwidth. Hence, Gaussian kernel may not always prove to be the best choice for a kernel function. To overcome this drawback, a Cauchy kernel function is constructed which gives reasonable estimation accuracy for a wide range of kernel bandwidth [26,27].

This paper deals with angles only target tracking problem in 2D and 3D. The literature presents with many variations of this tracking problem such as when the target is a curvilinear manoeuvring target [28,29]. However, as is common in passive sonar target tracking applications, the objective here is to estimate the states of a moving constant velocity target with the help of angles-only measurements, but corrupted with non Gaussian noise. The observer continuously monitors for the signals, that are generated due to the sound radiated by the target. The AoT can also be performed with other measurement sources like IRST sensor [4], radar [30] and also through video tracking [31]. Any irregularities in these signals received by the observer can be considered as glint noise. A mixture model of two zero-mean Gaussian for glint noise has been proposed in [32]. This consists of one Gaussian density with high probability and small variance while the other has small probability of occurrence and large variance. Alternatively, it is also modelled in [33] as a mixture of zero mean with small variance. In this work, the non Gaussian noise in angular measurements is modelled as a mixture of Gaussian densities plus shot noise.

The main contribution of this paper is the development of three new nonlinear filters for AoT problem, MC-UKF-CK, MC-NSKF-GK and MC-NSKF-CK, and their performance evaluation in the context of angles-only tracking. Accordingly, conventional filters UKF and NSKF have been reformulated based on maximum correntropy criterion. MC-UKF and MC-NSKF based on Gaussian kernel (MC-UKF-GK, MC-NSKF-GK) and Cauchy kernel (MC-UKF-CK, MC-NSKF-CK) functions have been derived. The performance evaluation of these estimators are conducted considering RMSE in position and track loss as the two performance metrics and a comparative discussion is presented. The simulation results highlight that the existing solutions behave poorly in comparison to the proposed algorithms.

The rest of the paper is organised as follows. Section 2 describes the problem formulation for AoT in 2D as well as 3D. Section 3 illustrates the correntropy, its properties for two random variables and MCC. In Section 4, the already existing Gaussian kernel based MC state estimation framework is revisited. In Section 5, the Cauchy kernel based MC state estimation framework for nonlinear systems is derived. Section 6 briefly discuss about the state estimators on which the developed MCC framework is incorporated. Section 7 describes the realization of non Gaussian noise, followed by simulation study in Section 8. Finally, the concluding remarks are given in Section 9.

2. Problem Formulation

The aim of the angles only tracking problem is to track the target trajectory using the noise corrupted angular measurements. The dynamics of the target is assumed to be a constant velocity motion. The observer motion is deterministic, implying that the position and velocity of the observer is known to us. The 2D and 3D target observer dynamics is illustrated below.

2.1. Process Model

The target and observer state vector with position and velocity as its states is given as

$$\mathbf{X}_k^t = \begin{bmatrix} x_k^t & y_k^t & \dot{x}_k^t & \dot{y}_k^t \end{bmatrix}' \qquad \mathbf{X}_k^o = \begin{bmatrix} x_k^o & y_k^o & \dot{x}_k^o & \dot{y}_k^o \end{bmatrix}'.$$

The discrete time linear process model representing the target motion is given as

$$\mathbf{X}_k^t = \mathbf{F}\mathbf{X}_{k-1}^t + w_{k-1}. \tag{1}$$

Now, the relative state vector dynamics is

$$\mathbf{X}_k = \mathbf{F}\mathbf{X}_{k-1} + w_{k-1} - \mathbf{X}_{k-1}^o + \mathbf{F}\mathbf{X}_{k-1}^o. \tag{2}$$

where $\mathbf{X}_k$, the relative vector is defined as

$$\begin{aligned}
\mathbf{X}_k &= \mathbf{X}_k^t - \mathbf{X}_k^o \\
&= \begin{bmatrix} x_k^t - x_k^o & y_k^t - y_k^o & \dot{x}_k^t - \dot{x}_k^o & \dot{y}_k^t - \dot{y}_k^o \end{bmatrix}' \\
&= \begin{bmatrix} x_k & y_k & \dot{x}_k & \dot{y}_k \end{bmatrix}'.
\end{aligned} \tag{3}$$

$\mathbf{F}$ is the state transition matrix and w_{k-1} is zero mean Gaussian process noise with $\mathbf{Q}$ as the covariance matrix. For problem formulation in the two dimensional space (let $n = 2$), $\mathbf{F}, \mathbf{Q}$ matrices are defined as,

$$\mathbf{F} = \begin{bmatrix} 1 & 0 & T & 0 \\ 0 & 1 & 0 & T \\ 0 & 0 & 1 & 0 \\ 0 & 0 & 0 & 1 \end{bmatrix} \quad \text{and} \quad \mathbf{Q} = \begin{bmatrix} \frac{T^3}{3}q_x & 0 & \frac{T^2}{2}q_x & 0 \\ 0 & \frac{T^3}{3}q_y & 0 & \frac{T^2}{2}q_y \\ \frac{T^2}{2}q_x & 0 & Tq_x & 0 \\ 0 & \frac{T^2}{2}q_y & 0 & Tq_y \end{bmatrix}.$$

The target observer dynamics in 2D for a moderately nonlinear scenario, is shown in Figure 1. Similarly for $n = 3$, the state and the associated matrices are

$$\mathbf{X}_k^t = \begin{bmatrix} x_k^t & y_k^t & z_k^t & \dot{x}_k^t & \dot{y}_k^t & \dot{z}_k^t \end{bmatrix}' \qquad \mathbf{X}^o = \begin{bmatrix} x_k^o & y_k^o & z_k^o & \dot{x}_k^o & \dot{y}_k^o & \dot{z}_k^o \end{bmatrix}'$$

$$\mathbf{F}_{k-1} = \begin{bmatrix} 1 & 0 & 0 & T & 0 & 0 \\ 0 & 1 & 0 & 0 & T & 0 \\ 0 & 0 & 1 & 0 & 0 & T \\ 0 & 0 & 0 & 1 & 0 & 0 \\ 0 & 0 & 0 & 0 & 1 & 0 \\ 0 & 0 & 0 & 0 & 0 & 1 \end{bmatrix}, \quad \mathbf{Q}_{k-1} = \begin{bmatrix} \frac{T^3}{3}q_x & 0 & 0 & \frac{T^2}{2}q_x & 0 & 0 \\ 0 & \frac{T^3}{3}q_y & 0 & 0 & \frac{T^2}{2}q_y & 0 \\ 0 & 0 & \frac{T^3}{3}q_z & 0 & 0 & \frac{T^2}{2}q_z \\ \frac{T^2}{2}q_x & 0 & 0 & Tq_x & 0 & 0 \\ 0 & \frac{T^2}{2}q_y & 0 & 0 & Tq_y & 0 \\ 0 & 0 & \frac{T^2}{2}q_z & 0 & 0 & Tq_z \end{bmatrix},$$

where T is the sampling time interval and q_x, q_y, q_z are the power spectral densities of the process noise along the $X, Y,$ and Z axes respectively.

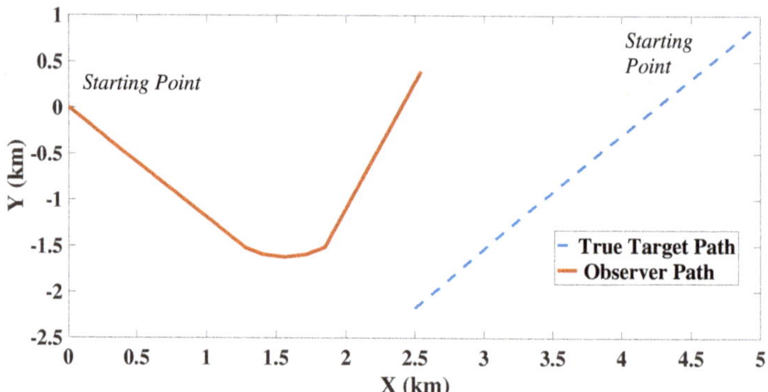

Figure 1. Target Observer Dynamics in 2D.

The 3D target observer trajectory referred in the problem is given by Figure 2.

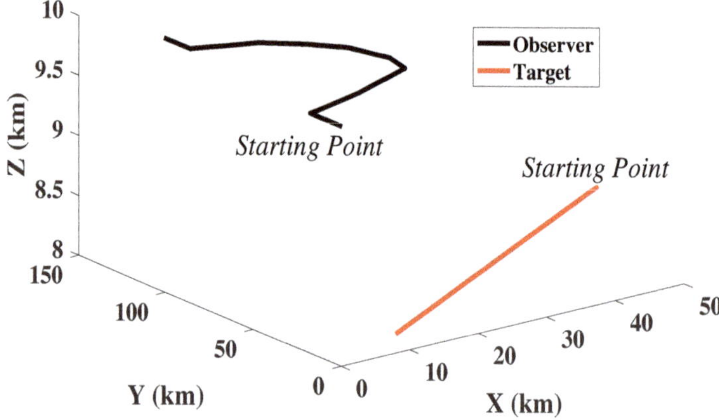

Figure 2. Target Observer Dynamics in 3D.

2.2. Measurement Model

2D AoT problem: The only available measurements are the bearing angles through which the states of the relative state vector can be estimated. The measurement model and the true angle measurements for the problem can be represented as

$$z_k = h(\mathbf{X}_k) + v_k \qquad\qquad h(\mathbf{X}_k) = \beta_k = \tan^{-1}(x_k, y_k). \qquad (4)$$

where v_k shall be modelled as the non Gaussian noise.

3D AoT problem: Figure 3 represents the target observer dynamics in Cartesian coordinate.

The range vector $\mathbf{r}$ is defined as

$$\mathbf{r} = \begin{bmatrix} x_k^t - x_k^o & y_k^t - y_k^o & z_k^t - z_k^o \end{bmatrix}'$$
$$= \begin{bmatrix} x_k & y_k & z_k \end{bmatrix}'.$$

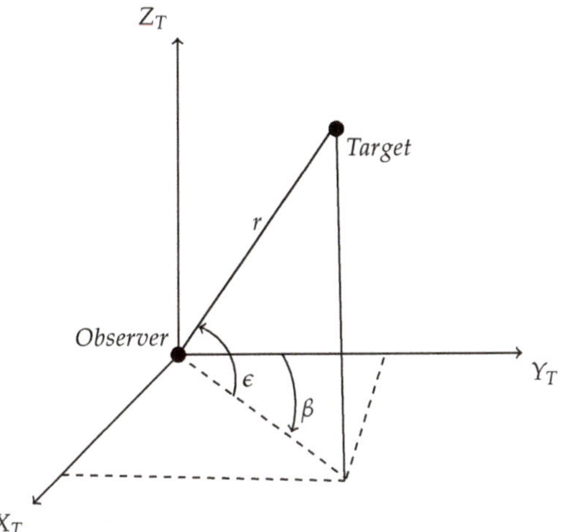

Figure 3. Target Observer in Cartesian Coordinate Frame.

From Figure 3, $\mathbf{r}$ can be expressed in terms of bearing (β) and elevation (ϵ) as

$$\mathbf{r} = [\, r\cos\epsilon\sin\beta \quad r\cos\epsilon\cos\beta \quad r\sin\epsilon \,]',$$

and the actual range is $r_k = \sqrt{x_k^2 + y_k^2 + z_k^2}$. Here, the nonlinear noise corrupted measurements are bearing (β) and elevation (ϵ) angles respectively, where $\beta \in [0, 2\pi]$ and $\epsilon \in [-\frac{\pi}{2}, \frac{\pi}{2}]$. The measurement model involving the bearing and elevation angle is

$$\mathbf{z}_k = \mathbf{h}(\mathbf{X}_k) + \mathbf{v}_k, \tag{5}$$

where,

$$\mathbf{h}(\mathbf{X}_k) = \begin{bmatrix} \beta_k \\ \epsilon_k \end{bmatrix} = \begin{bmatrix} \tan^{-1}(x_k, y_k) \\ \tan^{-1}\left(\dfrac{z_k}{\sqrt{x_k^2 + y_k^2}} \right) \end{bmatrix}.$$

Here, $\mathbf{v}_k$ is to be modelled as the non Gaussian noise.

3. Correntropy Measure

Correntropy is directly related to the probability of how similar two random variables are in the joint space controlled by the kernel bandwidth. The kernel bandwidth controls the window in which the similarity has to be assessed, and hence provides a way to eliminate the detrimental effect of outliers [16]. If X and Y are random variables, correntropy is defined as

$$V_\sigma(X, Y) = E[k_\sigma(X, Y)] = \iint k_\sigma(x, y) p_{XY}(x, y)\, dx\, dy,$$

where k_σ denotes a positive definite kernel function, $p_{XY}(.)$ denotes the joint PDF of X and Y and E is the expectation operator. Since the joint density is not accessible and if only a finite number of data points N are available, a sample estimator can be defined as

$$\widehat{V}_\sigma(X, Y) = \frac{1}{N} \sum_{i=1}^{N} G_\sigma(x_i - y_i).$$

Here $G_\sigma(\cdot)$ is the Gaussian kernel, defined as

$$G_\sigma(x_i - y_i) = \exp\left(-\frac{\|x_i - y_i\|^2}{2\sigma^2}\right),$$

(6)

which is bounded, positive and reaches its maximum only when $X = Y$, leading to the maximum correntropy criterion (MCC). By taking the Taylor series expansion of the Gaussian kernel, correntropy can also be expressed as a weighted sum of all even order moments of $(x_i - y_i)$, i.e.,

$$V_\sigma(X, Y) = \sum_{k=0}^{\infty} \frac{(-1)^k}{2^k \sigma^{2k} k!} E[(X - Y)^{2k}].$$

On the other hand, Cauchy kernel based non-linear state estimators can be developed using Cauchy kernel instead of Gaussian kernel function. It is defined as [34]

$$C_\delta(x_i - y_i) = \frac{1}{1 + \dfrac{\|x_i - y_i\|^2}{\delta}}.$$

Here δ is a positive scalar, representing the Cauchy kernel bandwidth. Similar to the Gaussian kernel, it can be shown that the Cauchy kernel also incorporates the higher order moments, given as

$$V_\delta(X, Y) = \sum_{k=0}^{\infty} \frac{(-1)^k}{\delta^k}\binom{N+k-1}{k} E\left[(X - Y)^{2k}\right].$$

A detailed derivation of the above equation is given in Appendix A.

4. Gaussian Kernel Based Maximum Correntropy Estimation Framework

Let us consider the process model described in Equations (1) and (5). To accommodate for the large outliers in measurements, the noise $\mathbf{v}_k$ is considered non-Gaussian. Hence for MC based estimation framework [24], the Gaussian assumption of $\mathbf{v}_k$ is relaxed.

In order to deal with the non Gaussian noises in the measurement update step, a statistical linearisation approach is employed. Consider that the nonlinear function $\mathbf{h}(\cdot)$, operating on vector random variables $\mathbf{X}_k$ is evaluated at N-points χ_k, $k = 1, \cdots, \mathrm{N}$, with $\mathbf{z}_k = \mathbf{h}(\chi_k) + \mathbf{v}_k$. Suppose that the weighted mean of χ_k is given by $\hat{\mathbf{X}}_{k|k-1} = \sum_{k=1}^{N} W_k \chi_k$, with $\sum_{k=1}^{N} W_k = 1$. Similarly, $\hat{\mathbf{z}}_{k|k-1} = \sum_{k=1}^{N} W_k \mathbf{z}_k$. Then the prior and cross covariance $\mathbf{P}_{k|k-1}$ and $\mathbf{P}_{Xz}$ are given as

$$\mathbf{P}_{k|k-1} = \sum_{k=1}^{N} W_k\left[(\chi_k - \hat{\mathbf{X}}_{k|k-1})(\chi_k - \hat{\mathbf{X}}_{k|k-1})'\right] \quad \text{and}$$

$$\mathbf{P}_{Xz} = \sum_{k=1}^{N} W_k\left[(\chi_k - \hat{\mathbf{X}}_{k|k-1})(\mathbf{z}_k - \hat{\mathbf{z}}_{k|k-1})'\right].$$

The nonlinear measurement function is represented in terms of measurement slope matrix $\overline{\mathbf{H}}_k$, and a constant term $\bar{c}_k$ as $\mathbf{h}(X_k) \approx \overline{\mathbf{H}}_k \mathbf{X}_k + \bar{c}_k$. Here $\overline{\mathbf{H}}_k$ and $\bar{c}_k$ are computed by minimizing the weighted least squares,

$$\underset{\overline{H}_k, \bar{c}_k}{\arg\min}\, W_k\|\bar{\mathbf{v}}_k\|^2, \quad \text{where} \quad \bar{\mathbf{v}}_k = \mathbf{z}_k - \overline{\mathbf{H}}_k \mathbf{X}_k - \bar{c}_k.$$

Then the solutions are $\overline{\mathbf{H}}_k = \left(\mathbf{P}_{k|k-1}^{-1}\mathbf{P}_{\mathbf{Xz}}\right)'$ and $\overline{c}_k = \widehat{\mathbf{z}}_{k|k-1} - \overline{\mathbf{H}}_k\widehat{\mathbf{X}}_{k|k-1}$. As the mean of $\overline{\mathbf{v}}_k$ is zero, that is $E[\overline{\mathbf{v}}_k] = \widehat{\mathbf{z}}_{k|k-1} - \overline{\mathbf{H}}_k\widehat{\mathbf{X}}_{k|k-1} - \overline{c}_k = 0$, the covariance matrix $\mathbf{R}_k$ can be calculated as

$$
\begin{aligned}
\mathbf{R}_k &= \sum_{k=1}^{N} W_k[\overline{\mathbf{v}}_k\overline{\mathbf{v}}_k'] \\
&= \sum_{k=1}^{N} W_k\left[(\mathbf{z}_k - \widehat{\mathbf{z}}_{k|k-1}) - \overline{\mathbf{H}}_k(\mathbf{X}_k - \widehat{\mathbf{X}}_{k|k-1})\right]\left[(\mathbf{z}_k - \widehat{\mathbf{z}}_{k|k-1}) - \overline{\mathbf{H}}_k(\mathbf{X}_k - \widehat{\mathbf{X}}_{k|k-1})\right]' \quad (7)\\
&= \mathbf{P}_{\mathbf{zz}} - \overline{\mathbf{H}}_k\mathbf{P}_{\mathbf{Xz}} - \mathbf{P}_{\mathbf{Xz}}'\overline{\mathbf{H}}_k' + \overline{\mathbf{H}}_k\mathbf{P}_{k|k-1}\overline{\mathbf{H}}_k' \\
&= \mathbf{P}_{\mathbf{zz}} - \overline{\mathbf{H}}_k\mathbf{P}_{k|k-1}\overline{\mathbf{H}}_k'.
\end{aligned}
$$

Thus, the linearised measurement equation is given as

$$
\mathbf{z}_k = \widehat{\mathbf{z}}_{k|k-1} + \overline{\mathbf{H}}_k\left(\mathbf{X}_k - \widehat{\mathbf{X}}_{k|k-1}\right) + \overline{\mathbf{v}}_k \quad \text{with} \quad \overline{\mathbf{v}}_k \sim \mathcal{N}(0, \mathbf{R}_k). \quad (8)
$$

Accordingly, a cost function is formulated with the help of weighted least squares (WLS) to handle Gaussian process noise. To handle non-Gaussian measurement noise, statistical linearisation approach was used to define WLS function which in turn is used in MCC. Hence the cost function can be defined as

$$
\mathbf{J} = \wp\|\mathbf{X}_k - \widehat{\mathbf{X}}_{k|k-1}\|_{\mathbf{P}_{k|k-1}^{-1}}^2 - \varrho\exp\left(-\frac{\aleph'\mathbf{R}^{-1}\aleph}{2\sigma^2}\right),
$$

where $\aleph = \mathbf{z}_k - \widehat{\mathbf{z}}_{k|k-1} - \overline{\mathbf{H}}_k(\mathbf{X}_k - \widehat{\mathbf{X}}_{k|k-1})$, $\wp$ and ϱ are adjustable weights. In order to find the optimal estimate of $\mathbf{X}_k$, the cost function has to be minimized i.e.,

$$
\widehat{\mathbf{X}}_k = \arg\min_{\mathbf{X}_k} \mathbf{J},
$$

and the solution can be obtained as $\frac{\partial \mathbf{J}}{\partial \mathbf{X}_k} = 0$. This implies that

$$
\begin{aligned}
\frac{\partial \mathbf{J}}{\partial \mathbf{X}_k} &= \wp\mathbf{P}_{k|k-1}^{-1}(\mathbf{X}_k - \widehat{\mathbf{X}}_{k|k-1}) + \frac{\varrho}{2\sigma^2}G_\sigma(\aleph_{\mathbf{R}})\overline{\mathbf{H}}_k'\mathbf{R}_k^{-1}\aleph \\
&= \wp\mathbf{P}_{k|k-1}^{-1}(\mathbf{X}_k - \widehat{\mathbf{X}}_{k|k-1}) + \frac{\varrho\mathbf{L}_k^G\overline{\mathbf{H}}_k'\mathbf{R}_k^{-1}}{2\sigma^2}\aleph = 0, \quad (9)
\end{aligned}
$$

where

$$
\mathbf{L}_k^G = G_\sigma(\aleph_{\mathbf{R}}) = \exp\left(-\frac{\aleph'\mathbf{R}^{-1}\aleph}{2\sigma^2}\right). \quad (10)
$$

In order to guarantee the convergence of the algorithm to a corresponding state estimator that follows a complete Gaussian assumption (when the kernel bandwidth σ becomes infinity), the values for weights in $\mathbf{J}$ are taken as $\wp = 1$ and $\varrho = -2\sigma^2$. Then Equation (9) becomes

$$
\mathbf{P}_{k|k-1}^{-1}(\mathbf{X}_k - \widehat{\mathbf{X}}_{k|k-1}) = \mathbf{L}_k^G\overline{\mathbf{H}}_k'\mathbf{R}_k^{-1}\aleph.
$$

Rearranging, we get

$$
\left(\mathbf{P}_{k|k-1}^{-1} + \mathbf{L}_k^G\overline{\mathbf{H}}_k'\mathbf{R}_k^{-1}\overline{\mathbf{H}}_k\right)\mathbf{X}_k = \mathbf{L}_k^G\overline{\mathbf{H}}_k'\mathbf{R}_k^{-1}\left(\mathbf{z}_k - \widehat{\mathbf{z}}_{k|k-1}\right) + \left(\mathbf{L}_k^G\overline{\mathbf{H}}_k'\mathbf{R}_k^{-1}\overline{\mathbf{H}}_k + \mathbf{P}_{k|k-1}^{-1}\right)\widehat{\mathbf{X}}_{k|k-1}. \quad (11)
$$

Since $\mathbf{L}_k^G$ is related to $\mathbf{X}_k$, Equation (11) represents a fixed point equation that can be solved using the fixed point iteration algorithm considering $\mathbf{X}_k$ equal to $\widehat{\mathbf{X}}_{k|k-1}$ in

Equation (10). But as mentioned in [19,22,24], for a satisfactory estimation performance, a single iteration is sufficient. Hence, adopting the same approach leads to the modification of Equation (11) as

$$\widehat{\mathbf{X}}_{k|k} = \widehat{\mathbf{X}}_{k|k-1} + \mathbf{K}_k^G \left(\mathbf{z}_k - \widehat{\mathbf{z}}_{k|k-1} \right),$$

where

$$\mathbf{K}_k^G = \left(\mathbf{P}_{k|k-1}^{-1} + L_k^G \overline{\mathbf{H}}_k' \mathbf{R}_k^{-1} \overline{\mathbf{H}}_k \right)^{-1} L_k^G \overline{\mathbf{H}}_k' \mathbf{R}_k^{-1} \quad \text{and}$$

$$L_k^G = \exp \left(- \frac{(\mathbf{z}_k - \widehat{\mathbf{z}}_{k|k-1})' \mathbf{R}_k^{-1} (\mathbf{z}_k - \widehat{\mathbf{z}}_{k|k-1})}{2\sigma^2} \right).$$

A more appropriate form for $\mathbf{K}_k^G$, in terms of reduced computational complexity, can be derived using the matrix inversion lemma (detailed derivation is given in Appendix B) as

$$\mathbf{K}_k^G = \mathbf{P}_{k|k-1} L_k^G \overline{\mathbf{H}}_k' \left(\mathbf{R}_k + \overline{\mathbf{H}}_k \mathbf{P}_{k|k-1} L_k^G \overline{\mathbf{H}}_k' \right)^{-1}. \tag{12}$$

Now, the corresponding posterior error covariance matrix is given as

$$\mathbf{P}_{k|k} = \left(\mathbf{I} - \mathbf{K}_k^G \overline{\mathbf{H}}_k \right) \mathbf{P}_{k|k-1} \left(\mathbf{I} - \mathbf{K}_k^G \overline{\mathbf{H}}_k \right)' + \mathbf{K}_k^G \mathbf{R}_k \mathbf{K}_k^{G'}.$$

5. Cauchy Kernel Based Maximum Correntropy Estimation Framework

In this section, we derive a maximum correntropy estimation framework using Cauchy kernel for potential improvement in estimation accuracy, in the presence of large multi dimensional non Gaussian noise. Hence the cost function becomes

$$\mathbf{J}_C = \wp^C \| \mathbf{X}_k - \widehat{\mathbf{X}}_{k|k-1} \|_{\mathbf{P}_{k|k-1}^{-1}}^2 - \varrho^C C_\delta \left(\aleph_\mathbf{R} \right)$$

where $\wp^C$ and ϱ^C are adjustable weights, and

$$C_\delta \left(\aleph_\mathbf{R} \right) = \frac{1}{1 + \dfrac{\aleph' \mathbf{R}_k^{-1} \aleph}{\delta}},$$

with $\aleph$ being the same as that mentioned in Section 4. To obtain the optimal estimate of $\mathbf{X}_k$, we equate $\dfrac{\partial \mathbf{J}_C}{\partial \mathbf{X}_k} = 0$, giving $\wp^C \mathbf{P}_{k|k-1}^{-1} \left(\mathbf{X}_k - \widehat{\mathbf{X}}_{k|k-1} \right) - \dfrac{\varrho^C L_k^C}{\delta} \overline{\mathbf{H}}' \mathbf{R}_k^{-1} \aleph = 0$, where

$$L_k^C = C_\delta^2(\aleph_\mathbf{R}) = C_\delta^2 \left(\| \mathbf{z}_k - \overline{\mathbf{H}}_k \mathbf{X}_k - \widehat{\mathbf{z}}_{k|k-1} + \overline{\mathbf{H}}_k \widehat{\mathbf{X}}_{k|k-1} \|_{\mathbf{R}_k^{-1}} \right).$$

We set $\wp^C = 1$ and $\varrho^C = \delta$ so as to guarantee the convergence of the estimator when kernel bandwidth δ tends to ∞. Rearranging,

$$\left(\mathbf{P}_{k|k-1}^{-1} + L_k^C \overline{\mathbf{H}}_k' \mathbf{R}_k^{-1} \overline{\mathbf{H}}_k \right) \mathbf{X}_k = L_k^C \overline{\mathbf{H}}_k' \mathbf{R}_k^{-1} \left(\mathbf{z}_k - \widehat{\mathbf{z}}_{k|k-1} \right) + \left(L_k^C \overline{\mathbf{H}}_k' \mathbf{R}_k^{-1} \overline{\mathbf{H}}_k + \mathbf{P}_{k|k-1}^{-1} \right) \widehat{\mathbf{X}}_{k|k-1}. \tag{13}$$

Here also, L_k^C is related to $\mathbf{X}_k$ and hence Equation (13) is a fixed point equation that is to be solved using fixed point iteration algorithm, considering $\mathbf{X}_k$ equal to $\widehat{\mathbf{X}}_{k|k-1}$. Using the same justification that was adopted in Gaussian kernel case that only a single iteration is required, the expression for posterior mean is obtained as

$$\widehat{\mathbf{X}}_{k|k} = \widehat{\mathbf{X}}_{k|k-1} + \mathbf{K}_k^C \left(\mathbf{z}_k - \widehat{\mathbf{z}}_{k|k-1} \right),$$

where

$$\mathbf{K}_k^C = \left(\mathbf{P}_{k|k-1}^{-1} + \mathbf{L}_k^C \overline{\mathbf{H}}_k' \mathbf{R}_k^{-1} \overline{\mathbf{H}}_k\right)^{-1} \mathbf{L}_k^C \overline{\mathbf{H}}_k' \mathbf{R}_k^{-1} \quad \text{and} \quad \mathbf{L}_k^C = C_\delta^2\left(\|\mathbf{z}_k - \hat{\mathbf{z}}_{k|k-1}\|_{\mathbf{R}_k^{-1}}\right).$$

As per the proof given in Appendix B, Kalman gain can be modified as

$$\mathbf{K}_k^C = \mathbf{P}_{k|k-1} \mathbf{L}_k^C \overline{\mathbf{H}}_k'\left(\mathbf{R}_k + \overline{\mathbf{H}}_k \mathbf{P}_{k|k-1} \mathbf{L}_k^C \overline{\mathbf{H}}_k'\right)^{-1}. \tag{14}$$

Then the posterior error covariance matrix shall be calculated as

$$\mathbf{P}_{k|k} = \left(\mathbf{I} - \mathbf{K}_k^C \overline{\mathbf{H}}_k\right)\mathbf{P}_{k|k-1}\left(\mathbf{I} - \mathbf{K}_k^C \overline{\mathbf{H}}_k\right)' + \mathbf{K}_k^C \mathbf{R}_k \mathbf{K}_k^{C'}. \tag{15}$$

Theorem 1. *As the kernel bandwidth $\delta \to \infty$, the Cauchy kernel based MC estimator reduces to the standard nonlinear state estimation algorithm.*

Proof. As the time update is the same for the developed algorithms with respect to the standard nonlinear state estimators, the prior mean and covariance is unchanged. Hence the focus shall be on the posterior mean and covariance. This implies that the Kalman gain equation has to be revisited. When $\delta \to \infty$,

$$\lim_{\delta \to \infty} \mathbf{L}_k^C = \lim_{\delta \to \infty} C_\delta^2\left(\|\mathbf{z}_k - \hat{\mathbf{z}}_{k|k-1}\|_{\mathbf{R}_k^{-1}}\right) = \lim_{\delta \to \infty} \frac{1}{\left(1 + \dfrac{\aleph' \mathbf{R}_k^{-1} \aleph}{\delta}\right)^2} = 1. \tag{16}$$

Substituting the Equations (7) and (16) and $\overline{\mathbf{H}}_k$ in $\mathbf{K}_k^C$, we have

$$\mathbf{K}_k^C = \mathbf{P}_{k|k-1}(\mathbf{P}_{k|k-1}^{-1}\mathbf{P}_{\mathbf{Xz}})(\mathbf{P}_{\mathbf{zz}} - \overline{\mathbf{H}}_k \mathbf{P}_{k|k-1}\overline{\mathbf{H}}_k' + \overline{\mathbf{H}}_k \mathbf{P}_{k|k-1}\overline{\mathbf{H}}_k')^{-1} = \mathbf{P}_{\mathbf{Xz}}\mathbf{P}_{\mathbf{zz}}^{-1}.$$

Since the expression of $\mathbf{K}_k^C$ is similar to the Kalman gain of standard nonlinear state estimator, posterior mean is also the same.

Now, for the posterior covariance $\mathbf{P}_{k|k}$, consider Equation (15),

$$\mathbf{P}_{k|k} = \mathbf{P}_{k|k-1} - \mathbf{P}_{k|k-1}\overline{\mathbf{H}}_k' \mathbf{K}_k^{C'} - \mathbf{K}_k^C \overline{\mathbf{H}}_k \mathbf{P}_{k|k-1} + \mathbf{K}_k^C\left(\mathbf{R}_k + \overline{\mathbf{H}}_k \mathbf{P}_{k|k-1} \mathbf{L}_k^C \overline{\mathbf{H}}_k'\right)\mathbf{K}_k^{C'}. \tag{17}$$

Post multiplying Equation (14) by $(\mathbf{R}_k + \overline{\mathbf{H}}_k \mathbf{P}_{k|k-1} \mathbf{L}_k^C \overline{\mathbf{H}}_k')$ on both sides give

$$\mathbf{K}_k^C(\mathbf{R}_k + \overline{\mathbf{H}}_k \mathbf{P}_{k|k-1} \mathbf{L}_k^C \overline{\mathbf{H}}_k') = \mathbf{P}_{k|k-1}\mathbf{L}_k^C \overline{\mathbf{H}}_k'. \tag{18}$$

Using Equations (16) and (18)

$$\mathbf{P}_{k|k} = \mathbf{P}_{k|k-1} - \mathbf{P}_{k|k-1}\overline{\mathbf{H}}_k' \mathbf{K}_k^{C'} - \mathbf{K}_k^C \overline{\mathbf{H}}_k \mathbf{P}_{k|k-1} + \mathbf{P}_{k|k-1}\overline{\mathbf{H}}_k' \mathbf{K}_k^{C'} = \mathbf{P}_{k|k-1} - \mathbf{K}_k^C \overline{\mathbf{H}}_k \mathbf{P}_{k|k-1}.$$

Substituting $\overline{\mathbf{H}}_k$, we get $\mathbf{P}_{k|k} = \mathbf{P}_{k|k-1} - \mathbf{K}_k^C \mathbf{P}_{\mathbf{Xz}}'$. For the given condition, $\mathbf{K}_k^C = \mathbf{P}_{\mathbf{Xz}}\mathbf{P}_{\mathbf{zz}}^{-1}$, then $\mathbf{P}_{\mathbf{Xz}}' = \mathbf{P}_{\mathbf{zz}}'\mathbf{K}_k^{C'}$. Thus $\mathbf{P}_{k|k}$ will become $\mathbf{P}_{k|k} = \mathbf{P}_{k|k-1} - \mathbf{K}_k^C \mathbf{P}_{\mathbf{zz}}'\mathbf{K}_k^{C'}$, which matches with the posterior error covariance of standard nonlinear estimator. $\square$

Remark 1. *For systems with non-Gaussian noise with large probability of abnormal noise, small value of δ is likely to provide more robustness. If the occurrence of abnormal noise is less, large value of δ could be considered.*

Remark 2. *Cauchy kernel based nonlinear estimator with different δ performs with more estimation accuracy than Gaussian kernel based nonlinear estimator with different σ. Hence it is easier to select a value for δ that can provide accurate and robust estimates in the presence of abnormal noise.*

6. Nonlinear State Estimators

This section deals with the nonlinear state estimators UKF and NSKF with a generalized algorithm for kernel based MC estimator.

6.1. Unscented Kalman Filter (UKF)

In the Bayesian framework, when the functions are nonlinear, the integrals encountered are intractable in nature and has to be evaluated using suitable numerical approximation methods. The UKF, through its unscented transformation, provides a way to numerically evaluate these integrals. Assuming that the integral to be approximated is

$$\mathrm{I}(\mathbf{X}) = \int \mathbf{h}(\mathbf{X}) p_{\mathbf{X}}(\mathbf{X}) d\mathbf{X},$$

and $\mathbf{X} \sim \mathcal{N}(\hat{\mathbf{x}}, \mathbf{P})$, the unscented transformation defines a set of sigma points $(\hat{\mathbf{X}}_i)$ and weights (W_i) such that [35]

$$\mathrm{I}(\mathbf{X}) \approx \sum_{i=1}^{N} W_i \mathbf{h}(\hat{\mathbf{X}}_i), \text{ where } n \text{ is the dimension of the state space and } \mathbf{N} = 2n + 1.$$

The sigma points and weights are defined as

$$\hat{\mathbf{X}}_1 = \hat{\mathbf{x}}, \quad W_1 = \frac{\kappa}{n + \kappa},$$
$$\hat{\mathbf{X}}_i = \hat{\mathbf{x}} + \left(\sqrt{(n+\kappa)\mathbf{P}} \right)_i, \quad W_i = \frac{1}{2(n+\kappa)}, \quad i = 1, \cdots, n$$
$$\hat{\mathbf{X}}_i = \hat{\mathbf{x}} - \left(\sqrt{(n+\kappa)\mathbf{P}} \right)_i, \quad W_i = \frac{1}{2(n+\kappa)}, \quad i = 1, \cdots, n, \tag{19}$$

with κ being the tuning parameter and $\hat{\mathbf{x}}$ is the mean.

6.2. New Sigma Point Kalman Filter (NSKF)

From Equation (19), it can observed that in the unscented transformation, the maximum weight is assigned to the mean value. All the other sigma points are assigned equal weights, i.e., same probability of occurrence. In NSKF, a new approach was considered such that the sigma points closer to the mean will have more probability of occurrence. To realize this, a new method was formulated for defining the sigma points and weights, stated as [8]

$$\hat{\mathbf{X}}_1 = \hat{\mathbf{x}}, \quad W_1 = 1 - \frac{\sum_{i=1}^{n} \alpha_i}{2(\sum_{i=1}^{n} \alpha_i + b)}$$
$$\hat{\mathbf{X}}_{i+1} = \hat{\mathbf{x}} + \sqrt{\frac{\sum_{i=1}^{n} \alpha_i + b}{m\alpha_i}} S_i, \quad W_{i+1} = \frac{m\alpha_i}{4(\sum_{i=1}^{n} \alpha_i + b)}, \quad i = 1, \cdots, n$$
$$\hat{\mathbf{X}}_{i+1} = \hat{\mathbf{x}} - \sqrt{\frac{\sum_{i=1}^{n} \alpha_i + b}{m\alpha_{i-n}}} S_{i-n}, \quad W_{i+1} = \frac{m\alpha_{i-n}}{4(\sum_{i=1}^{n} \alpha_i + b)}, \quad i = n+1, \cdots, 2n \tag{20}$$
$$\hat{\mathbf{X}}_{i+1} = \hat{\mathbf{x}} + \sqrt{\frac{\sum_{i=1}^{n} \alpha_i + b}{(1-m)\alpha_{i-2n}}} S_{i-2n}, \quad W_{i+1} = \frac{(1-m)\alpha_{i-2n}}{4(\sum_{i=1}^{n} \alpha_i + b)}, \quad i = 2n+1, \cdots, 3n$$
$$\hat{\mathbf{X}}_{i+1} = \hat{\mathbf{x}} - \sqrt{\frac{\sum_{i=1}^{n} \alpha_i + b}{(1-m)\alpha_{i-3n}}} S_{i-3n}, \quad W_{i+1} = \frac{(1-m)\alpha_{i-3n}}{4(\sum_{i=1}^{n} \alpha_i + b)}, \quad i = 3n+1, \cdots, 4n.$$

Now the total number of sigma points $\mathbf{N} = 4n + 1$, $\mathbf{P}_i$ and S_i denote the ith column of $\mathbf{P}$ and S respectively, and $SS' = P$. The scalar variables are defined as $b > \{\frac{1}{4}\max(m\alpha_i) - \frac{1}{2}\sum_{i=1}^{n}\alpha_i\}$, $m \in (0.5, 1)$ and $\alpha_i = \frac{|<\hat{x}, P_i>|}{\|\hat{x}\|_2 \|P_i\|_2}$.

The Algorithm 1 for the developed estimators, both Gaussian kernel and Cauchy kernel based is given below. In this algorithm, $\mathbf{K}_k$ and $\mathbf{L}_k$ can be defined as per the chosen kernel function. R_k is the noise covariance matrix which is assumed to be known in case there are no measurement outliers.

Algorithm 1: For MC-UKF-CK and MC-NSKF-CK

Initialise $\widehat{\mathbf{X}}_{k-1|k-1}$ and $\mathbf{P}_{k-1|k-1}$

$$\widehat{\mathbf{X}}_{k|k-1} = \mathbf{F}_{k-1}\widehat{\mathbf{X}}_{k-1|k-1} - \mathbf{X}_k^o + \mathbf{F}_{k-1}\mathbf{X}_{k-1}^o$$

$$\mathbf{P}_{k|k-1} = \mathbf{F}_{k-1}\mathbf{P}_{k-1|k-1}\mathbf{F}_{k-1}' + \mathbf{Q}$$

Calculate $\widehat{\mathbf{X}}_i$ and W_i using (19) or (20), $i = 1, \cdots, \mathbf{N}$

$$Z_{i,k|k-1} = \mathbf{h}(\widehat{\mathbf{X}}_i)$$

$$\widehat{\mathbf{z}}_k = \sum_{i=1}^{N} W_i Z_{i,k|k-1}$$

$$\mathbf{P_{zz}} = \sum_{i=1}^{N} W_i [Z_{i,k|k-1} - \widehat{\mathbf{z}}_k][Z_{i,k|k-1} - \widehat{\mathbf{z}}_k]' + \mathbf{R}_k$$

$$\mathbf{P_{Xz}} = \sum_{i=1}^{N} W_i [\widehat{\mathbf{X}}_{i,k|k-1} - \widehat{\mathbf{X}}_{k|k-1}][Z_{i,k|k-1} - \widehat{\mathbf{z}}_k]'$$

$$\overline{\mathbf{H}}_k = (\mathbf{P}_{k|k-1}^{-1}\mathbf{P_{Xz}})'$$

$$\mathbf{R}_k = \mathbf{P_{zz}} - \overline{\mathbf{H}}_k\mathbf{P}_{k|k-1}\overline{\mathbf{H}}_k'$$

$$\mathbf{K}_k = \mathbf{P}_{k|k-1}\mathbf{L}_k\overline{\mathbf{H}}_k'(\mathbf{R}_k + \overline{\mathbf{H}}_k\mathbf{P}_{k|k-1}\mathbf{L}_k\overline{\mathbf{H}}_k')^{-1}.$$

Posterior mean: $\widehat{\mathbf{X}}_{k|k} = \widehat{\mathbf{X}}_{k|k-1} + \mathbf{K}_k\left(\mathbf{z}_k - \widehat{\mathbf{z}}_k\right).$

Posterior covariance: $\mathbf{P}_{k|k} = (\mathbf{I} - \mathbf{K}_k\overline{\mathbf{H}}_k)\mathbf{P}_{k|k-1}(\mathbf{I} - \mathbf{K}_k\overline{\mathbf{H}}_k)' + \mathbf{K}_k\mathbf{R}_k\mathbf{K}_k'.$

---

7. Modelling of Non Gaussian Noise in Angular Measurements

As mentioned in [36], a suitable way of modelling glint noise is to assume a Gaussian mixture. It is observed that the glint is more like Gaussian around the mean but has a non-Gaussian nature towards the tail region. The tail region represents the outliers, termed as glint spikes [32]. But shot noise, on the other hand, is modelled as an impulse with fixed amplitude at specific time steps. The mixture density of glint noise is modelled as $f(x) = (1 - \mu)f_{g1}(x) + \mu f_{g2}(x)$, where μ is the glint probability and $f_{g1}(x) \sim \mathcal{N}(0, \sigma_1^2)$, $f_{g2}(x) \sim \mathcal{N}(0, \sigma_2^2)$ with $\sigma_1 \neq \sigma_2$. The non Gaussian noises for angular measurements have been modelled by taking appropriate values for μ, σ_1 and σ_2.

8. Simulation Results

The scenario for angles-only tracking problem in 2D as well as 3D Cartesian coordinate frame is considered in this section. The parameters required for generating the target-observer dynamics, and simulation results are discussed. For simulations, moderately nonlinear tracking scenario for 2D as well as for 3D is considered. The simulation is carried for 1000 Monte Carlo runs with sampling time interval denoted as T. The entire tracking scenario is implemented and simulated in MATLAB software.

8.1. 2D Scenario and Filter Initialisation

Figure 4 shows the tracking performance of MC-NSKF-CK, where the estimated target path is plotted along with the truth target path, and the observer path for a single Monte Carlo run. It should be noted that for each run, observer path remains the same where as the target path varies due to the process noise. Further, the filter initialisation is also changing because of the randomness introduced in each run, as mentioned in Equation (21).

The filter is initialised as given in [10]. It is to be noted that for filter initialisation, we need an initial guess for speed, initial course and range of the target. Considering the problem at hand, these estimates have to be obtained from the initial angle measurement received. From these initial guess for parameters, the initial estimate for the states are obtained which are the positions and velocities. Accordingly, the initial range, target course and speed values are considered and mentioned in the Table 1. They are defined as $\bar{s} = \mathcal{N}(s, \sigma_s^2)$, $c_r = \mathcal{N}(\bar{c}_r, \sigma_c^2)$ and $\bar{r} = \mathcal{N}(r, \sigma_r^2)$ where $\bar{c}_r$ can be defined as $\bar{c}_r = z_0 + \pi$ with z_0 as the first bearing measurement. Finally the initial state vector $\hat{\mathbf{X}}_{0|0}$ and the initial covariance $\mathbf{P}_{0|0}$ is calculated as

$$
\hat{\mathbf{X}}_{0|0} = \begin{bmatrix} r\sin(z_0) \\ r\cos(z_0) \\ s\sin(\bar{c}_r) - \dot{x}_0^o \\ s\cos(\bar{c}_r) - \dot{y}_0^o \end{bmatrix} \qquad
\mathbf{P}_{0|0} = \begin{bmatrix} \mathrm{P_{xx}} & \mathrm{P_{xy}} & 0 & 0 \\ \mathrm{P_{yx}} & \mathrm{P_{yy}} & 0 & 0 \\ 0 & 0 & \mathrm{P_{\dot{x}\dot{x}}} & \mathrm{P_{\dot{x}\dot{y}}} \\ 0 & 0 & \mathrm{P_{\dot{y}\dot{x}}} & \mathrm{P_{\dot{y}\dot{y}}} \end{bmatrix} \tag{21}
$$

where

$$
\mathrm{P_{xx}} = r^2\sigma_\beta^2\cos^2(z_0) + \sigma_r^2\sin^2(z_0) \qquad \mathrm{P_{yy}} = r^2\sigma_\beta^2\sin^2(z_0) + \sigma_r^2\cos^2(z_0)
$$

$$
\mathrm{P_{xy}} = \mathrm{P_{yx}} = (\sigma_r^2 - r^2\sigma_\beta^2)\sin(z_0)\cos(z_0) \qquad \mathrm{P_{\dot{x}\dot{x}}} = s^2\sigma_c^2\cos^2(\bar{c}_r) + \sigma_s^2\sin^2(\bar{c}_r)
$$

$$
\mathrm{P_{\dot{y}\dot{y}}} = s^2\sigma_c^2\sin^2(\bar{c}_r) + \sigma_s^2\cos^2(\bar{c}_r) \qquad \mathrm{P_{\dot{x}\dot{y}}} = \mathrm{P_{\dot{y}\dot{x}}} = (\sigma_s^2 - s^2\sigma_c^2)\sin(\bar{c}_r)\cos(\bar{c}_r).
$$

Table 1. Tracking parameters for 2D scenario.

Parameters	Values
Initial Target Position	$\begin{bmatrix} 4.9286 & 0.8420 \end{bmatrix}$ (km)
Initial Observer Position	$\begin{bmatrix} 0 & 0 \end{bmatrix}$ (km)
Initial Target Speed (s)	4 (knots)
Initial Observer Speed	5 (knots)
Target Course	$-135.4°$
Observer manoeuvre	From 780 to 1020 (s)
Initial Range (r)	5 (km)
Observation time	1800 (s)
q_x, q_y	9×10^{-12} (km^2/s^3)
σ_β	$1.5°$
σ_r	2 (km)
σ_s	2 (knots)
Sampling time	$T = 10$ (s)
Initial Observer Course	$140°$
Final Observer Course	$20°$
σ_c	$\pi/\sqrt{12}$

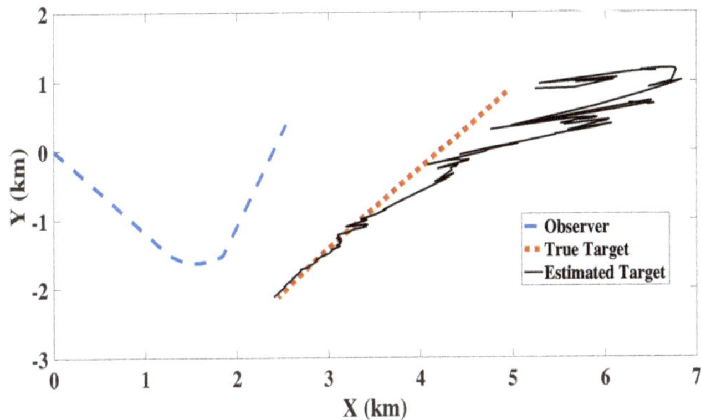

Figure 4. Target truth and estimated path obtained from MC-NSKF-CK.

8.2. 3D Scenario and Filter Initialisation

Figure 5 shows the estimated target path obtained from MC-NSKF-CK and truth target path with observer trajectory. The initial parameter values required for generating the 3D scenario is given in the Table 2. Assuming that there are no outliers in the measurement, R_k is defined as $R_k = diag(\sigma_\beta, \sigma_\epsilon)$. The bearing angle β is calculated with reference to the true North.

Table 2. Target & Observer Initial Parameters.

Parameters	Values
Initial Target Position	$\begin{bmatrix} 138/\sqrt{2} & 138/\sqrt{2} & 9 \end{bmatrix}$ (km)
Initial Observer Position	$\begin{bmatrix} 0 & 0 & 10 \end{bmatrix}$ (km)
Initial Target Speed (s)	0.297 (km/s)
Initial Observer Speed (s)	0.297 (km/s)
Target Course	$-135°$
Observer manoeuvre	From 70 to 370 (s)
Initial Range (r)	150 (km)
Observation time	420 (s)
q_x, q_y	10^{-8} km^2/s^3
q_z	10^{-10} km^2/s^3
$\sigma_\beta, \sigma_\epsilon$	0.057°
σ_r	13.6 (km)
σ_s	41.6 (m/s)
Elevation Angle	0.415°
Sampling time	$T = 10$ (s)

For each Monte Carlo run, according to the new measurement received, initial range r and speed of the target s is assumed as mentioned in Table 2. According to these values, the relative state is initialised using the range estimate $\bar{r} \sim \mathcal{N}(r, \sigma_r^2)$, initial bearing and elevation estimate $\hat{\beta}_1$ and $\hat{\epsilon}_1$ with headings $\bar{\alpha}_1 = \beta_1 + \pi$ rad/s and $\bar{\gamma}_1 = 0$ rad/s, and the initial speed estimate $\bar{s} \sim \mathcal{N}(s, \sigma_s^2)$ with s as 0.258 km/s. The $\sigma_{\bar{\alpha}_1} = \pi/\sqrt{12}$ and $\sigma_{\bar{\gamma}_1} = \pi/60$ respectively [37].

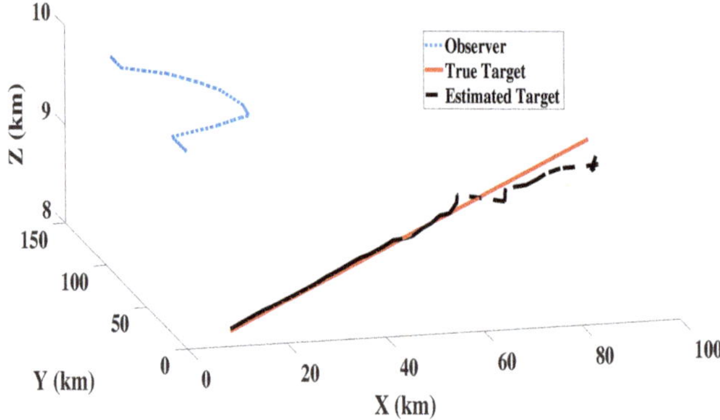

Figure 5. Target truth and estimated path obtained from MC-NSKF-CK.

The initial relative state vector $\widehat{\mathbf{X}}_{0|0}$ is given as [38]

$$\widehat{\mathbf{X}}_{0|0} = \begin{bmatrix} \bar{r}\,\zeta_{1,0}(\hat{\epsilon}_1,\sigma_\epsilon{}^2)\,\zeta_{0,1}(\hat{\beta}_1,\sigma_\beta{}^2) \\ \bar{r}\,\zeta_{1,0}(\hat{\epsilon}_1,\sigma_\epsilon{}^2)\,\zeta_{1,0}(\hat{\beta}_1,\sigma_\beta{}^2) \\ \bar{r}\,\zeta_{0,1}(\hat{\epsilon}_1,\sigma_\epsilon{}^2) \\ \bar{s}\,\zeta_{1,0}(\bar{\gamma}_1,\sigma_\gamma{}^2)\,\zeta_{0,1}(\bar{\alpha}_1,\sigma_\alpha{}^2) - \dot{x}_1^{\mathrm{o}} \\ \bar{s}\,\zeta_{1,0}(\bar{\gamma}_1,\sigma_\gamma{}^2)\,\zeta_{1,0}(\bar{\alpha}_1,\sigma_\alpha{}^2) - \dot{y}_1^{\mathrm{o}} \\ \bar{s}\,\zeta_{0,1}(\bar{\gamma}_1,\sigma_\gamma{}^2) - \dot{z}_1^{\mathrm{o}} \end{bmatrix},$$

where $\zeta_{1,0}(\mu,\sigma^2) = \cos\mu\,\exp(-\sigma^2/2)$ and $\zeta_{0,1}(\mu,\sigma^2) = \sin\mu\,\exp(-\sigma^2/2)$.

The initial covariance matrix $\mathbf{P}_{0|0}$, whose entries are considered as mentioned in [38], is defined as

$$\mathbf{P}_{0|0} = \begin{bmatrix} P_{xx} & P_{xy} & P_{xz} & 0 & 0 & 0 \\ P_{xy} & P_{yy} & P_{yz} & 0 & 0 & 0 \\ P_{xz} & P_{yz} & P_{zz} & 0 & 0 & 0 \\ 0 & 0 & 0 & P_{\dot{x}\dot{x}} & P_{\dot{x}\dot{y}} & P_{\dot{x}\dot{z}} \\ 0 & 0 & 0 & P_{\dot{x}\dot{y}} & P_{\dot{y}\dot{y}} & P_{\dot{y}\dot{z}} \\ 0 & 0 & 0 & P_{\dot{x}\dot{z}} & P_{\dot{y}\dot{z}} & P_{\dot{z}\dot{z}} \end{bmatrix}.$$

8.3. Performance Metrics

Performance analysis of the estimators formulated is evaluated by considering the below mentioned error statistics.

1. **RMSE:** Root-mean-square error in resultant target position is computed as follows

$$\mathrm{RMSE}_k = \sqrt{\frac{1}{M}\sum_{j=1}^{M}[(x_{j,k}^{\mathrm{t}} - \widehat{x}_{j,k}^{\mathrm{t}})^2 + (y_{j,k}^{\mathrm{t}} - \widehat{y}_{j,k}^{\mathrm{t}})^2]} \qquad {\scriptstyle n=2}$$

$$\mathrm{RMSE}_k = \sqrt{\frac{1}{M}\sum_{j=1}^{M}[(x_{j,k}^{\mathrm{t}} - \widehat{x}_{j,k}^{\mathrm{t}})^2 + (y_{j,k}^{\mathrm{t}} - \widehat{y}_{j,k}^{\mathrm{t}})^2 + (z_{j,k}^{\mathrm{t}} - \widehat{z}_{j,k}^{\mathrm{t}})^2]} \qquad {\scriptstyle n=3}$$

where k denotes the time steps and M the total number of Monte Carlo runs.

2. **Track Divergence:** In order to identify if a track is divergent or not, a certain threshold value ($\mathcal{T}_b$) is set according to the position error computed at the final time instant of observation (k_{max}) as

$$\text{pos}_{err} = \sqrt{\left(x^t_{j,k_{max}} - \widehat{x}^t_{j,k_{max}}\right)^2 + \left(y^t_{j,k_{max}} - \widehat{y}^t_{j,k_{max}}\right)^2}\Bigg|_{n=2}$$

$$\text{pos}_{err} = \sqrt{\left(x^t_{j,k_{max}} - \widehat{x}^t_{j,k_{max}}\right)^2 + \left(y^t_{j,k_{max}} - \widehat{y}^t_{j,k_{max}}\right)^2 + \left(z^t_{j,k_{max}} - \widehat{z}^t_{j,k_{max}}\right)^2}\Bigg|_{n=3}$$

$$\text{for } j = 1, 2, \cdots, M.$$

So, if the difference between estimated and truth target position is more than the threshold value ($\mathcal{T}_b$), then we can say that the estimated path is moving away from the truth path. Thus, the track is considered to be divergent, and the number of such tracks are counted over M Monte Carlo runs.

8.4. Performance Analysis

The performance analysis of the developed filters is evaluated in the presence of glint plus shot noise in angle measurements. The accuracy of the estimators are evaluated by computing root mean square error (RMSE) in position at the end of the simulation period by imposing a track loss condition of 1 km.

$$v_k = \begin{cases} 0.2\mathcal{N}(0,\sigma_{\theta_1}{}^2) + 0.8\mathcal{N}(0,\sigma_{\theta_2}{}^2) + 10°, & \text{when } k = 1200 \text{ and } 900 \text{ s} \\ 0.2\mathcal{N}(0,\sigma_{\theta_1}{}^2) + 0.8\mathcal{N}(0,\sigma_{\theta_2}{}^2), & \text{otherwise.} \end{cases} \tag{22}$$

The measurement noise v_k for both the scenarios are given in Equations (22) and (23), respectively. Here, $\sigma_{\theta_1} = 0.5°$, $\sigma_{\theta_2} = 5°$, $(\sigma_{\beta_1}, \sigma_{\epsilon_1})$ as 0.0001 rad and $(\sigma_{\beta_2}, \sigma_{\epsilon_2})$ as 0.01 rad. The noise corrupted angle measurement for 2D is plotted as Figure 6. For 3D, the noise corrupted bearing and elevation angle are as shown in Figures 7 and 8 respectively. For illustration, in the figures, we have also plotted the angle measurements with Gaussian noise.

$$v_k = \begin{cases} 0.8\mathcal{N}(0, diag([\sigma_{\beta_1}{}^2 \ \sigma_{\epsilon_1}{}^2])) + 0.2\mathcal{N}(0, diag([\sigma_{\beta_2}{}^2 \ \sigma_{\epsilon_2}{}^2])) + [10° \ 1°]^T, \\ \qquad\qquad\qquad\qquad\qquad \text{when } k = 270 \text{ and } 390 \text{ s} \\ 0.8\mathcal{N}(0, diag([\sigma_{\beta_1}{}^2 \ \sigma_{\epsilon_1}{}^2])) + 0.2\mathcal{N}(0, diag([\sigma_{\beta_2}{}^2 \ \sigma_{\epsilon_2}{}^2])), \quad \text{otherwise.} \end{cases} \tag{23}$$

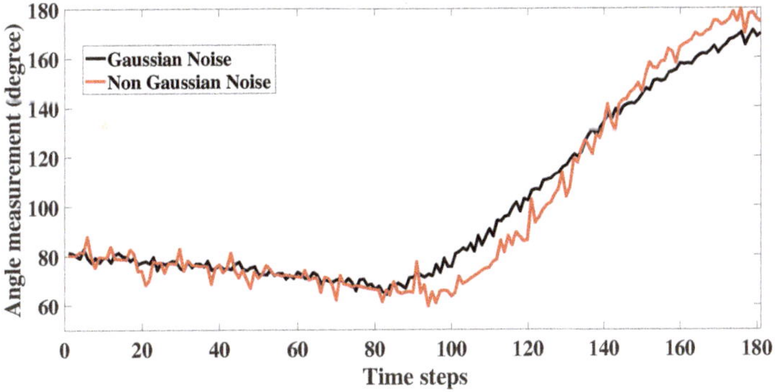

Figure 6. 2D: Angle measurement with glint plus shot noise.

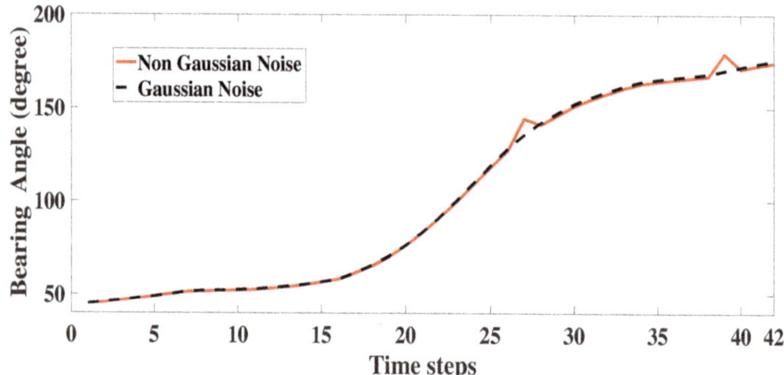

Figure 7. 3D: Bearing angle measurement with glint plus shot noise.

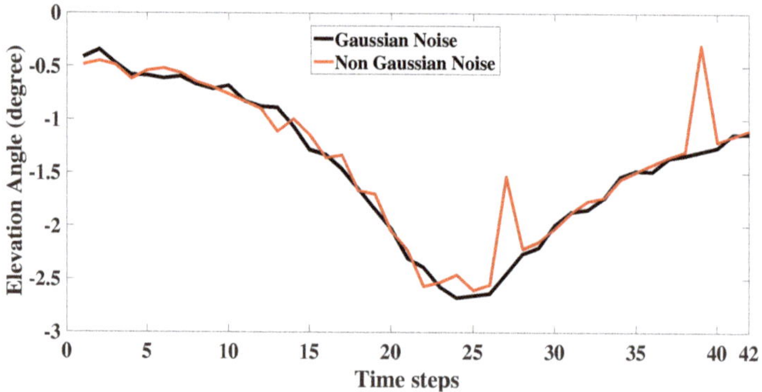

Figure 8. 3D: Elevation angle measurement with glint plus shot noise.

With track loss condition of less than 1 km, the observed RMSE in position at the last time instant and percentage track loss for 2D as well as 3D is given in the Tables 3 and 4 respectively. Also, RMSE in resultant position (after excluding the diverged tracks) is evaluated and plotted in Figures 9 and 10. From these figures it can be inferred that in the presence of non Gaussian noise the estimation accuracy of UKF deteriorates, whereas filters based on MC framework performed with superior estimation accuracy. From the tabulation results it is evident that the Cauchy kernel based MC-UKF and MC-NSKF gives 108.9 m, 108.8 m RMSE and 1.1 and 0.5% track loss which is much less than that of the conventional UKF and NSKF which gives 152.8 m and 151.1 m RMSE with 4.4 and 2.8% track loss in 2D scenario. Similar observations can be made with respect to Gaussian kernel based MC-UKF and MC-NSKF giving much better accuracy but slightly less than Cauchy kernel MC framework. However, in 3D scenario, the MC based filters gave even superior estimation efficiency than that of the 2D scenario. UKF and NSKF in 3D with non Gaussian noise resulted in 100% track loss. Hence it can be inferred that for the given problem set up and noise statistics, UKF and NSKF failed to give estimates that met the track loss condition set, where as the Gaussian and Cauchy kernel based maximum correntropy filters gave more robust and accurate estimates. This can be inferred from the simulation results where the developed filters incurred only 13 to 14% track loss, with a final error in range of around 500 m. All these simulations are carried out by assuming the bandwidth (σ, δ) for 2D as (9,70) and for 3D as (11,75) such that the estimators can achieve maximum estimation accuracy. Also, the tuning parameter value of NSKF, $m = 0.6$ is assumed for simulation.

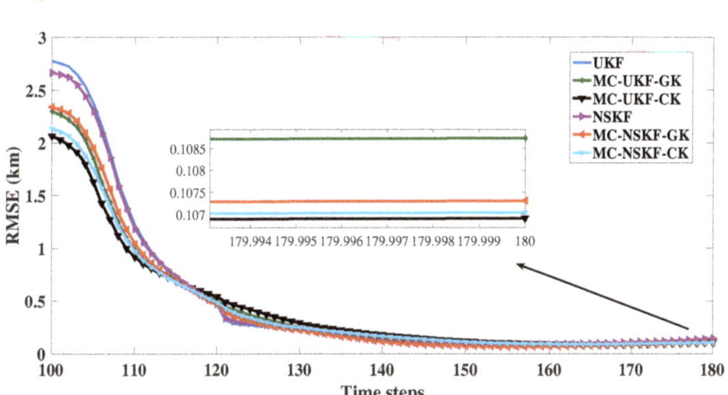

Figure 9. 2D: RMSE in position.

Table 3. 2D: RMSE in position and % Track Loss.

Filters	% Track Loss	RMSE (m)
UKF	4.4	152.8
MC-UKF-GK	1.1	111.0
MC-UKF-CK	1.1	108.9
NSKF	2.8	151.1
MC-NSKF-GK	1.2	109.6
MC-NSKF-CK	0.5	108.8

Table 4. 3D: RMSE in position and % Track Loss.

Filters	% Track Loss	RMSE (m)
UKF	100	-
MC-UKF-GK	14	496.1
MC-UKF-CK	13.5	499.8
NSKF	100	-
MC-NSKF-GK	14	496.8
MC-NSKF-CK	13.5	498.9

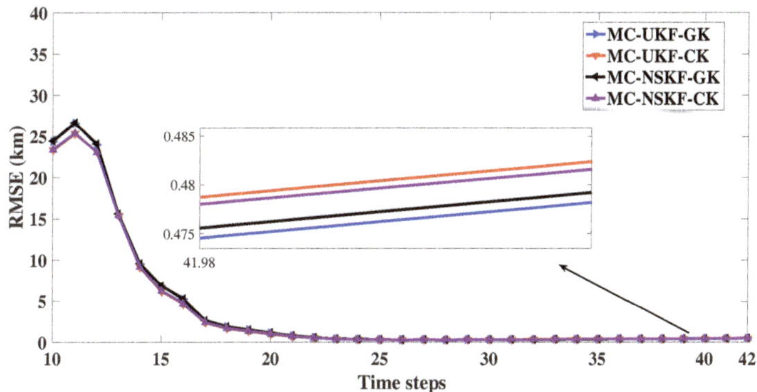

Figure 10. 3D: RMSE in position.

9. Conclusions

Since, measurements obtained in target tracking scenarios are corrupted with non Gaussian noise, this paper presents a maximum correntropy framework for 2D as well as 3D angles-only target tracking problem. The reformulation of UKF and NSKF in terms of Gaussian and Cauchy kernel based MC framework was realized. The non Gaussian noise is modelled as a Gaussian mixture (glint noise) plus shot noise. Finally, the performance of the estimators were evaluated and a comparative analysis is presented on the basis of RMSE in position and % track loss. From the comparative analysis, it can be concluded that the Gaussian and Cauchy kernel based MC framework provides improved estimation accuracy than UKF and NSKF in non Gaussian noise environments. Thus, it can be inferred that MC based estimators have the potential to give accurate and robust state estimates in the presence of non Gaussian noises in angle measurements. As a future work, the proposed estimation framework can be extended to track a manoeuvring target in the presence of angles-only measurements corrupted with non Gaussian noise.

Author Contributions: Conceptualization, R.R. and B.R.; methodology, A.U., A.D. and R.R.; software, A.U., A.D. and R.R.; validation, R.R. and B.R.; writing—original draft preparation, A.U.; writing—review and editing, R.R. and B.R.; supervision, R.R. and B.R. All authors have read and agreed to the published version of the manuscript.

Funding: This research was funded by SVNIT Surat Project No. 2020-21/seed money/30.

Institutional Review Board Statement: Not applicable.

Informed Consent Statement: Not applicable.

Data Availability Statement: Not applicable.

Conflicts of Interest: The authors declare no conflict of interest.

Abbreviations

The following abbreviations are used in this manuscript:

$\mathbf{X^t}_k$	Target state vector at sample k
$\mathbf{X^o}_k$	Observer state vector at sample k
$\mathbf{X}_k$	Relative state vector at sample k
$\mathbf{w}$	Zero mean Gaussian process noise
$\mathbf{Q}$	Process covariance
$\mathbf{F}$	State transition matrix
T	Sampling time
q_x, q_y, q_z	Power spectral densities of the process noise along the X, Y, and Z axes
$\mathbf{z}_k$	Measurement vector at sample k
$\mathbf{v}_k$	Non Gaussian measurement noise at sample k
$\mathbf{r}$	Range vector
β and ϵ	Bearing and Elevation angle measurement
$\mathbf{R}_k$	Measurement noise covariance matrix at sample k
σ_β and σ_ϵ	Standard deviations of error in bearing and elevation angles
k_σ	Kernel function
G_σ and σ	Gaussian kernel and Gaussian bandwidth
C_δ and δ	Cauchy kernel and Cauchy bandwidth
$\mathbf{X}_{k\|k-1}$	Prior mean at sample k
$\mathbf{P}_{k\|k-1}$	Prior covariance at sample k
W_k	Weights at sample k
$\mathbf{\overline{H}}_k$	Measurement slope matrix
$\mathbf{P_{Xz}}$	Cross covariance
$\mathbf{P_{zz}}$	Measurement covariance
J	Cost function

$\wp$ and ϱ	Adjustable weights
$\mathbf{L}_i^G$	Gaussian scalar term
$\mathbf{K}_i^G$	Gaussian Kalman gain
$\mathbf{X}_{k|k}$	Posterior mean at sample k
$\mathbf{P}_{k|k}$	Posterior covariance at sample k
$\bar{c}_r$	Initial course estimate
$\hat{\beta}_1$ and $\hat{\epsilon}_1$	True initial bearing and elevation measurement estimate
$\bar{\alpha}_1$ and $\bar{\gamma}_1$	Bearing and Elevation angle heading
$\mathcal{T}_b$	Threshold
RMSE	Root Mean Square Error
MMSE	Minimum Mean Square Error
MC-UKF-GK	Maximum correntropy unscented Kalman filter Gaussian kernel
MC-UKF-CK	Maximum correntropy unscented Kalman filter Cauchy kernel
MC-NSKF-GK	Maximum correntropy new sigma point Kalman filter Gaussian kernel
MC-NSKF-CK	Maximum correntropy new sigma point Kalman filter Cauchy kernel

Appendix A. Power Series Expansion of Cauchy Kernel Function

The binomial expansion of $(1+x)^{-N}$ for negative integer $-N$ is given as follows:

$$\left(1+x\right)^{-N} = 1 + (-N)x + \frac{(-N)(-N-1)}{2!}x^2 + \frac{(-N)(-N-1)(-N-2)}{3!}x^3 + \cdots,$$

$$= \sum_{k=0}^{\infty} (-1)^k \binom{N+k-1}{k} x^k, \quad \text{for } |x| < 1.$$

Now the correntropy measure, by taking the binomial series expansion of Cauchy kernel with $x = \frac{(X-Y)^2}{\delta}$ is

$$V_\delta(X,Y) = \sum_{k=0}^{\infty} (-1)^k \binom{N+k-1}{k} \left(\frac{(X-Y)^2}{\delta}\right)^k$$

$$= \sum_{k=0}^{\infty} \frac{(-1)^k}{\delta^k} \binom{N+k-1}{k} E\left[(X-Y)^{2k}\right], \quad \text{for } \left|\frac{(X-Y)^2}{\delta}\right| < 1.$$

Appendix B. Derivation of Kalman Gain

The Kalman gain for Gaussian kernel based nonlinear estimator is $\mathbf{K}_k^G$ with $\mathbf{L}_k^G$ as a scalar term. Similarly, for Cauchy kernel, it is $\mathbf{K}_k^C$ and $\mathbf{L}_k^C$ respectively. A general expression for Kalman gain is given as $\mathbf{K}_k = (\mathbf{P}_{k|k-1}^{-1} + \mathbf{L}_k \overline{\mathbf{H}}' \mathbf{R}_k^{-1} \overline{\mathbf{H}})^{-1} \mathbf{L}_k \overline{\mathbf{H}}' \mathbf{R}_k^{-1}$. Applying matrix inversion lemma

$$\mathbf{K}_k = \left(\mathbf{P}_{k|k-1} - \mathbf{P}_{k|k-1} \mathbf{L}_k \overline{\mathbf{H}}' (\mathbf{R}_k + \overline{\mathbf{H}} \mathbf{P}_{k|k-1} \mathbf{L}_k \overline{\mathbf{H}}')^{-1} \overline{\mathbf{H}} \mathbf{P}_{k|k-1}\right) \mathbf{L}_k \overline{\mathbf{H}}' \mathbf{R}_k^{-1}$$

$$= \mathbf{P}_{k|k-1} \mathbf{L}_k \overline{\mathbf{H}}' \mathbf{R}_k^{-1} - \mathbf{P}_{k|k-1} \mathbf{L}_k \overline{\mathbf{H}}' (\mathbf{R}_k + \overline{\mathbf{H}} \mathbf{P}_{k|k-1} \mathbf{L}_k \overline{\mathbf{H}}')^{-1} \overline{\mathbf{H}} \mathbf{P}_{k|k-1} \mathbf{L}_k \overline{\mathbf{H}}' \mathbf{R}_k^{-1}$$

$$= \mathbf{P}_{k|k-1} \mathbf{L}_k \overline{\mathbf{H}}' \left(\mathbf{R}_k^{-1} - (\mathbf{R}_k + \overline{\mathbf{H}} \mathbf{P}_{k|k-1} \mathbf{L}_k \overline{\mathbf{H}}')^{-1} \overline{\mathbf{H}} \mathbf{P}_{k|k-1} \mathbf{L}_k \overline{\mathbf{H}}' \mathbf{R}_k^{-1}\right)$$

$$= \mathbf{P}_{k|k-1} \mathbf{L}_k \overline{\mathbf{H}}' \left(\mathbf{I} - (\mathbf{R}_k + \overline{\mathbf{H}} \mathbf{P}_{k|k-1} \mathbf{L}_k \overline{\mathbf{H}}')^{-1} \overline{\mathbf{H}} \mathbf{P}_{k|k-1} \mathbf{L}_k \overline{\mathbf{H}}'\right) \mathbf{R}_k^{-1}.$$

After certain algebraic manipulations, we get

$$\mathbf{K}_k = \mathbf{P}_{k|k-1} \mathbf{L}_k \overline{\mathbf{H}}' (\mathbf{I} + \mathbf{R}_k^{-1} \overline{\mathbf{H}} \mathbf{P}_{k|k-1} \mathbf{L}_k \overline{\mathbf{H}}')^{-1} \mathbf{R}_k^{-1} = \mathbf{P}_{k|k-1} \mathbf{L}_k \overline{\mathbf{H}}_k' (\mathbf{R}_k + \overline{\mathbf{H}}_k \mathbf{P}_{k|k-1} \mathbf{L}_k \overline{\mathbf{H}}_k')^{-1}. \tag{A1}$$

From the above equation, $\mathbf{K}_k^G$ and $\mathbf{K}_k^C$ can be defined by making necessary substitution for $\mathbf{L}_k^G$ and $\mathbf{L}_k^C$.

References

1. Auger, F.; Hilairet, M.; Guerrero, J.M.; Monmasson, E.; Orlowska-Kowalska, T.; Katsura, S. Industrial applications of the Kalman filter: A review. *IEEE Trans. Ind. Electron.* **2013**, *60*, 5458–5471. [CrossRef]
2. Welch, G.; Bishop, G. *An Introduction to the Kalman Filter*; University of North Carolina: Chapel Hill, NC, USA, 1995.
3. Ristic, B.; Arulampalam, S.; Gordon, N. *Beyond the Kalman Filter: Particle Filters for Tracking Applications*; Artech house: Norwood, MA, USA, 2003.
4. Mallick, M.; Tian, X.; Zhu, Y.; Morelande, M. Angle-Only Filtering of a Maneuvering Target in 3D. *Sensors* **2022**, *22*, 1422. [CrossRef]
5. Reif, K.; Gunther, S.; Yaz, E.; Unbehauen, R. Stochastic stability of the discrete-time extended Kalman filter. *IEEE Trans. Autom. Control.* **1999**, *44*, 714–728. [CrossRef]
6. Julier, S.J.; Uhlmann, J.K. Unscented filtering and nonlinear estimation. *Proc. IEEE* **2004**, *92*, 401–422. [CrossRef]
7. Haykin, S.; Arasaratnam, I. Cubature kalman filters. *IEEE Trans. Autom. Control.* **2009**, *54*, 1254–1269.
8. Radhakrishnan, R.; Yadav, A.; Date, P.; Bhaumik, S. A new method for generating sigma points and weights for nonlinear filtering. *IEEE Control. Syst. Lett.* **2018**, *2*, 519–524. [CrossRef]
9. Wu, Z.; Shi, J.; Zhang, X.; Ma, W.; Chen, B.; Senior Member, I. Kernel recursive maximum correntropy. *Signal Process.* **2015**, *117*, 11–16. [CrossRef]
10. Radhakrishnan, R.; Bhaumik, S.; Tomar, N.K. Gaussian sum shifted Rayleigh filter for underwater bearings-only target tracking problems. *IEEE J. Ocean. Eng.* **2018**, *44*, 492–501. [CrossRef]
11. Radhakrishnan, R.; Asfia, U.; Sharma, S. Gaussian sum state estimators for three dimensional angles-only underwater target tracking problems. *IFAC-PapersOnLine* **2022**, *55*, 333–338. [CrossRef]
12. Kotecha, J.H.; Djuric, P.M. Gaussian sum particle filtering. *IEEE Trans. Signal Process.* **2003**, *51*, 2602–2612. [CrossRef]
13. Karlgaard, C.D. Nonlinear regression Huber–Kalman filtering and fixed-interval smoothing. *J. Guid. Control. Dyn.* **2015**, *38*, 322–330. [CrossRef]
14. Li, W.; Jia, Y. H-infinity filtering for a class of nonlinear discrete-time systems based on unscented transform. *Signal Process.* **2010**, *90*, 3301–3307. [CrossRef]
15. Liu, W.; Pokharel, P.P.; Principe, J.C. Correntropy: A localized similarity measure. In Proceedings of the 2006 IEEE International Joint Conference on Neural Network Proceedings, Vancouver, BC, Canada, 16–21 July 2006; pp. 4919–4924.
16. Liu, W.; Pokharel, P.P.; Principe, J.C. Correntropy: Properties and applications in non-Gaussian signal processing. *IEEE Trans. Signal Process.* **2007**, *55*, 5286–5298. [CrossRef]
17. Cinar, G.T.; Príncipe, J.C. Hidden state estimation using the correntropy filter with fixed point update and adaptive kernel size. In Proceedings of the 2012 International Joint Conference on Neural Networks (IJCNN), Brisbane, Australia, 10–15 June 2012; pp. 1–6.
18. Cinar, G.T.; Príncipe, J.C. Adaptive background estimation using an information theoretic cost for hidden state estimation. In Proceedings of the 2011 International Joint Conference on Neural Networks, San Jose, CA, USA, 31 July–5 August 2011; pp. 489–494.
19. Chen, B.; Liu, X.; Zhao, H.; Principe, J.C. Maximum correntropy Kalman filter. *Automatica* **2017**, *76*, 70–77. [CrossRef]
20. Izanloo, R.; Fakoorian, S.A.; Yazdi, H.S.; Simon, D. Kalman filtering based on the maximum correntropy criterion in the presence of non-Gaussian noise. In Proceedings of the 2016 Annual Conference on Information Science and Systems (CISS), Princeton, NJ, USA, 16–18 March 2016; pp. 500–505.
21. Liu, X.; Qu, H.; Zhao, J.; Chen, B. Extended Kalman filter under maximum correntropy criterion. In Proceedings of the 2016 International Joint Conference on Neural Networks (IJCNN), Vancouver, BC, Canada, 24–29 July 2016; pp. 1733–1737.
22. Liu, X.; Qu, H.; Zhao, J.; Yue, P.; Wang, M. Maximum correntropy unscented Kalman filter for spacecraft relative state estimation. *Sensors* **2016**, *16*, 1530. [CrossRef]
23. Qin, W.; Wang, X.; Cui, N. Maximum correntropy sparse Gauss–Hermite quadrature filter and its application in tracking ballistic missile. *IET Radar Sonar Navig.* **2017**, *11*, 1388–1396. [CrossRef]
24. Wang, G.; Li, N.; Zhang, Y. Maximum correntropy unscented Kalman and information filters for non-Gaussian measurement noise. *J. Frankl. Inst.* **2017**, *354*, 8659–8677. [CrossRef]
25. Wang, G.; Gao, Z.; Zhang, Y.; Ma, B. Adaptive maximum correntropy Gaussian filter based on variational Bayes. *Sensors* **2018**, *18*, 1960. [CrossRef]
26. Wang, J.; Lyu, D.; He, Z.; Zhou, H.; Wang, D. Cauchy kernel-based maximum correntropy Kalman filter. *Int. J. Syst. Sci.* **2020**, *51*, 3523–3538. [CrossRef]
27. Song, H.; Ding, D.; Dong, H.; Yi, X. Distributed filtering based on Cauchy-kernel-based maximum correntropy subject to randomly occurring cyber-attacks. *Automatica* **2022**, *135*, 110004. [CrossRef]
28. Best, R.A.; Norton, J. A new model and efficient tracker for a target with curvilinear motion. *IEEE Trans. Aerosp. Electron. Syst.* **1997**, *33*, 1030–1037. [CrossRef]
29. Cai, X.; Sun, F. Two-layer IMM tracker with variable structure for curvilinear maneuvering targets. *Wirel. Pers. Commun.* **2018**, *103*, 727–740. [CrossRef]
30. Blackman, S.; Popoli, R. *Design and Analysis of Modern Tracking Systems*; Artech House: Norwood, MA, USA, 1999.
31. Cavallaro, A.; Maggio, E. *Video Tracking: Theory and Practice*; John Wiley & Sons: Hoboken, NJ, USA, 2011.

32. Hewer, G.; Martin, R.; Zeh, J. Robust preprocessing for Kalman filtering of glint noise. *IEEE Trans. Aerosp. Electron. Syst.* **1987**, 120–128. [CrossRef]
33. Wu, W.R.; Cheng, P.P. A nonlinear IMM algorithm for maneuvering target tracking. *IEEE Trans. Aerosp. Electron. Syst.* **1994**, *30*, 875–886.
34. Souza, C.R. Kernel functions for machine learning applications. *Creat. Commons Attrib.-Noncommercial-Share Alike* **2010**, *3*, 1–1.
35. Julier, S.; Uhlmann, J.; Durrant-Whyte, H.F. A new method for the nonlinear transformation of means and covariances in filters and estimators. *IEEE Trans. Autom. Control.* **2000**, *45*, 477–482. [CrossRef]
36. Wu, W.R. Target racking with glint noise. *IEEE Trans. Aerosp. Electron. Syst.* **1993**, *29*, 174–185.
37. Mallick, M.; Morelande, M.; Mihaylova, L.; Arulampalam, S.; Yan, Y. Angle-only filtering in three dimensions. In *Integrated Tracking, Classification, and Sensor Management: Theory and Applications*; Wiley-IEEE Press: Hoboken, NJ, USA, 2012; Volume 1, pp. 3–42.
38. Asfia, U.; Radhakrishnan, R.; Sharma, S. Three-Dimensional Bearings-Only Target Tracking: Comparison of Few Sigma Point Kalman Filters. In *Communication and Control for Robotic Systems*; Springer: Berlin/Heidelberg, Germany, 2022; pp. 273–289.

Article

Optimal Geometries for AOA Localization in the Bayesian Sense

Kutluyil Dogancay

UniSA STEM, University of South Australia, Mawson Lakes, SA 5095, Australia; kutluyil.dogancay@unisa.edu.au

Abstract: This paper considers the optimal sensor placement problem for angle-of-arrival (AOA) target localization in the 2D plane with a Gaussian prior. Optimal sensor locations are analytically determined for a single AOA sensor using the D- and A-optimality criteria and an approximation of the Bayesian Fisher information matrix (BFIM). Optimal sensor placement is shown to align with the minor axis of the prior covariance error ellipse for both optimality criteria. The approximate BFIM is argued to be valid for a sufficiently small prior covariance compared with the target range. Optimal sensor placement results obtained for Bayesian target localization are extended to manoeuvring target tracking. For sensor trajectory optimization subject to turn-rate constraints, numerical search methods based on the D- and A-optimality criteria as well as a new closed-form projection algorithm that aims to achieve alignment with the minor axis of the prior error ellipse are proposed. It is observed that the two optimality criteria generate significantly different optimal sensor trajectories despite having the same optimal sensor placement for the localization of a stationary target. Analysis results and the performance of the sensor trajectory optimization methods are demonstrated with simulation examples. It is observed that the new closed-form projection algorithm achieves superior tracking performance compared with the two numerical search methods.

Keywords: optimal target–sensor geometries; bearings-only localization; Fisher information matrix; Bayesian estimation

Citation: Dogancay, K. Optimal Geometries for AOA Localization in the Bayesian Sense. *Sensors* **2022**, *22*, 9802. https://doi.org/10.3390/s22249802

Academic Editors: Mahendra Mallick and Ratnasingham Tharmarasa

Received: 6 November 2022
Accepted: 12 December 2022
Published: 14 December 2022

Publisher's Note: MDPI stays neutral with regard to jurisdictional claims in published maps and institutional affiliations.

1. Introduction

In target tracking and localization problems, target–sensor geometries are known to play a significant role in determining the localization and tracking performance. In this paper, we focus on optimal target–sensor geometries for angle-of-arrival (AOA) localization in the 2D plane using a single moving sensor. First, the localization problem is cast as a Bayesian estimation problem, which assumes the availability of prior information in the form of a Gaussian prior for the unknown target location. For this problem, optimal sensor placement results are developed using approximate estimation bounds. Next, the Bayesian estimation problem is extended to target tracking using the Kalman filter, and optimal sensor trajectories are developed to track a manoeuvring target.

Optimal sensor placement has been researched for several decades. Early works included [1–3], where the performance of the extended Kalman filter (EKF) [4] and several deterministic (non-Bayesian) estimators was reported for different bearings-only sensor manoeuvres, mostly in sonar applications. In [5], optimal bearings-only sensor manoeuvres for tracking a constant-velocity target were derived using optimal control theory. The sensor trajectory optimization problem was formulated as a partially observable Markov decision problem (POMDP) for a manoeuvring target using the trace of FIM, which is similar to the A-optimality criterion (minimizing trace of inverse FIM), as the reward function in [6]. In [7], the D-optimality criterion, whereby the sensor location is determined to maximize the determinant of the Fisher information matrix (FIM), was adopted to determine optimal sensor trajectories for the localization of a stationary target. Optimal sensor manoeuvres necessary to make a constant-velocity target observable were discussed in [8]. The sensor trajectory optimization problem in the Bayesian sense was considered in [9], where the posterior Cramer–Rao lower bound (PCRLB) [10] was employed to minimize the largest root-mean-square-error (RMSE)-bound approximated by the reciprocal of measurement data contributions to target location information.

A comprehensive analysis of the 2D optimal AOA sensor placement problem for a stationary target was presented in [11,12]. Optimal 3D AOA target–sensor geometries for a stationary target were analysed in [13]. A gradient-descent algorithm for sensor path optimization to minimize the mean-square error of predicted EKF target location estimates was proposed in [14]. A UAV path optimization algorithm that solves a nonlinear programming problem based on the D-optimality criterion was developed in [15] to geolocate a stationary target using a heterogeneous mix of passive payload sensors. In [16], a unified framework was proposed for AOA, range-only, and received signal strength localization when the target was stationary. Optimal target–sensor geometries for maximum a posteriori (MAP) target localization with a Gaussian prior were investigated in [17]. In [18], the optimality criteria for target–sensor geometries in a Kalman filtering setting were analysed for several sensor types using an approximation of the Bayesian FIM. A unified 2D target–sensor geometry optimization framework was proposed in [19] for stationary target localization with a Gaussian prior, reducing the optimization problem to minimization of the modulus of a vector sum, akin to [11]. In [20], optimal sensor placement in 3D space was studied for AOA target localization with a Gaussian prior, employing rotational invariance arguments.

This paper develops optimal sensor placement results for a single AOA sensor at a fixed distance from the mean of the Gaussian prior. To do this, the Bayesian FIM is approximated by replacing the expectation of the contribution form measurement data with its instantaneous value calculated at the mean of the Gaussian prior. It is argued that this approximation is valid when the covariance of the Gaussian prior is relatively small compared with the target range. The optimal sensor placements for the D- and A-optimality criteria are shown to be identical and align with the minor axis of the error ellipse of the prior covariance. In the context of bearings-only manoeuvring target tracking, numerical search methods based on the D- and A-optimality criteria and a new closed-form projection algorithm that attempts to achieve alignment with the minor axis of the prior covariance error ellipse are proposed for sensor trajectory optimization subject to turn-rate constraints. It is observed that the D- and A-optimality criteria yield markedly different optimal sensor trajectories even though they produce identical optimal sensor placement for the localization of a stationary target. The projection algorithm is shown to outperform the other two methods in simulation studies.

This paper is organized as follows. Section 2 investigates the optimal sensor placement problem for a stationary target with a Gaussian prior using the D- and A-optimality criteria. Section 3 extends the results of Section 2 to Kalman filter tracking of a manoeuvring target, proposing two sensor trajectory optimization methods and a new closed-from projection algorithm. Section 4 presents simulation examples to verify the optimal sensor placement results derived in Section 2 and to compare and demonstrate the effectiveness of the sensor trajectory optimization algorithms proposed in Section 3. Concluding remarks are made in Section 5.

2. Optimal Target-Sensor Geometry with Gaussian Prior

In tracking problems with a state space that can be modelled or approximated as a Gauss–Markov process, the Kalman filter has been extensively used to compute the Gaussian prior for state estimates from the noisy sensor measurements available in each recursion in the form of the predicted state estimate and predicted state covariance. Starting with the Gaussian prior $\mathcal{N}(x_0, P_0)$, where x_0 is the mean and P_0 is the covariance of the prior, the objective of optimal sensor placement is to determine a sensor location s at a fixed distance of $d = \|d\|$, where $d = x_0 - s$, from the mean of the Gaussian prior (or predicted target location estimate) x_0 so that sensor measurements collected at the new sensor location will optimize a well-defined objective function that is related to the Bayesian localization performance. The new measurements together with the Gaussian prior are then used to compute the filtered state estimate and covariance, which are optimized in terms of target–sensor geometry. We consider two optimality criteria for sensor placement: namely, the D-optimality and A-optimality criteria [21–23], which are commonly used in practice. Referring to Figure 1, the task of geometry optimization is reduced to finding a range vector d or bearing angle $\theta(x_0)$ pivoted at the mean of the Gaussian prior with $d = \|d\|$ fixed, which gives the location of the sensor to satisfy the chosen optimization criterion.

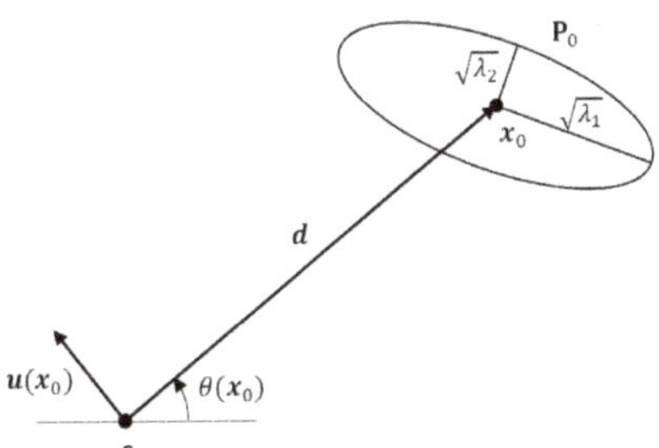

Figure 1. The 2D AOA geometry and 1-σ error ellipse for Gaussian prior with mean x_0 and covariance P_0.

The bearing measurements collected by the sensor located at s are given by

$$\theta = \theta(x) + w \tag{1}$$

where $x \sim \mathcal{N}(x_0, P_0)$ is the Bayesian prior, $w \sim \mathcal{N}(0, \sigma^2)$ is the bearing angle noise, and

$$\theta(x) = \tan^{-1}(x_2 - s_2, x_1 - s_1), \quad -\pi \leq h(x) \leq \pi \tag{2}$$

is the true bearing angle with $x = [x_1, x_2]^T$ and $s = [s_1, s_2]^T$. In (2), $\tan^{-1}(\cdot)$ is the four-quadrant arc-tangent. The objective of Bayesian estimation is to determine an estimate for the unknown random target location x from the bearing measurement θ and the knowledge of the Gaussian prior with mean x_0 and covariance P_0.

The optimality criteria considered in this paper employ estimation bounds obtained from the FIM or CRLB. In Bayesian estimation problems that involve random unknown parameters, these bounds are replaced by the Bayesian FIM (BFIM) or Bayesian CRLB (BCRLB). The BFIM for the single-sensor AOA localization problem is defined as [24]

$$\mathbf{\Phi} = K_0 + E_x\left\{ \frac{1}{\sigma^2 d^2(x)} u(x) u^T(x) \right\} \tag{3}$$

where $E_x\{\cdot\}$ denotes the expectation over x,

$$u(x) = \begin{bmatrix} -\sin\theta(x) \\ \cos\theta(x) \end{bmatrix} \tag{4}$$

is the unit vector orthogonal to the range vector, and $d(x) = \|x - s\|$ is the target range from the sensor positioned at s. The matrix $K_0 = P_0^{-1}$ represents the contribution of the a priori information, and

$$E_x\left\{ \frac{1}{\sigma^2 d^2(x)} u(x) u^T(x) \right\} \tag{5}$$

is the contribution of data. The inverse of the BFIM gives the BCRLB.

Equation (5) can be rewritten as

$$E_x\left\{\frac{1}{\sigma^2 d^2(x)}u(x)u^T(x)\right\} = \frac{1}{\sigma^2}E_x\left\{\frac{1}{d^2(x)}u(x)u^T(x)\right\} \tag{6}$$

$$= \frac{1}{\sigma^2}\begin{bmatrix} E\left\{\frac{\sin^2\theta(x)}{\|x-s\|^2}\right\} & -E\left\{\frac{\sin\theta(x)\cos\theta(x)}{\|x-s\|^2}\right\} \\ -E\left\{\frac{\sin\theta(x)\cos\theta(x)}{\|x-s\|^2}\right\} & E\left\{\frac{\cos^2\theta(x)}{\|x-s\|^2}\right\} \end{bmatrix} \tag{7}$$

$$\approx \frac{1}{\sigma^2 d^2}u(x_0)u^T(x_0), \quad u(x_0) = \begin{bmatrix} -\sin\theta(x_0) \\ \cos\theta(x_0) \end{bmatrix} \tag{8}$$

which gives

$$\boldsymbol{\Phi} \approx K_0 + \frac{1}{\sigma^2 d^2}u(x_0)u^T(x_0). \tag{9}$$

The approximation in (9) is valid for a sufficiently "small" prior covariance compared with the target range. Here, we measure the size of prior covariance by its trace (see (16)). Referring to Figure 2, we have

$$\sin\theta(x) = \sin\theta(x_0 + \eta) \tag{10}$$

$$= \frac{x_{0,2} - s_2 + \eta_2}{\sqrt{(x_{0,1} - s_1 + \eta_1)^2 + (x_{0,2} - s_2 + \eta_2)^2}} \tag{11}$$

where $\eta = [\eta_1, \eta_2]^T \sim \mathcal{N}(0, P_0)$, and $x_0 = [x_{0,1}, x_{0,2}]^T$. Taking the expectation of the squared sine function yields

$$E\{\sin^2\theta(x)\} \approx \frac{E\{(x_{0,2} - s_2 + \eta_2)^2\}}{E\{(x_{0,1} - s_1 + \eta_1)^2 + (x_{0,2} - s_2 + \eta_2)^2\}} \tag{12}$$

$$\approx \frac{(x_{0,2} - s_2)^2 + E\{\eta_2^2\}}{d^2 + E\{\eta_1^2\} + E\{\eta_2^2\}} \tag{13}$$

$$\approx \left(\frac{x_{0,2} - s_2}{d}\right)^2 \tag{14}$$

$$\approx \sin^2\theta(x_0) \tag{15}$$

if

$$d^2 \gg E\{\eta_1^2\} + E\{\eta_2^2\} = \operatorname{tr} P_0. \tag{16}$$

Here, tr denotes trace. Applying the same line of reasoning to $E\{\sin\theta(x)\cos\theta(x)\}$ and $E\{\cos^2\theta(x)\}$, we conclude that (16) is the condition that must be met to justify (9).

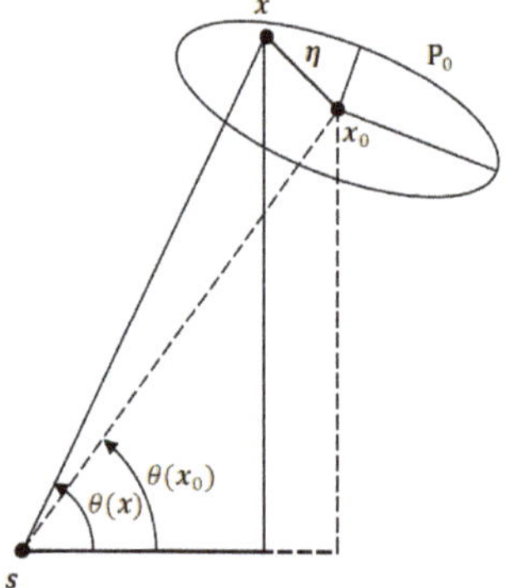

Figure 2. Geometric interpretation of instantaneous angle measurements.

2.1. D-Optimality Criterion

The D-optimality criterion aims to maximize the determinant of the Fisher information matrix (FIM). For the Bayesian estimation problem considered here, the FIM is replaced by the Bayesian FIM (BFIM). Considering the optimization problem described in Figure 1, the optimal placement for the sensor is obtained from the solution of

$$\max_{s} |\mathbf{\Phi}| \tag{17}$$

where $|\cdot|$ denotes the determinant. The optimization problem in (17) determines the sensor location s that maximizes the determinant of BFIM for a given Gaussian prior.

Noting that in (9) K_0 is a square matrix and $u(x_0)$ is a column vector, the determinant of the BFIM can be rewritten as a sum of two terms [25]

$$|\mathbf{\Phi}| \approx \left| K_0 + \frac{1}{\sigma^2 d^2} u(x_0) u^T(x_0) \right| \tag{18}$$

$$\approx |K_0| + \frac{1}{\sigma^2 d^2} u^T(x_0) K_0^* u(x_0) \tag{19}$$

where K_0^* is the adjoint of K_0, defined by

$$K_0^* = |K_0| K_0^{-1} = |K_0| P_0. \tag{20}$$

Thus, for a given Gaussian prior P_0 and fixed d, the optimization problem in (17) reduces to

$$\max_{\theta(x_0)} u^T(x_0) P_0 u(x_0). \tag{21}$$

Since $u(x_0)$ is a unit vector, (21) is a problem of quadratic form maximization over the unit circle. Using the eigenvalues of P_0, denoted by λ_1, λ_2 with $\lambda_1 \geq \lambda_2$, the solution of (21) is given by [26]:

$$\lambda_1 = \max_{\theta(x_0)} u^T(x_0) P_0 u(x_0) \tag{22}$$

$$v_1 = \arg\max_{u(x_0)} u^T(x_0) P_0 u(x_0) \tag{23}$$

where v_1 is an orthonormal eigenvector of P_0 corresponding to λ_1. The optimal bearing angle for the sensor, θ_{opt}, is easily obtained from the optimal unit vector v_1 by noting that it is orthogonal to the range vector (see Figure 1). In other words, the optimal range vector d_{opt} must be aligned with the minor axis of the error ellipsoid of the Gaussian prior, as shown in Figure 3.

Some remarks are in order here:

- If the Gaussian prior has a circular error ellipse with P_0 given by a scaled identity matrix, (21) becomes

$$\max_{\theta(x_0)} u^T(x_0) u(x_0) \quad \text{or} \quad \max_{\theta(x_0)} 1 \tag{24}$$

which means that optimality is achieved by any bearing angle $\theta(x_0)$.
- In all other cases, there are two optimal bearing angles aligned with the minor axis of the error ellipse, producing two possible optimal sensor locations with range vectors $\pm d_{\text{opt}}$ symmetric about x_0.

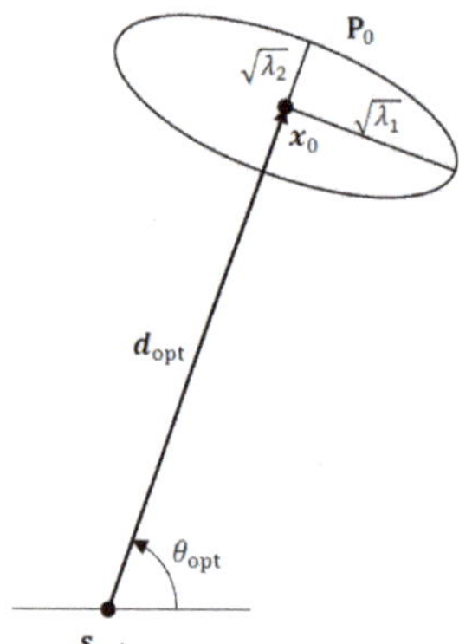

Figure 3. Optimal sensor placement in 2D using the D-optimality criterion with Gaussian prior $\mathcal{N}(x_0, P_0)$.

2.2. A-Optimality Criterion

The objective of the A-optimality criterion is to minimize the trace of the BCRLB or the inverse BFIM. In this case, the optimal sensor placement is obtained from

$$\min_{s} \ \mathrm{tr}(\mathbf{\Phi}^{-1}).$$ (25)

Applying the matrix inversion lemma [27] to the approximate BFIM in (9), we get

$$\mathbf{\Phi}^{-1} \approx P_0 \left(I - \frac{u(x_0)u^T(x_0)P_0}{\sigma^2 d^2 + u^T(x_0)P_0u(x_0)} \right)$$ (26)

It is clear that to solve (25), we need to maximize the trace of the second term on the right-hand side of (26); i.e.,

$$\max_{s} \mathrm{tr} \left(\frac{P_0 u(x_0)u^T(x_0)P_0}{\sigma^2 d^2 + u^T(x_0)P_0u(x_0)} \right)$$ (27)

or

$$\max_{u(x_0)} \frac{u^T(x_0)P_0^2 u(x_0)}{\sigma^2 d^2 + u^T(x_0)P_0u(x_0)}.$$ (28)

To solve (28) for $u(x_0)$, let $R = \frac{u^T(x_0)P_0^2 u(x_0)}{\sigma^2 d^2 + u^T(x_0)P_0u(x_0)}$ and set its gradient equal to zero,

$$\frac{\partial}{\partial u(x_0)} R = 0$$ (29)

which results in

$$\frac{2P_0^2 u(x_0)(\sigma^2 d^2 + u^T(x_0)P_0u(x_0)) - 2(u^T(x_0)P_0^2 u(x_0))P_0u(x_0)}{(\sigma^2 d^2 + u^T(x_0)P_0u(x_0))^2} = 0$$ (30)

$$P_0^2 u(x_0)(\sigma^2 d^2 + u^T(x_0)P_0u(x_0)) = (u^T(x_0)P_0^2 u(x_0))P_0u(x_0)$$ (31)

$$P_0^2 u(x_0) = R P_0 u(x_0).$$ (32)

Left-multiplying both sides of the above equation with P_0^{-1} finally gives

$$P_0 u(x_0) = R u(x_0)$$ (33)

where the nonzero scalar R is an eigenvalue of P_0 and the unit vector $u(x_0)$ is the corresponding eigenvector. We therefore conclude that R is maximized when $u(x_0) = v_1$, which is the eigenvector of P_0 associated with its largest eigenvalue λ_1. Note that this optimality result is identical to that for the A-optimality criterion derived in Section 2.1.

3. Application to Tracking

In this section, sensor waypoint optimization algorithms are devised to embed the optimal sensor placement results derived in Section 2 into the Kalman filter. As a specific application, bearings-only manoeuvring target tracking is considered. When the target is moving, it is often the case that the target dynamics and constraints on the motion of a single sensor, such as turn-rates and distances between successive waypoints, do not allow strictly optimal sensor placement geometries to be achieved from one Kalman filter recursion to the next. We develop sensor trajectory optimization methods that respect dynamic sensor constraints.

The principle we follow is based on the treatment of each Kalman filter recursion as solving a Bayesian target localization problem with a Gaussian prior available from the previous recursion and measurements taken at a new optimized sensor location to compute filtered state estimates and an updated prior for the next Kalman filter recursion. Figure 4 captures the computational steps of a Kalman filter recursion with sensor waypoint optimization embedded into it. The details of how optimal sensor waypoints are computed from Kalman filter parameters are discussed later in the section.

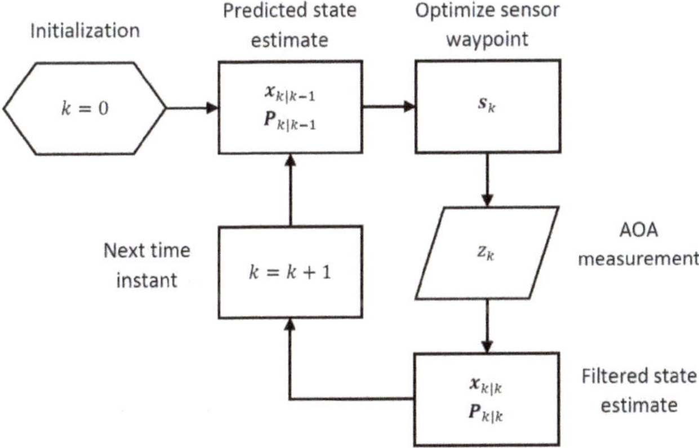

Figure 4. Optimal sensor waypoint computation in a Kalman filter recursion.

The single moving sensor collects AOA measurements from a manoeuvring target at time instants $k = 0, 1, 2, \ldots$. The sensor location at time k is denoted by s_k. The process equation for the target is

$$x_{k+1} = Fx_k + n_k, \quad k = 0, 1, \ldots \tag{34}$$

where $x_k = [x_k, \dot{x}_k, y_k, \dot{y}_k]^T$ is the target state vector with $[x_k, y_k]^T$ and $[\dot{x}_k, \dot{y}_k]^T$ denoting the target location and velocity, respectively, at time k. In (34) the dynamical constraint (the state transition matrix) is given by

$$F = \begin{bmatrix} A & 0 \\ 0 & A \end{bmatrix}, \quad A = \begin{bmatrix} 1 & T \\ 0 & 1 \end{bmatrix} \tag{35}$$

where T denotes the time interval between discrete-time instants k. The process noise n_k accounts for unknown target manoeuvres and is zero-mean white Gaussian with covariance

$$Q = \begin{bmatrix} q_x B & 0 \\ 0 & q_y B \end{bmatrix}, \quad B = \begin{bmatrix} T^4/4 & T^3/2 \\ T^3/2 & T^2 \end{bmatrix} \tag{36}$$

where q_x and q_y are often determined from maximum target acceleration [28].

The AOA measurement equation is

$$z_k = h(x_k) + w_k \tag{37}$$

where $h(\cdot)$ is the bearing angle of the target from the sensor location [see (2)], and $w_k \sim \mathcal{N}(0, \sigma^2)$ is the bearing measurement noise. As the measurement Equation (37) is nonlinear, the extended Kalman filter (EKF) is often used to estimate the target state vector, which is given by the recursion:

State Prediction:

$$x_{k|k-1} = F x_{k-1|k-1} \qquad \text{mean of prior} \tag{38}$$
$$P_{k|k-1} = F P_{k-1|k-1} F^T + Q \qquad \text{covariance of prior} \tag{39}$$

State Update:

$$\tilde{z}_k = z_k - h(x_{k|k-1}) \tag{40}$$
$$G_k = P_{k|k-1} h_k^T (h_k P_{k|k-1} h_k^T + \sigma^2)^{-1} \tag{41}$$
$$x_{k|k} = x_{k|k-1} + G_k \tilde{z}_k \qquad \text{state estimate} \tag{42}$$
$$P_{k|k} = (I - G_k h_k) P_{k|k-1} \qquad \text{covariance of state estimate} \tag{43}$$

where $x_{k|k-1}$ is the state prediction at time k given all measurements up to time $k-1$, and $x_{k|k}$ is the filtered state estimate at time k. The EKF replaces the nonlinear measurement Equation (37) with

$$z_k = h_k x_k + w_k \tag{44}$$

where h_k is the Jacobian of $h(x_k)$ evaluated at $x_{k|k-1}$:

$$h_k = \frac{1}{d_{k|k-1}} \begin{bmatrix} u_1(x_{k|k-1}) & 0 & u_2(x_{k|k-1}) & 0 \end{bmatrix}. \tag{45}$$

Here, $d_{k|k-1}$ is the target–sensor range estimate computed from $x_{k|k-1}$ and s_k, and $u(\cdot) = [u_1(\cdot), u_2(\cdot)]^T$ is the unit vector defined in (4).

The moving AOA sensor is assumed to travel with a constant velocity, which means that the distance between successive waypoints is constant, i.e., $\|s_k - s_{k-1}\| = s$. Assuming a maximum turn-rate of $\pm \vartheta_{\max}$ in azimuth, the next waypoint is constrained to lie on an arc defined by

$$s_k = s_{k-1} + s\, v(\vartheta_k), \quad |\vartheta_k - \vartheta_{k-1}| < \vartheta_{\max} \tag{46}$$

where $v(\vartheta_k) = [\cos \vartheta_k, \sin \vartheta_k]^T$ is the sensor heading vector with heading angle ϑ_k at time instant k.

The recursive BFIM for the Kalman filter tracking problem is given by [24]

$$\Phi_k = (F \Phi_{k-1} F^T + Q)^{-1} + \frac{1}{\sigma^2} E_{x_k}\{\tilde{h}_k^T \tilde{h}_k\} \tag{47}$$

where $\tilde{h}_k$ is the Jacobian matrix in (45) calculated at target state x_k. Equation (47) has the same structure as (3) in that it is the sum of prior information $(F \Phi_{k-1} F^T + Q)^{-1}$ and contribution from measurements $\frac{1}{\sigma^2} E_{x_k} \{\tilde{h}_k^T \tilde{h}_k\}$. It is necessary to simplify (47) so that readily available Kalman filter estimates can be used rather than resorting to computationally expensive Monte Carlo simulations to calculate the expectation.

The contribution from measurements is approximated by

$$\frac{1}{\sigma^2} E_{x_k} \{\tilde{h}_k^T \tilde{h}_k\} \approx \frac{1}{\sigma^2} h_k^T h_k \tag{48}$$

where h_k is the Jacobian in (45). Thus, using $P_{k|k-1}^{-1}$ as the prior information and (48) as the contribution from measurements, we have

$$\Phi_k \approx P_{k|k-1}^{-1} + \frac{1}{\sigma^2} h_k^T h_k \tag{49}$$

where both $P_{k|k-1}$ and h_k are calculated by the EKF. In the following subsections, we show how to apply the D- and A-optimality criteria to the approximate recursive BFIM expression in (49) to derive trajectory optimization algorithms.

3.1. Sensor Trajectory Optimization Using D-Optimality

Referring to (21) and (49), the optimal waypoint for the sensor at time k using the D-optimality for the EKF takes the following form:

$$\max_{s_k \in \mathcal{S}_k} h_k P_{k|k-1} h_k^T \tag{50}$$

which can be rewritten as

$$\max_{s_k \in \mathcal{S}_k} \frac{1}{d_{k|k-1}^2} u^T(x_{k|k-1}) P_{\mathrm{loc},k|k-1} u(x_{k|k-1}) \tag{51}$$

where $\mathcal{S}_k$ is the set of permissible waypoints compliant with the turn-rate

$$\mathcal{S}_k = \{s_k \mid s_k - s_{k-1} = sv(\vartheta_k),\ |\vartheta_k - \vartheta_{k-1}| < \vartheta_{\max}\} \tag{52}$$

and

$$P_{\mathrm{loc},k|k-1} = \begin{bmatrix} p_{\mathrm{loc},k|k-1}(1,1) & p_{\mathrm{loc},k|k-1}(1,2) \\ p_{\mathrm{loc},k|k-1}(2,1) & p_{\mathrm{loc},k|k-1}(2,2) \end{bmatrix} \tag{53}$$

is the 2×2 covariance matrix for predicted target location, which is extracted from $P_{k|k-1}$ as shown below:

$$P_{k|k-1} = \begin{bmatrix} p_{\mathrm{loc},k|k-1}(1,1) & * & p_{\mathrm{loc},k|k-1}(1,2) & * \\ * & * & * & * \\ p_{\mathrm{loc},k|k-1}(2,1) & * & p_{\mathrm{loc},k|k-1}(2,2) & * \\ * & * & * & * \end{bmatrix}. \tag{54}$$

Note that as $d_{k|k-1}$ also depends on s_k (i.e., the fixed range constraint does not apply), (51) does not have a simple closed-form solution. This means that it must be solved by a numerical search over a finite number of permissible waypoints contained in the set $\mathcal{S}_k$.

3.2. Sensor Trajectory Optimization Using A-Optimality

Using (28), the A-optimality criterion for sensor waypoints is given by

$$\max_{s_k \in \mathcal{S}_k} \frac{h_k^T P_{k|k-1}^2 h_k}{\sigma^2 + h_k^T P_{k|k-1} h_k} \tag{55}$$

which is obtained by substituting h_k for $u^T(x_0)/d$ and $P_{k|k-1}$ for P_0 into (28).

Different from the D-optimality criterion in (51), (55) depends not only on the target range $d_{k|k-1}$ through the Jacobian $\mathbf{h}_k$, but also the bearing noise variance σ^2. Again, it is not straightforward to solve (55) for the optimal $\mathbf{s}_k$. A numerical search over the members of the set $\mathcal{S}_k$ is necessary.

3.3. Projection Algorithm: A Closed-Form Solution

The D- and A-optimality solutions for determining optimal sensor waypoints described above can be computationally expensive, especially if the numerical search must be carried out over a large number of candidate waypoints in the set $\mathcal{S}_k$. In this subsection, we present an alternative closed-form solution, called the projection algorithm, inspired by the ultimate objective of aligning the sensor with the minor axis of the prior covariance error ellipse.

The idea behind the projection algorithm is illustrated in Figure 5. The next waypoint $\mathbf{s}_k$ is chosen to guide the sensor towards the closest point $\boldsymbol{\psi}_k$ that is aligned with the minor axis of the target location prior $\mathbf{P}_{\text{loc},k|k-1}$ and is at the same distance from the mean of the prior $\mathbf{x}_{\text{loc},k|k-1} = [x_{k|k-1}, y_{k|k-1}]^T$ as the estimated target–sensor range $d_{k|k-1}$. The next waypoint is found by projecting the waypoint vector $\mathbf{s}_k - \mathbf{s}_{k-1}$ to $\boldsymbol{\psi}_k - \mathbf{s}_{k-1}$ subject to the turn-rate constraint. If the projection causes the sensor heading angle to exceed the turn-rate, the next waypoint is chosen to have the maximum turn-rate. This projection also brings the sensor closer to the target with $d_k < d_{k|k-1}$. The reduction in d_k is proportional to how far $\boldsymbol{\psi}_k$ is from $\mathbf{s}_{k-1}$. If $\mathbf{x}_{\text{loc},k|k-1} - \mathbf{s}_{k-1}$ is aligned with the major axis of the error ellipse, which represents the worst geometry, the distance between $\mathbf{s}_{k-1}$ and $\boldsymbol{\psi}_k$ is maximized and d_k will have the maximum reduction.

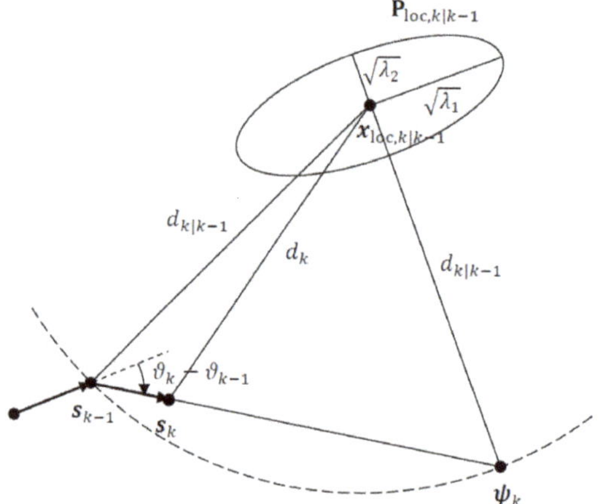

Figure 5. Closed-form projection algorithm to determine the next waypoint $\mathbf{s}_k$. The heading angle ϑ_k is constrained by the turn-rate $|\vartheta_k - \vartheta_{k-1}| < \vartheta_{\text{max}}$.

This behaviour makes sense because the optimality of a target–sensor geometry is improved at the maximum rate by moving the sensor directly towards the optimal sensor location for the given Gaussian prior with mean $x_{\text{loc},k|k-1}$ and covariance $\mathbf{P}_{\text{loc},k|k-1}$.

A detailed description of the projection algorithm is provided in Algorithm 1.

Algorithm 1 Projection algorithm.

Input: $s_{k-1}, x_{\mathrm{loc},k|k-1}, P_{\mathrm{loc},k|k-1}, \vartheta_{\max}, \vartheta_{k-1}, s$
Output: s_k
Calculate v_2, eigenvector of $P_{\mathrm{loc},k|k-1}$ associated with its smallest eigenvalue
Calculate $d_{k|k-1} = \|x_{\mathrm{loc},k|k-1} - s_{k-1}\|$
Calculate $\psi_k^{\pm} = x_{\mathrm{loc},k|k-1} \pm d_{k|k-1} v_2$
If $\|\psi_k^{+} - s_{k-1}\| < \|\psi_k^{-} - s_{k-1}\|$
$\qquad \psi_k = \psi_k^{+}$
Else
$\qquad \psi_k = \psi_k^{-}$
End
If $s_{k-1} = \psi_k$ (sensor is already aligned with prior covariance minor axis)
$\qquad$ Return $s_k = s_{k-1} + s[\cos\vartheta_{k-1}, \sin\vartheta_{k-1}]^T$ (no change in sensor heading)
Else
$\qquad$ Calculate $\Delta s = s\dfrac{\psi_k - s_{k-1}}{\|\psi_k - s_{k-1}\|}$
$\qquad$ Calculate $\Delta\vartheta = \angle\Delta s - \vartheta_{k-1}, -\pi \le \Delta\vartheta \le \pi$ ($\angle$ denotes bearing angle)
$\qquad$ If $|\Delta\vartheta| < \vartheta_{\max}$
$\qquad\qquad$ Return $s_k = s_{k-1} + \Delta s$
$\qquad$ Else
$\qquad\qquad$ Calculate $\vartheta_k = \vartheta_{k-1} + \mathrm{sign}(\Delta\vartheta)\vartheta_{\max}$
$\qquad\qquad$ Return $s_k = s_{k-1} + s[\cos\vartheta_k, \sin\vartheta_k]^T$
$\qquad$ End
End

---

4. Simulation Examples

This section presents simulation examples to verify the optimization results and to demonstrate the performance of the sensor trajectory optimization algorithms developed in Sections 2 and 3. In the first set of simulations, the focus is on optimal target–sensor geometries using the D- and A-optimality criteria for Bayesian target localization. The Gaussian prior has zero mean $x_0 = [0,0]^T$ and covariance

$$P_0 = \begin{bmatrix} 27.6047 & -14.7721 \\ -14.7721 & 22.3953 \end{bmatrix} \tag{56}$$

with eigenvalues $\lambda_1 = 40$ and $\lambda_2 = 10$. The minor axis of the error ellipse make an angle of $50°$ with the positive x-axis. The AOA sensor is allowed to be located on a circle of radius $d = 50$ km centred at the mean of the Gaussian prior x_0, and the bearing angle noise standard deviation is $\sigma = 5°$.

Figure 6 shows the D- and A-optimality measures, $|\Phi|$ and $\mathrm{tr}\,\Phi^{-1}$, respectively, versus the bearing angle θ in the range $0 \le \theta < \pi$ for the approximate and exact BFIM given by (9) and (3), respectively. The exact BFIM was calculated using 50,000 Monte Carlo runs for each bearing angle. As evident from Figure 6, the simulated optimal bearing angles are not significantly different for the approximate and exact BFIM. This is also backed up by the close proximity of the two curves in Figure 6. The optimal bearing angle obtained from the approximate BFIM aligns with the minor axis of the error ellipse at $\theta_{\mathrm{opt}} = 50°$ for both D- and A-optimality criteria, which is in agreement with the analytical results derived in Section 2.

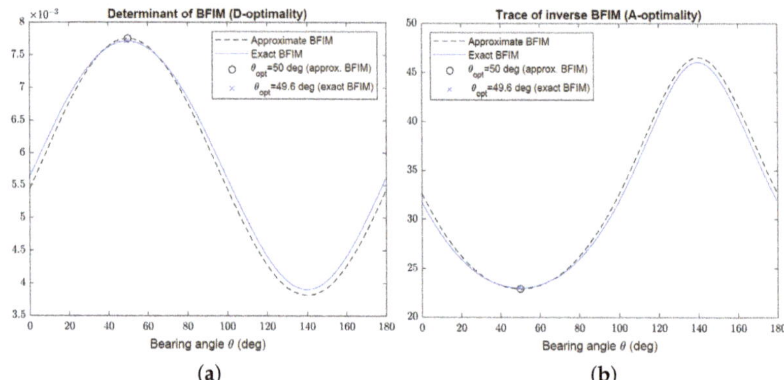

(a)   **(b)**

Figure 6. (**a**) Plot of determinant of Φ (D-optimality criterion) versus bearing angle using approximate and exact BFIM. (**b**) Plot of trace of Φ^{-1} (A-optimality criterion) versus bearing angle using approximate and exact BFIM.

To confirm that (9) is a valid approximation only for sufficiently small P_0 compared with the target range, we repeated the previous simulations for P_0 increased by a factor of two. Figure 7 shows the resulting D- and A-optimality measures for the approximate and exact BFIM. While the approximate and exact BFIM still yield almost the same optimal bearing angles, the curves corresponding to them exhibit significant discrepancy, in particular at bearing angles away from θ_{opt}.

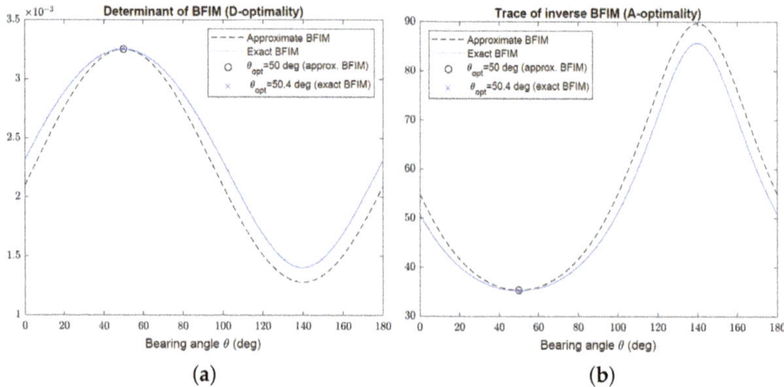

(a)   **(b)**

Figure 7. (**a**) Plot of determinant of Φ (D-optimality criterion) versus bearing angle for large P_0 using approximate and exact BFIM. (**b**) Plot of trace of Φ^{-1} (A-optimality criterion) versus bearing angle for large P_0 using approximate and exact BFIM.

The optimal target–sensor geometry for the simulated scenario is depicted in Figure 8. As expected, at the optimal bearing angle, the target–sensor range vector is perfectly aligned with the minor axis of the error ellipse of the Gaussian prior. Note that there are, in fact, two optimal sensor locations. The other one has the bearing angle $\theta_{\mathrm{opt}} - 180 = -130$ degrees.

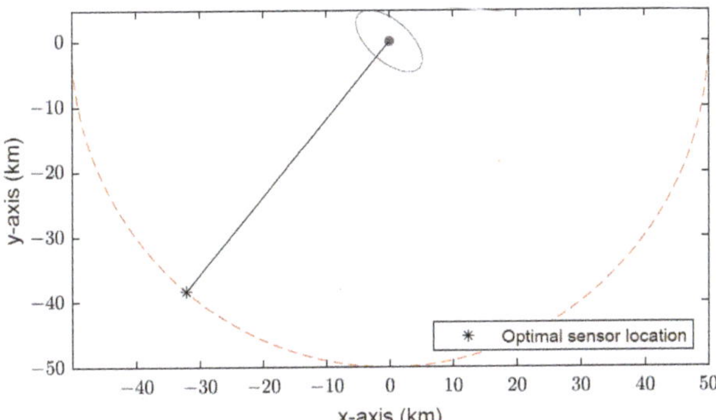

Figure 8. Optimal target–sensor geometry, where the target–sensor range vector is perfectly aligned with the minor axis of the error ellipse of the Gaussian prior covariance.

In the next set of simulations, we consider sensor trajectory optimization in a bearings-only manoeuvring target tracking problem. The algorithms in (51), (55) and Algorithm 1 are simulated for a single realization of target manoeuvres. The process noise parameter for the target is $q_x = q_y = 10^{-4}$ m^2/s^4. The bearing angle noise is assumed to be $\sigma = 4°$. The initial target dynamics are

$$x_{0|-1} = [0, 0, 0, 0]^T \tag{57}$$

and

$$P_{0|-1} = \begin{bmatrix} 19.7500 & 0 & 9.0933 & 0 \\ 0 & 0 & 0 & 0 \\ 9.0933 & 0 & 9.2500 & 0 \\ 0 & 0 & 0 & 0 \end{bmatrix}. \tag{58}$$

The time interval between successive EKF recursions is set equal to $T = 10$ s. The sensor is initially located at $s_0 = [-26.8116, -22.4976]$ km and moves with a constant speed of 25 m/s (90 km/h). The maximum turn-rate for the sensor is $30°$ per 10 s. The separation between successive sensor waypoints is $s = 0.25$ km. For the sensor trajectory optimization methods based on the D- and A-optimality criteria, 10 uniformly spaced points are used for numerical search over sensor waypoints in the set $\mathcal{S}_k$.

The simulated sensor trajectory for the D-optimality criterion in (51) is depicted in Figure 9. The 2-σ error ellipses for predicted target location estimates are plotted every 50 time instants. The initial error ellipse is drawn in black and all others are in grey. The optimal sensor trajectory achieves a rapid reduction in the size of error ellipses. Following the initial approach, the sensor chases the target by circling around it.

Figure 10 shows the simulated sensor trajectory for the A-optimality criterion in (55). The optimal sensor trajectory has a markedly different behaviour to that observed for the D-optimality criterion (see Figure 9) in that it seems to favour circling the target more than getting close to it initially. As a consequence, it takes longer to achieve a significant reduction in error ellipses than the A-optimality criterion.

The simulation results for the projection algorithm in Algorithm 1 are shown in Figure 11. The sensor follows a more direct route towards the target than both the D- and A-optimality methods, followed by circling manoeuvres. This is expected to produce faster convergence to the minimum estimation error than the D- and A-optimality methods at the expense of somewhat larger estimation error initially. This observation is confirmed by the root-mean-square error (RMSE) of target location estimates shown in Figure 12, which were computed from 5000 Monte Carlo simulation runs.

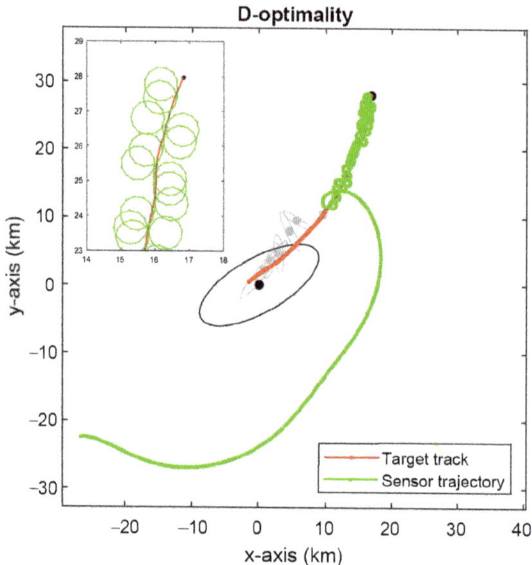

Figure 9. Optimal sensor trajectory for D-optimality criterion.

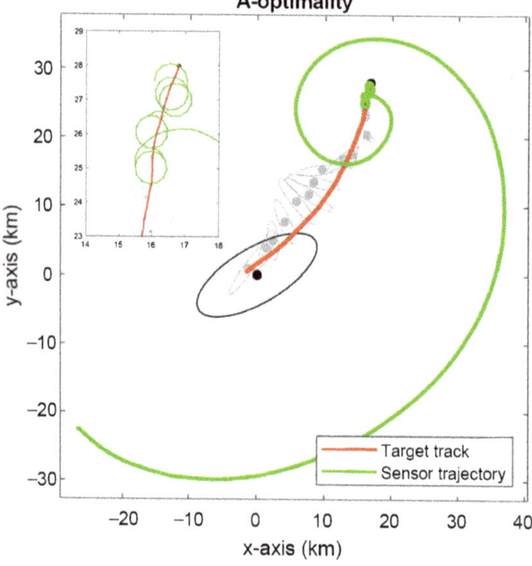

Figure 10. Optimal sensor trajectory for A-optimality criterion.

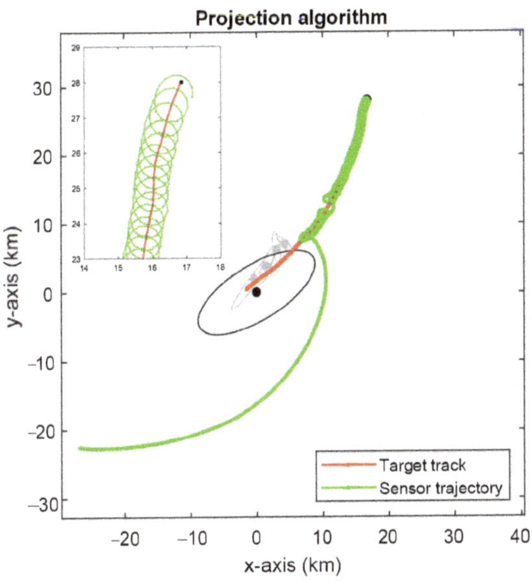

Figure 11. Optimal sensor trajectory using the projection algorithm.

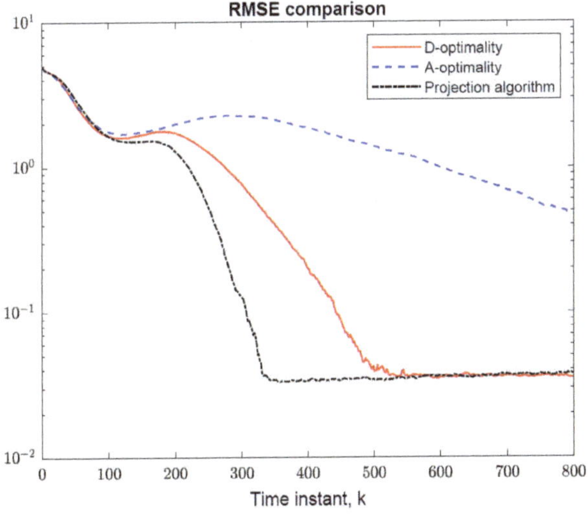

Figure 12. RMSE comparison of sensor trajectory optimization methods. The projection algorithm minimizes target tracking error fastest, followed by the D-optimality method.

5. Conclusions

Optimal sensor placement for the Bayesian AOA target localization problem was considered. The concept of updating prior information with data from measurements was extended to the Kalman filter tracking problem. Using an approximation of the BFIM, the optimal sensor placement for a given Gaussian prior was shown to be aligned with the minor axis of the prior error ellipse for both D- and A-optimality criteria. By way of simulations, this result was shown to match the optimal sensor placement result for the exact BFIM, which was numerically calculated using Monte Carlo simulations. The D- and A-optimality criteria were adopted for optimal sensor guidance in target tracking applications. Simple methods requiring a numerical search over sensor heading were developed and demonstrated in simulations. A new method, called the projection algorithm, was also developed predicated on the optimal sensor placement result for the D- and A-optimality criteria. The efficacy of the projection algorithm in achieving fast minimization of tracking error was confirmed by numerical simulations. Even though the D- and A-optimality criteria share the same optimal sensor placement result for a stationary target, they generate quite different sensor trajectories in a target tracking setting where optimal target–sensor geometries cannot be realized instantaneously because of constrained sensor dynamics. The A-optimality method was observed to favour circular motion around the target initially. This results in delays in minimizing the tracking error to a desirable level.

Funding: This research received no external funding.

Institutional Review Board Statement: Not applicable.

Informed Consent Statement: Not applicable.

Data Availability Statement: Not applicable.

Conflicts of Interest: The author declares no conflict of interest.

References

1. Aidala, V.J. Kalman filter behavior in bearings-only tracking applications. *IEEE Trans. Aerosp. Electron. Syst.* **1979**, *15*, 29–39. [CrossRef]
2. Nardone, S.C.; Aidala, V.J. Observability criteria for bearings-only target motion analysis. *IEEE Trans. Aerosp. Electron. Syst.* **1981**, *17*, 162–166. [CrossRef]
3. Nardone, S.C.; Lindgren, A.G.; Gong, K.F. Fundamental properties and performance of conventional bearings-only target motion analysis. *IEEE Trans. Autom. Control.* **1984**, *29*, 775–787. [CrossRef]
4. Anderson, B.D.O.; Moore, J.B. *Optimal Filtering*; Prentice Hall: Hoboke, NJ, USA, 1979.
5. Passerieux, J.M.; Van Cappel, D. Optimal observer maneuver for bearings-only tracking. *IEEE Trans. Aerosp. Electron. Syst.* **1998**, *34*, 777–788. [CrossRef]
6. Tremois, O.; Le Cadre, J.P. Optimal observer trajectory in bearings-only tracking for manoeuvring sources. *IEE Proc. Radar Sonar Navig.* **1999**, *146*, 31–39. [CrossRef]
7. Oshman, Y.; Davidson, P. Optimization of observer trajectories for bearings-only target localization. *IEEE Trans. Aerosp. Electron. Syst.* **1999**, *35*, 892–902. [CrossRef]
8. Farina, A. Target tracking with bearings-only measurements. *Signal Process.* **1999**, *78*, 61–78. [CrossRef]
9. Hernandez, M.L. Optimal sensor trajectories in bearings-only tracking. In Proceedings of the Seventh International Conference on Information Fusion (FUSION 2004), Stockholm, Sweden, 28 June–1 July 2004; Volume 2, pp. 893–900.
10. Tichavský, P.; Muravchik, C.H.; Nehorai, A. Posterior Cramér-Rao bounds for discrete-time nonlinear filtering. *IEEE Trans. Signal Process.* **1998**, *46*, 1386–1396. [CrossRef]
11. Doğançay, K.; Hmam, H. Optimal angular sensor separation for AOA localization. *Signal Process.* **2008**, *88*, 1248–1260. [CrossRef]
12. Bishop, A.N.; Fidan, B.; Anderson, B.D.; Dogancay, K.; Pathirana, P.N. Optimality analysis of sensor-target localization geometries. *Automatica* **2010**, *46*, 479–492. [CrossRef]
13. Xu, S.; Dogancay, K. Optimal Sensor Placement for 3-D Angle-of-Arrival Target Localization. *IEEE Trans. Aerosp. Electron. Syst.* **2017**, *53*, 1196–1211. [CrossRef]
14. Dogancay, K. Single- and multi-platform constrained sensor path optimization for angle-of-arrival target tracking. In Proceedings of the 2010 18th European Signal Processing Conference, Aalborg, Denmark, 23–27 August 2010; pp. 835–839.
15. Dogancay, K. UAV Path Planning for Passive Emitter Localization. *IEEE Trans. Aerosp. Electron. Syst.* **2012**, *48*, 1150–1166. [CrossRef]
16. Zhao, S.; Chen, B.M.; Lee, T.H. Optimal sensor placement for target localisation and tracking in 2D and 3D. *Int. J. Control.* **2013**, *86*, 1687–1704. [CrossRef]
17. Yang, C.; Kaplan, L.; Blasch, E.; Bakich, M. Optimal Placement of Heterogeneous Sensors for Targets with Gaussian Priors. *IEEE Trans. Aerosp. Electron. Syst.* **2013**, *49*, 1637–1653. [CrossRef]
18. Yang, C.; Kaplan, L.; Blasch, E. Performance Measures of Covariance and Information Matrices in Resource Management for Target State Estimation. *IEEE Trans. Aerosp. Electron. Syst.* **2012**, *48*, 2594–2613. [CrossRef]
19. Nguyen, N.H. Optimal Geometry Analysis for Target Localization With Bayesian Priors. *IEEE Access* **2021**, *9*, 33419–33437. [CrossRef]
20. Zhou, R.; Chen, J.; Tan, W.; Yan, Q.; Cai, C. Optimal 3D Angle of Arrival Sensor Placement with Gaussian Priors. *Entropy* **2021**, *23*, 1379. [CrossRef] [PubMed]

21. Fedorov, V.V. *Theory of Optimal Experiments*; Academic Press: New York, NY, USA, 1972.
22. Goodwin, G.C.; Payne, R.L. *Dynamic System Identification: Experiment Design and Data Analysis*; Academic Press: New York, NY, USA, 1977.
23. Uciński, D. *Optimal Measurement Methods for Distributed Parameter System Identification*; CRC Press: Boca Radon, FL, USA, 2005.
24. Van Trees, H.L.; Bell, K.L.; Tian, Z. *Detection, Estimation, and Modulation Theory, Part I: Detection, Estimation, and Filtering Theory*, 2nd ed.; Wiley: Hoboken, NJ, USA, 2013.
25. Balestra, P. *Determinant and Inverse of a Sum of Matrices with Applications in Economics and Statistics*; Technical Report hal-01527161; Laboratoire d'Analyse et de Techniques Economiques (LATEC), Institut de Mathematiques Economiques, Universite de Dijon: Dijon, France, 1978.
26. Hager, W.W. Minimizing a Quadratic Over a Sphere. *SIAM J. Optim.* **2001**, *12*, 188–208. [CrossRef]
27. Sayed, A.H. *Fundamentals of Adaptive Filtering*; Wiley: Hoboken, NJ, USA, 2003.
28. Bar-Shalom, Y.; Blair, W.D. (Eds.) *Multitarget-Multisensor Tracking: Applications and Advances*; Artech House: Boston, MA, USA, 2000; Volume III.

Article

Signal Source Positioning Based on Angle-Only Measurements in Passive Sensor Networks

Yidi Chen [1,†], **Linhai Wang** [2,†], **Shenghua Zhou** [2,*] and **Renwen Chen** [1]

[1] State Key Laboratory of Mechanics and Control of Mechanical Structures, Nanjing University of Aeronautics and Astronautics, No. 29, Yudao Street, Nanjing 210016, China; 18001296740@189.cn (Y.C.); rwchen@nuaa.edu.cn (R.C.)

[2] National Laboratory of Radar Signal Processing, Xidian University, No. 2 Taibai Road, Xi'an 710071, China; xd2020lhwang@163.com

[*] Correspondence: shzhou@mail.xidian.edu.cn; Tel.: +86-29-88201220

[†] The authors contributed equally to this work.

Abstract: Some passive sensors can measure only directions of arrival of signals, but the real positions of signal sources are often desirable, which can be estimated by combining distributed passive sensors as a network. However, passive observations should be correctly associated first. This paper studies the multi-target data association and signal localization problem in distributed passive sensor networks. With angle-only measurements from distributed passive sensors, multiple lines in a 3-dimensional (3D) scenario can be built and then those that will intersect in a small volume in 3D are classified into the same source. The center of the small volume is taken as an estimate of the signal source position, whose statistical distributions are formulated. If the minimum distance is less than an association threshold, then two lines are considered to be from the same signal source. In numerical results, the impacts of angle measurement accuracy and platform self-positioning accuracy are analyzed, indicating that this method can achieve a prescribed data association rate and a high positioning performance with a low computation cost.

Keywords: passive sensor network; signal localization; data association; angle-only measurements; accuracy analysis

Citation: Chen, Y.; Wang, L.; Zhou, S.; Chen, R. Signal Source Positioning Based on Angle-Only Measurements in Passive Sensor Networks. *Sensors* **2022**, *22*, 1554. https://doi.org/10.3390/s22041554

Academic Editors: Ratnasingham Tharmarasa and Mahendra Mallick

Received: 10 January 2022
Accepted: 14 February 2022
Published: 17 February 2022

Publisher's Note: MDPI stays neutral with regard to jurisdictional claims in published maps and institutional affiliations.

1. Introduction

Unlike active sensors such as radars, passive sensors do not transmit signals and thus have no anti-jamming problem [1,2]. However, some passive sensors, such as infrared sensors, photoelectric sensors, electronic counter measurement (ECM) and cameras, can estimate only angles of signal sources. Therefore, their signal source positioning performances are typically poor since they have no accurate range information of signal sources. In order to estimate the positions of signal sources, passive sensors with angle-only observations can be connected with communication links into a network to measure signals sources from different spatial locations. In this case, an algorithm to combine the angle-only observations is needed [3–7]. Compared with the time of arrival (TOA) [8] and the time difference of arrival (TDOA) localization , angle-only localization does not require accurate time synchronization between distributed passive sensors for signal sources with low speeds [6].

In a passive sensor network, there may be multiple signal sources, and before accurate localization, one should first correctly associate observations regarding the same sources [9]. The multi-dimension assignment model is a classical method for data association in passive sensor networks [10–12], but it needs the locations of signal sources, which is unavailable before correct data association. A geometry-based localization algorithm for a distributed sensor network is presented in [13], which constructs a test statistic based on the minimal distance between the lines of sight for data association. The measurement errors are considered, but the platform self-positioning errors are not considered.

In this paper, how to perform data association and signal source localization in a 3-dimensional (3D) scenario for distributed passive sensors with only angle measurements is studied. We improve the intersection localization algorithm for passive sensor networks in a multi-target scenario. We first consider the data association problem and then the signal source position estimation problem. The basic concept is to construct a set of lines in a 3D scenario according to angle measurements of signal sources. In data association, measurement lines that will intersect within a small space volume are categorized into the same group. The statistical distribution of the minimal distances of the lines are formulated and the minimum distance between any two observation lines is a random variable, proved to follow the Chi-square distribution. The threshold for correct association is formulated by the misassociation probability; namely, two observations are from the same signal source but are classified into two groups. In the test statistics, not only the measurement errors but also the platform positioning errors are considered, which makes the association performance robust when the platform positioning errors exist.

After data association, observations regarding the same signal sources are grouped, based on which the location of signal sources can be estimated. Three positioning algorithms are considered. It is known that angle measurements are nonlinear functions of coordinates of signal sources. With the Taylor expansion, the least square (LS) algorithm linearizes the nonlinear angle measurements about the target position and then uses the LS method to obtain the target position estimate [6,14–19]. In real applications, different passive sensors may obtain observations of different signal-to-noise ratios (SNRs) and then the weighted least squares algorithm (WLS) [6,14] and total least square (TLS) algorithm [20] can be used to obtain a better estimate. Another source location method is the intersection localization algorithm [1,4,5,21–23]. The basic concept is that if multiple passive sensors simultaneously measure the signal sources without measurement error, these measurement lines of sight will intersect to the target position. The geometric method and algebraic solution method use this property to estimate the positions of signal sources.

The data association process and target location process of this method are closely combined, which ensures a lower algorithm complexity and a better positioning performance. In numerical results, the improvement of data association and signal-source positioning are analyzed. The impact of the target-sensor geometry on the localization accuracy is also studied, indicating that the localization performance will be better if the lines associated with different observations are perpendicular to each other.

We follow the convention that bold lower and upper case letters denote column vectors and matrices, respectively. A symbol with an upper script o denotes the true value. For instance, $\mathbf{a}^o$ denotes the true value of $\mathbf{a}$. $\mathrm{diag}(\cdot)$ with a vector entry denotes a diagonal matrix with the entry vector as diagonal elements. The notation $\mathrm{diag}(\mathbf{A}_1, \mathbf{A}_2, \ldots, \mathbf{A}_N)$ stands for the block-diagonal matrix formed by the matrices $\mathbf{A}_1, \mathbf{A}_2, \ldots, \mathbf{A}_N$.

2. Localization with Angle-Only Passive Sensors

2.1. Signal Model of Passive Observations

Consider a passive sensor network with N widely separated sensors and M targets in the surveillance volume. Assume that a coordinate system is available for all the sensors, such as earth-centered earth-fixed (ECEF) of the World Geodetic System 84 (WGS84). The real position of the nth sensor at instant t is denoted as $\mathbf{s}_n^o(t) = [x_{n,s}^o(t), y_{n,s}^o(t), z_{n,s}^o(t)]^{\mathrm{T}}$, $n = 1, 2, \ldots, N$, where $(\cdot)^{\mathrm{T}}$ denotes the transpose operation, t denotes time, and $x_{n,s}^o(t), y_{n,s}^o(t), z_{n,s}^o(t)$ denote the x, y, z coordinates of the nth sensor at instant t, respectively. The real position of the mth target at instant t is denoted by $\mathbf{g}_m^o(t) = [x_{m,g}^o(t), y_{m,g}^o(t), z_{m,g}^o(t)]^{\mathrm{T}}$, $m = 1, \cdots, M$, where $x_{m,g}^o(t), y_{m,g}^o(t), z_{m,g}^o(t)$ denote the x, y, z coordinates of the mth target at instant t, respectively. Assume that there is a self-positioning device to measure the position of each sensor. There are different kinds of instruments that can measure the position of a platform, such as the Global Positioning System (GPS) and inertial sensors. The topology of the passive sensors and targets are shown in Figure 1.

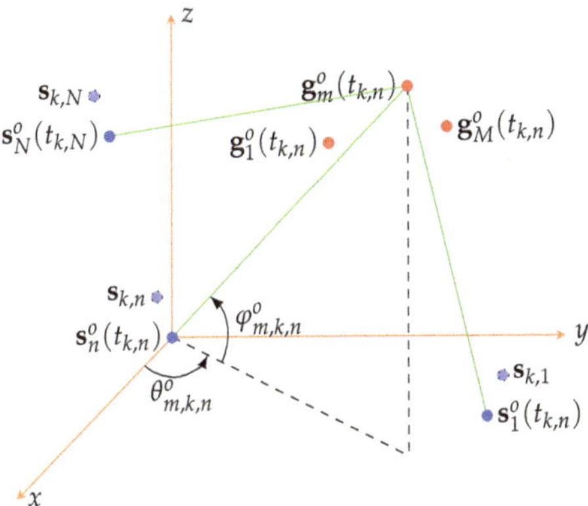

Figure 1. Measurement scenario of the passive sensors.

For the nth sensor, signals are detected at instants denoted by $t_{k,n}, k = 1, \cdots, N_n$, where N_n denotes the number of observations of the nth sensor. At instant $t_{k,n}$, assume that the position of the nth sensors is measured as

$$\mathbf{s}_{k,n} = \mathbf{s}_n^o(t_{k,n}) + \Delta\mathbf{s}_n(t_{k,n}) = [x_{n,s}(t_{k,n}), y_{n,s}(t_{k,n}), z_{n,s}(t_{k,n})]^T, k = 1, \cdots, N_n \tag{1}$$

where $\Delta\mathbf{s}_n(t_{k,n})$ denotes the sensor self-positioning error. For simplicity, we assume that the positioning errors follow zero mean Gaussian distributions with covariance matrices $\mathbf{C}_{k,n} = \mathbb{E}(\Delta\mathbf{s}_n(t_{k,n})\Delta\mathbf{s}_n^T(t_{k,n}))$, where $\mathbb{E}$ denotes the expectation operation. In practice, the self-positioning error of the sensor is approximately subject to zero-mean Gaussian distribution, so this assumption is reasonable and widely used.

For the angle-only sensors, the observations are directions of the signals and the lth signal at instant $t_{k,n}$ is denoted by $\boldsymbol{\theta}_{l,k,n} = [\theta_{l,k,n}, \varphi_{l,k,n}]^T$, where $(l, k, n) \in \mathcal{M}_{k,n}$, $k = 1, \cdots, N_n$ and $n = 1, \cdots, N$, and $\mathcal{M}_{k,n}$ denotes a set of triples of signals detected at the kth measurement by the nth sensor. Assume that $|\mathcal{M}_{k,n}| = M_{k,n}$, where $|\cdot|$ over a set denotes the cardinality of the set. As the existence of miss detection, false alarms and overlapping of signal sources, $M_{k,n}$ may not be equal to M. Denote $\mathcal{M}_n = \cup_{k=1}^{N_n}\mathcal{M}_{k,n}$ and $\mathcal{A} = \cup_{n=1}^{N}\mathcal{M}_n$, where $\cup$ denotes the union operation. The total number of observations by N sensors is denoted by

$$N_s = |\mathcal{A}| = \sum_{n=1}^{N}\sum_{k=1}^{N_n} M_{k,n}. \tag{2}$$

At instant $t_{k,n}$, the real position of the mth signal source is denoted by

$$\mathbf{g}_m^o(t_{k,n}) = [x_{m,g}(t_{k,n}), y_{m,g}(t_{k,n}), z_{m,g}(t_{k,n})]^T, m = 1, \cdots, M. \tag{3}$$

For the mth signal source, the real azimuth angle and elevation angle regarding the nth sensor can be expressed by

$$\theta_{m,k,n}^o = \tan^{-1}\left(y_{m,g}^o(t_{k,n}) - y_{n,s}^o(t_{k,n}), x_{m,g}^o(t_{k,n}) - x_{n,s}^o(t_{k,n})\right) \tag{4}$$

$$\varphi_{m,k,n}^o = \arctan\left((z_{m,g}^o(t_{k,n}) - z_{n,s}^o(t_{k,n})) / \sqrt{(x_{m,g}^o(t_{k,n}) - x_{n,s}^o(t_{k,n}))^2 + (y_{m,g}^o(t_{k,n}) - y_{n,s}^o(t_{k,n}))^2}\right)$$

respectively, where $\theta_{m,k,n}^o \in (-\pi, \pi)$, $\varphi_{m,k,n}^o \in (-\frac{\pi}{2}, \frac{\pi}{2})$, $\tan^{-1}(*)$ is called the two-argument inverse tangent function [24,25] and $\arctan(*)$ is the inverse tangent function. Denote $\boldsymbol{\theta}_{m,k,n}^o = [\theta_{m,k,n}^o, \varphi_{m,k,n}^o]^T$.

Each observation is associated with one of M targets or the false alarm indexed by 0, represented by a set $\mathcal{M} = \{0, 1, \cdots, M\}$. It can be considered as a mapping $\psi : \mathcal{A} \to \mathcal{M}$.

According to our setting, the index set $\mathcal{A}$ can be partitioned into $M+1$ disjoint sets $\mathcal{A}_0, \mathcal{A}_1, \cdots, \mathcal{A}_M$ defined by

$$\mathcal{A}_i = \{x \mid \psi(x) = i, x \in \mathcal{A}\}, \tag{5}$$

where $\mathcal{A}_0$ denotes the index of observations corresponding to false alarms, and $\mathcal{A}_m$ denotes the index set of observations from the mth signal source. As a partition of $\mathcal{A}$, we have $\mathcal{A}_i \cap \mathcal{A}_j = \varnothing, i, j \in \mathcal{M}, i \neq j$, and $\mathcal{A} = \cup_{i=0}^{M} \mathcal{A}_i$, where $\cap$ denotes the intersection operation of sets.

If the mth signal source is detected and indexed as the (l, k, n) observation, then the azimuth angle and elevation angle measurements can be written as

$$\boldsymbol{\theta}_{l,k,n} = [\theta_{l,k,n}, \varphi_{l,k,n}]^{\mathrm{T}} \tag{6}$$

$$= \boldsymbol{\theta}_{m,k,n}^{o} + \Delta\boldsymbol{\theta}_{l,k,n} \tag{7}$$

$$\theta_{l,k,n} = \theta_{m,k,n}^{o} + \Delta\theta_{l,k,n} \tag{8}$$

$$\varphi_{l,k,n} = \varphi_{m,k,n}^{o} + \Delta\varphi_{l,k,n} \tag{9}$$

$$\Delta\boldsymbol{\theta}_{l,k,n} = [\Delta\theta_{l,k,n}, \Delta\theta_{l,k,n}]^{\mathrm{T}} \tag{10}$$

where $\Delta\theta_{l,k,n}$ and $\Delta\varphi_{l,k,n}$ represent the measurement noise of the azimuth angle and elevation angle, respectively. For simplicity, we assume that observation noises $\Delta\theta_{l,k,n}$ and $\Delta\varphi_{l,k,n}, l = 1, \cdots, M_{k,n}, k = 1, \cdots, N_n, n = 1, \cdots, N$ are statistically independent and follow zero-mean Gaussian distribution by assumption. The covariance matrices of $\Delta\boldsymbol{\theta}$ are denoted by $\mathbf{C}_{l,k,n} = \mathbb{E}(\Delta\boldsymbol{\theta}\Delta\boldsymbol{\theta}^{\mathrm{T}}) \in \mathbb{C}^{2\times2}$, namely $\Delta\boldsymbol{\theta} \sim \mathcal{N}(\mathbf{0}, \mathbf{C}_{l,k,n})$, which is typically affected by the SNR of the signal, where $\mathcal{N}(\mathbf{0}, \mathbf{C}_{l,k,n})$ denotes the zero-mean Gaussian distribution with mean $\mathbf{0}$ and covariance matrix $\mathbf{C}_{l,k,n}$. It should be noted that modeling the angle measurement noise as a zero-mean Gaussian distribution is a commonplace assumption.

2.2. The Distance of Observation Lines

The angle-only observations provide information on the directions of signal sources, and in theory, the directions can be expressed by 3D lines. Therefore, it is important to study the properties of the lines. Consider a simple scenario where the target can be deemed to be static when the observations are recorded and the location of the mth signal is simplify denoted by $\mathbf{g}_m^o$. For passive observations, each observation, say (l, k, n), will contribute a line in space and the line can be written as $L_{l,k,n}$

$$L_{l,k,n} : \mathbf{x} = \mathbf{s}_{k,n} + \alpha_{l,k,n}\mathbf{e}_{l,k,n}, \; \alpha_{l,k,n} \in \mathbb{R} \tag{11}$$

where $\alpha_{l,k,n}$, a parameter indicating the distance to the origin $\mathbf{s}_{k,n}$, $\mathbf{e}_{n,m} = [e_{l,k,n,x}, e_{l,k,n,y}, e_{l,k,n,z}]^{\mathrm{T}} \in \mathbb{R}^{3\times1}$, is the normalized direction vector associated with the angle observation $\boldsymbol{\theta}_{l,k,n}$, and

$$e_{l,k,n,x} = \cos(\theta_{l,k,n})\cos(\varphi_{l,k,n}) \tag{12}$$

$$e_{l,k,n,y} = \sin(\theta_{l,k,n})\cos(\varphi_{l,k,n}) \tag{13}$$

$$e_{l,k,n,y} = \sin(\varphi_{l,k,n}). \tag{14}$$

It can be seen that the subscripts of the denotations are complicated. Therefore, for two observations $i, j \in \mathcal{A}_m$, we simplify the expression of lines by

$$L_{i,m} : \mathbf{x} = \mathbf{s}_i + \alpha_{i,m}\mathbf{e}_{i,m} \tag{15}$$

$$L_{j,m} : \mathbf{x}' = \mathbf{s}_j + \alpha_{j,m}\mathbf{e}_{j,m} \tag{16}$$

where the locations of two sensors regarding the two observations are denoted by $\mathbf{s}_i, \mathbf{s}_j$, $\alpha_{i,m}, \alpha_{j,m}$ are two scalars indicating the distances to two origins, and $\mathbf{e}_{i,m}$ and $\mathbf{e}_{j,m}$ are two normalized vectors associated with two observations.

The difference between two points over the two lines are

$$\Rightarrow \quad \mathbf{x}' - \mathbf{x} = \mathbf{s}_j - \mathbf{s}_i + \alpha_{j,m}\mathbf{e}_{j,m} - \alpha_{i,m}\mathbf{e}_{i,m} \tag{17}$$

$$= \mathbf{s}_j - \mathbf{s}_i + (-\mathbf{e}_{i,m}, \mathbf{e}_{j,m})\boldsymbol{\alpha}, \tag{18}$$

where

$$\boldsymbol{\alpha} = \begin{bmatrix} \alpha_{i,m} \\ \alpha_{j,m} \end{bmatrix}. \tag{19}$$

Without measurement error, then there will be two $\alpha_{i,m}$ and $\alpha_{j,m}$ such that $\mathbf{x}' - \mathbf{x} = 0$, i.e.,

$$\mathbf{g}_m = \mathbf{s}_i + \alpha_{i,m}\mathbf{e}_{i,m} = \mathbf{s}_j + \alpha_{j,m}\mathbf{e}_{j,m}. \tag{20}$$

In general, due to inevitable measurement errors, the lines even regarding the same signal source may not coincide to each other. Therefore, we calculate minimal distance between those lines. The distance between two points over two lines can be expressed by

$$\begin{aligned} d &= \|\mathbf{x}' - \mathbf{x}\|^2 \\ &= \|\mathbf{s}_j - \mathbf{s}_i + (-\mathbf{e}_{i,m}, \mathbf{e}_{j,m})\boldsymbol{\alpha}\|^2 \\ &= \boldsymbol{\alpha}^{\mathrm{T}}\mathbf{E}_{i,j}^{\mathrm{T}}\mathbf{E}_{i,j}\boldsymbol{\alpha} + 2\boldsymbol{\alpha}^{\mathrm{T}}\mathbf{E}_{i,j}^{\mathrm{T}}(\mathbf{s}_j - \mathbf{s}_i) + (\mathbf{s}_j - \mathbf{s}_i)^{\mathrm{T}}(\mathbf{s}_j - \mathbf{s}_i) \end{aligned} \tag{21}$$

where $\|\cdot\|$ denotes the ℓ_2-norm, $\boldsymbol{\alpha} = [\alpha_{i,m}, \alpha_{j,m}]^{\mathrm{T}}$, and $\mathbf{E}_{i,j} = [-\mathbf{e}_{i,m}, \mathbf{e}_{j,m}]$. For simplicity, the distance d is actually the squared distance, instead of the distance.

In particular, if $\mathbf{e}_{i,m} = \mathbf{e}_{j,m}$, namely two lines are parallel to each other, then

$$d = \|\mathbf{s}_j - \mathbf{s}_i + (\alpha_{j,m} - \alpha_{i,m})\mathbf{e}_{i,m}\|^2, \tag{22}$$

and the minimal distance between points in two lines will be achieved if

$$\alpha_{j,m} - \alpha_{i,m} = \mathbf{e}_{i,m}^{\mathrm{T}}(\mathbf{s}_i - \mathbf{s}_j). \tag{23}$$

It can be proved that the minimal distance is

$$d = \|(\mathbf{e}_{i,m}\mathbf{e}_{i,m}^{\mathrm{T}} - \mathbf{I})(\mathbf{s}_i - \mathbf{s}_j)\|^2. \tag{24}$$

If $\mathbf{e}_{i,m} \neq \mathbf{e}_{j,m}$ or $|\mathbf{E}_{i,j}| \neq 0$, the distance is a second order function of $\alpha_{i,m}$ and $\alpha_{j,m}$, thus the minimal value of d is unique, where $|\cdot|$ over a matrix denotes the determinant of the input matrix. To obtain the minimal value, we take a derivative of d with respect to $\boldsymbol{\alpha}$,

$$\frac{\mathrm{d}d}{\mathrm{d}\boldsymbol{\alpha}} = \left[\frac{\mathrm{d}d}{\mathrm{d}\alpha_{i,m}}, \frac{\mathrm{d}d}{\mathrm{d}\alpha_{j,m}}\right]^{\mathrm{T}} \tag{25}$$

$$= 2\mathbf{E}_{i,j}^{\mathrm{T}}\mathbf{E}_{i,j}\boldsymbol{\alpha} + 2\mathbf{E}_{i,j}^{\mathrm{T}}(\mathbf{s}_j - \mathbf{s}_i) \tag{26}$$

where

$$\frac{\mathrm{d}d}{\mathrm{d}\alpha_{i,m}} = -2\mathbf{e}_{i,m}^{\mathrm{T}}(\mathbf{s}_j - \mathbf{s}_i + \mathbf{E}\boldsymbol{\alpha}) \tag{27}$$

$$\frac{\mathrm{d}d}{\mathrm{d}\alpha_{j,m}} = 2\mathbf{e}_{j,m}^{\mathrm{T}}(\mathbf{s}_j - \mathbf{s}_i + \mathbf{E}\boldsymbol{\alpha}). \tag{28}$$

Let the derivative be zero and then we obtain a solution

$$\frac{\mathrm{d}d}{\mathrm{d}\boldsymbol{\alpha}} = 0 \Rightarrow \quad \mathbf{E}_{i,j}^{\mathrm{T}}\mathbf{E}_{i,j}\boldsymbol{\alpha} = -\mathbf{E}_{i,j}^{\mathrm{T}}(\mathbf{s}_j - \mathbf{s}_i). \tag{29}$$

Under the assumption that $|\mathbf{E}_{i,j}| \neq 0$, $\mathbf{R}_{i,j} = \mathbf{E}_{i,j}^{\mathrm{T}}\mathbf{E}_{i,j}$ has a reverse matrix and then the solution can be immediately obtained by

$$\boldsymbol{\alpha}_{\mathrm{opt}} = -(\mathbf{E}_{i,j}^{\mathrm{T}}\mathbf{E}_{i,j})^{-1}\mathbf{E}_{i,j}^{\mathrm{T}}(\mathbf{s}_j - \mathbf{s}_i). \tag{30}$$

The minimal distance can be expressed by

$$d_{\min} = (\mathbf{s}_j - \mathbf{s}_i)^{\mathrm{T}}(\mathbf{I} - \mathbf{E}_{i,j}\mathbf{R}_{i,j}^{-1}\mathbf{E}_{i,j}^{\mathrm{T}})(\mathbf{s}_j - \mathbf{s}_i) \tag{31}$$

$$= \|(\mathbf{I} - \mathbf{E}_{i,j}\mathbf{R}_{i,j}^{-1}\mathbf{E}_{i,j}^{\mathrm{T}})(\mathbf{s}_j - \mathbf{s}_i)\|^2. \tag{32}$$

In particular, if

$$\mathbf{E}_{i,j}\mathbf{R}_{i,j}^{-1}\mathbf{E}_{i,j}^{\mathrm{T}}(\mathbf{s}_j - \mathbf{s}_i) = \mathbf{s}_j - \mathbf{s}_i, \tag{33}$$

namely, $\mathbf{s}_j - \mathbf{s}_i$ is an eigenvector of $\mathbf{E}_{i,j}\mathbf{R}_{i,j}^{-1}\mathbf{E}_{i,j}^{\mathrm{T}}$ and the eigenvalue is 1, then

$$d_{\min} = 0. \tag{34}$$

It means that two lines will intersect at a point. Without measurement errors, observations regarding the same target will form lines intersecting at the target position.

In practice, the real mapping ψ should be estimated through an association algorithm. Due to measurement errors, the estimated mapping may not be correct, and then, the positioning error may raise. Therefore, an accurate data association method is important, which will be studied subsequently.

3. Data Association Based on Minimal Distance

3.1. Data Association Model

In order to perform data association, we first need to build the statistical model of the minimal distance of the observation lines. Because of both the platform positioning error $\Delta \mathbf{s}_i, \Delta \mathbf{s}_j$ and the angle measurement error $\Delta \theta_{i,m}$ and $\Delta \theta_{j,m}, i, j \in \mathcal{A}_m$, the minimal distance $d_{\min}$ is not zero and follows a certain distribution depending on the measurement errors, where $\Delta \mathbf{s}_i, \Delta \mathbf{s}_j$ denote the sensor self-positioning errors of the ith and the jth observations, respectively, and $\Delta \theta_{i,m}, \Delta \theta_{j,m}$ denote the angle measurement error of the ith and the jth observations on the mth target.

In practice, the ith and the jth observations may belong to the same target or not, and the data association problem is treated as a test to make a decision here. For $i, j \in \mathcal{A}$, the data association problem for a target can be formulated as the following hypothesis problem

$$\begin{cases} H_0 : d_{\min} = 0 & \psi(i) = \psi(j) \\ H_1 : d_{\min} > 0 & \psi(i) \neq \psi(j). \end{cases} \tag{35}$$

To determine the statistical distribution of $d_{\min}$ under the H_0 hypothesis, we first define

$$\mathbf{K}_{i,j} = \mathbf{I} - \mathbf{E}_{i,j}\mathbf{R}_{i,j}^{-1}\mathbf{E}_{i,j}^{\mathrm{T}} \tag{36}$$

and then

$$d_{\min} = \|\mathbf{K}_{i,j}(\mathbf{s}_j - \mathbf{s}_i)\|^2. \tag{37}$$

Next, we discuss the eigenvalues of $\mathbf{K}_{i,j}$. It can be proved that $\mathbf{K}_{i,j}^{\mathrm{T}}\mathbf{K}_{i,j} = \mathbf{K}_{i,j}$ and $\mathbf{K}_{i,j}^{\mathrm{T}} = \mathbf{K}_{i,j}$. Therefore, $\mathbf{K}_{i,j}$ is a positive semi-definite matrix and its possible eigenvalues are either 0 or 1. As $\mathbf{K}_{i,j}$ is not an all-zero matrix, then there is at least an eigenvalue of 1.

The trace of the matrix $\mathbf{K}(i,j)$ satisfies

$$\mathrm{tr}(\mathbf{K}_{i,j}) = \mathrm{tr}(\mathbf{I} - \mathbf{E}_{i,j}\mathbf{R}_{i,j}^{-1}\mathbf{E}_{i,j}^{\mathrm{T}}) \tag{38}$$

$$= \mathrm{tr}(\mathbf{I}) - \mathrm{tr}(\mathbf{E}_{i,j}\mathbf{R}_{i,j}^{-1}\mathbf{E}_{i,j}^{\mathrm{T}}) \tag{39}$$

$$= 3 - \mathrm{tr}(\mathbf{E}_{i,j}^{\mathrm{T}}\mathbf{E}_{i,j}\mathbf{R}_{i,j}^{-1}) \tag{40}$$

$$= 3 - \mathrm{tr}(\mathbf{R}_{i,j}\mathbf{R}_{i,j}^{-1}) \tag{41}$$

$$= 3 - \mathrm{tr}(\mathbf{I}_2 \in \mathbb{R}^{2\times 2}) = 1 \tag{42}$$

where $\mathrm{tr}(\cdot)$ with a matrix input denotes the trace of the matrix.

Therefore, it can be inferred that three eigvenvalues of $\mathbf{K}_{i,j}$ are $1, 0, 0$ and the rank of $\mathbf{K}_{i,j}$ is 1. In other words, $\mathbf{K}_{i,j}$ can be written as

$$\mathbf{K}_{i,j} = \mathbf{e}_s\mathbf{e}_s^{\mathrm{T}} \tag{43}$$

where $\mathbf{e}_s$ is a unity direction vector perpendicular to both $\mathbf{e}_i$ and $\mathbf{e}_j$, defined by

$$\mathbf{e}_s = \mathbf{e}_i \times \mathbf{e}_j / \sqrt{1 - (\mathbf{e}_i^{\mathrm{T}}\mathbf{e}_j)^2} \tag{44}$$

and $\mathbf{e}_i$ and $\mathbf{e}_j$ denote the unity direction vectors associated with the ith and the jth observations, respectively.

With this fact, the minimal distance can be rewritten as

$$d_{\min} = |\mathbf{e}_s^{\mathrm{T}}(\mathbf{s}_j - \mathbf{s}_i)|^2. \tag{45}$$

In practice, both $\mathbf{s}_j - \mathbf{s}_i$ and $\mathbf{e}_s$ may be inaccurate. A statistical distribution is necessary to determine the impact of the measurement errors. In order to formulate the statistical distribution of $d_{\min}$ under the H_0 hypothesis, we define

$$r = \mathbf{e}_s^{\mathrm{T}}(\mathbf{s}_j - \mathbf{s}_i) \tag{46}$$

and its relationship to $d_{\min}$ is $d_{\min} = |r|^2$.

The minimal distance varies with a total of 10 parameters; namely $\theta_{i,m}$, $\theta_{j,m}$, $\mathbf{s}_{i,m}$ and $\mathbf{s}_{j,m}$, and the corresponding partial derivatives can be written as

$$\frac{\partial r}{\partial \boldsymbol{\theta}_{i,m}} = \left[\frac{\partial r}{\partial \theta_{i,m}}, \frac{\partial r}{\partial \varphi_{i,m}}\right]^{\mathrm{T}} \tag{47}$$

$$\frac{\partial r}{\partial \boldsymbol{\theta}_{j,m}} = \left[\frac{\partial r}{\partial \theta_{j,m}}, \frac{\partial r}{\partial \varphi_{j,m}}\right]^{\mathrm{T}} \tag{48}$$

$$\frac{\partial r}{\partial \mathbf{s}_i} = -\mathbf{e}_s \tag{49}$$

$$\frac{\partial r}{\partial \mathbf{s}_j} = \mathbf{e}_s. \tag{50}$$

Consequently, under the H_0 hypothesis, with denotation $\mathbf{v} = [\boldsymbol{\theta}_{i,m}^{\mathrm{T}}, \boldsymbol{\theta}_{j,m}^{\mathrm{T}}, \mathbf{s}_{i,m}^{\mathrm{T}}, \mathbf{s}_{j,m}^{\mathrm{T}}]^{\mathrm{T}}$, we have an approximation

$$r \approx \frac{\partial r}{\partial \boldsymbol{\theta}_{i,m}^{\mathrm{T}}}\Delta\boldsymbol{\theta}_{i,m} + \frac{\partial r}{\partial \boldsymbol{\theta}_{j,m}^{\mathrm{T}}}\Delta\boldsymbol{\theta}_{j,m} + \frac{\partial r}{\partial \mathbf{s}_i^{\mathrm{T}}}\Delta\mathbf{s}_{i,m} + \frac{\partial r}{\partial \mathbf{s}_j^{\mathrm{T}}}\Delta\mathbf{s}_{j,m} \tag{51}$$

$$= \frac{\partial r}{\partial \mathbf{v}^{\mathrm{T}}}\Delta\mathbf{v} \tag{52}$$

where $\Delta\mathbf{v} = [\Delta\boldsymbol{\theta}_{i,m}^{\mathrm{T}}, \Delta\boldsymbol{\theta}_{j,m}^{\mathrm{T}}, \Delta\mathbf{s}_{i,m}^{\mathrm{T}}, \Delta\mathbf{s}_{j,m}^{\mathrm{T}}]^{\mathrm{T}}$, and

$$\frac{\partial r}{\partial \mathbf{v}^{\mathrm{T}}} = \left[\frac{\partial r}{\partial \boldsymbol{\theta}_{i,m}^{\mathrm{T}}}, \frac{\partial r}{\partial \boldsymbol{\theta}_{j,m}^{\mathrm{T}}}, \frac{\partial r}{\partial \mathbf{s}_i^{\mathrm{T}}}, \frac{\partial r}{\partial \mathbf{s}_j^{\mathrm{T}}}\right]. \tag{53}$$

Under the assumption that the measurement errors $\Delta\boldsymbol{\theta}_{i,m}, \Delta\boldsymbol{\theta}_{j,m}, \Delta\mathbf{s}_{i,m}, \Delta\mathbf{s}_{j,m}$ are statistically independent of each other and follow a zero-mean Gaussian distribution, r follows the Gaussian distribution and then $d_{\min}$ follows the central weighted Chi-square distribution with 1 degree of freedom, namely $d_{\min} \sim \chi_1^2(\lambda)$, where λ is the variance

$$\lambda = \mathbb{E}(\frac{\partial r}{\partial \mathbf{v}^{\mathrm{T}}}\Delta\mathbf{v}\Delta\mathbf{v}^{\mathrm{T}}\frac{\partial r}{\partial \mathbf{v}}) = \frac{\partial r}{\partial \mathbf{v}^{\mathrm{T}}}\mathbf{C}_{\mathbf{v}}\frac{\partial r}{\partial \mathbf{v}} \tag{54}$$

and

$$\mathbf{C}_{\mathbf{v}} = \mathbb{E}(\Delta\mathbf{v}\Delta\mathbf{v}^{\mathrm{T}}). \tag{55}$$

The probability density function (PDF) and cumulative distribution function (CDF) can be written as

$$p_{d_{\min}}(d) = \frac{1}{2\lambda\sqrt{\pi}}(\frac{d}{\lambda})^{-\frac{1}{2}}\exp(-\frac{d}{2\lambda}), d \geq 0 \tag{56}$$

$$F_{d_{\min}}(d) = \frac{1}{\sqrt{\pi}}\gamma(\frac{d}{2\lambda}, \frac{1}{2}), d \geq 0 \tag{57}$$

respectively, where $\Gamma(\cdot)$ denotes the Gamma function and $\gamma(\cdot, \cdot)$ denotes the incomplete Gamma function. If the decision rule is to keep the misassociation probability $P(H_1|H_0)$ as a constant, say p_f, then the decision threshold can be obtained as

$$\rho = 2\lambda\gamma^{-1}(\sqrt{\pi}(1 - p_f), \frac{1}{2}) \tag{58}$$

where $\gamma^{-1}(\cdot, \frac{1}{2})$ denotes the inverse incomplete Gamma function with $\frac{1}{2}$ a degree of freedom.

Therefore, the key is to derive the variance λ. In practice, as the measurements are carried out by different sensors, it is reasonable to assume that the measurement errors are statistically independent of each other. In this case, $\mathbf{C}_{\mathbf{v}}$ is a block diagonal matrix and the variance can be formulated conveniently.

3.2. Measurement Errors and Association Threshold

For simplicity, we first consider the self-positioning error $\Delta\mathbf{s}_i, \Delta\mathbf{s}_j$, whose covariance matrices are assumed to be

$$\mathbf{C}_{i,s} = \mathbb{E}(\Delta\mathbf{s}_i\Delta\mathbf{s}_i^{\mathrm{T}}) \tag{59}$$

$$\mathbf{C}_{j,s} = \mathbb{E}(\Delta\mathbf{s}_j\Delta\mathbf{s}_j^{\mathrm{T}}). \tag{60}$$

Consequently, from (49) and (50), the variances due to the two terms are

$$\lambda_{i,s} = \mathbb{E}(\frac{\partial r}{\partial \mathbf{s}_i}\Delta\mathbf{s}_{i,m}\Delta\mathbf{s}_{i,m}^{\mathrm{T}}(\frac{\partial r}{\partial \mathbf{s}_i})^{\mathrm{T}}) = \mathbf{e}_s^{\mathrm{T}}\mathbf{C}_{i,s}\mathbf{e}_s \tag{61}$$

$$\lambda_{j,s} = \mathbb{E}(\frac{\partial r}{\partial \mathbf{s}_j}\Delta\mathbf{s}_{j,m}\Delta\mathbf{s}_{j,m}^{\mathrm{T}}(\frac{\partial r}{\partial \mathbf{s}_j})^{\mathrm{T}}) = \mathbf{e}_s^{\mathrm{T}}\mathbf{C}_{j,s}\mathbf{e}_s. \tag{62}$$

Next, we consider the variance caused by the angle measurement error $\Delta\theta_{i,m}$ and $\Delta\theta_{j,m}$. Assume that the covariance matrices of $\Delta\theta_{i,m}$ and $\Delta\theta_{j,m}$ are

$$\mathbf{C}_{i,\theta} = \mathbb{E}(\Delta\theta_{i,m}\Delta\theta_{i,m}^{\mathrm{T}}) \tag{63}$$

$$\mathbf{C}_{j,\theta} = \mathbb{E}(\Delta\theta_{j,m}\Delta\theta_{j,m}^{\mathrm{T}}) \tag{64}$$

respectively. The variances caused by the two terms are

$$\lambda_{i,\theta} = \mathbb{E}\left(\frac{\partial r}{\partial\theta_{i,m}^{\mathrm{T}}}\Delta\theta_{i,m}\Delta\theta_{i,m}^{\mathrm{T}}\frac{\partial r}{\partial\theta_{i,m}}\right) = \frac{\partial r}{\partial\theta_{i,m}^{\mathrm{T}}}\mathbb{E}(\Delta\theta_{i,m}\Delta\theta_{i,m}^{\mathrm{T}})\frac{\partial r}{\partial\theta_{i,m}} = \frac{\partial r}{\partial\theta_{i,m}^{\mathrm{T}}}\mathbf{C}_{i,\theta}\frac{\partial r}{\partial\theta_{i,m}} \tag{65}$$

$$\lambda_{j,\theta} = \mathbb{E}\left(\frac{\partial r}{\partial\theta_{j,m}^{\mathrm{T}}}\Delta\theta_{j,m}\Delta\theta_{j,m}^{\mathrm{T}}\frac{\partial r}{\partial\theta_{j,m}}\right) = \frac{\partial r}{\partial\theta_{j,m}^{\mathrm{T}}}\mathbb{E}(\Delta\theta_{j,m}\Delta\theta_{j,m}^{\mathrm{T}})\frac{\partial r}{\partial\theta_{j,m}} = \frac{\partial r}{\partial\theta_{j,m}^{\mathrm{T}}}\mathbf{C}_{j,\theta}\frac{\partial r}{\partial\theta_{j,m}} \tag{66}$$

It can be seen that the distance is not a linear function of θ. As a cross product of $\mathbf{e}_{i,m}$ and $\mathbf{e}_{j,m}$, $\mathbf{e}_s$ is a complicated function. From (47) and (48), $\frac{\partial r}{\partial\theta_{i,m}}$ and $\frac{\partial r}{\partial\theta_{j,m}}$ are proved in Appendix A.

Under these assumptions, $\mathbf{C}_v = \mathrm{diag}(\mathbf{C}_{i,\theta}, \mathbf{C}_{j,\theta}, \mathbf{C}_{i,s}, \mathbf{C}_{i,s})$. Consequently, the variance λ can be expressed by

$$\lambda = \lambda_{i,s} + \lambda_{j,s} + \lambda_{i,\theta} + \lambda_{j,\theta}. \tag{67}$$

It should be noted that under the H_1 hypothesis, the distance may be arbitrary, and for simplicity, we assume that the distance is uniformly distributed over the surveillance volume. In this case, we can determine whether two observations are from the same target by the following decision rule

$$d_{\min} \underset{H_0}{\overset{H_1}{\gtrless}} \rho. \tag{68}$$

The association algorithm is a mapping $\psi' : \mathcal{A} \to \mathcal{M}$, which partitions $\mathcal{A}$ into $M+1$ groups and this mapping may disagree with real ψ due to inevitable errors. In practice, some knowledge about the targets of interest may be available and thus can be used for better performance. For instance, if only targets on the ground are of interest, then we may use this information and discard observations in which the cross points are obviously apart from the ground, which will be studied in the future.

3.3. Localization Algorithms

With mapping ψ', we obtain another partition $\mathcal{A} = \cup_{m=0}^{M}\mathcal{A}'_m$, where $\mathcal{A}'_m = \{x|\psi'(x) = m, x \in \mathcal{A}\}$. For observations in $\mathcal{A}'_m$, we can conduct signal source positioning, and three target positioning estimation methods will be introduced subsequently.

We first consider the intersection method. In presence of measurement errors, we often have $d_{\min} \neq 0$ even if two observations are from the same target. In this case, over two lines, the two points with minimal distance are

$$\mathbf{x}_i = \mathbf{s}_i + \alpha_{\mathrm{opt}}(1)\mathbf{e}_{i,m} \tag{69}$$

$$\mathbf{x}_j = \mathbf{s}_j + \alpha_{\mathrm{opt}}(2)\mathbf{e}_{j,m}. \tag{70}$$

In this case, it is reasonable to take the middle of two points as the estimate of the signal position, namely

$$\begin{aligned}
\hat{\mathbf{g}}_{i,j} &= \frac{1}{2}(\mathbf{x}_i + \mathbf{x}_j) \\
&= \frac{1}{2}(\mathbf{s}_i + \mathbf{s}_j) - \frac{1}{2}(\mathbf{e}_{i,m}, \mathbf{e}_{j,m})\mathbf{R}_{i,j}^{-1}\mathbf{E}_{i,j}^{\mathrm{T}}(\mathbf{s}_j - \mathbf{s}_i).
\end{aligned} \tag{71}$$

With more observations available, there will be an estimate of the target location for each observation pair , and a simple estimate of the target location can be expressed by their average, namely

$$\hat{\mathbf{g}}_m = \frac{1}{N_m(N_m - 1)} \sum_{i,j \in \mathcal{A}_m} \hat{\mathbf{g}}_{i,j} \tag{72}$$

where N_m denotes the number of observations associated with the mth signal source.

In practice, another widely used localization algorithm is the LS algorithm. It stems from the delta method concerned in [26]. From (4), the ith angle observation denoted by $\theta_{i,m}$ actually contributes a geometric relationship, which can be expressed by

$$\mathbf{G}_i \mathbf{g}_m = \mathbf{G}_i \mathbf{s}_i, i \in \mathcal{A}_m \tag{73}$$

where

$$\mathbf{G}_i = \begin{bmatrix} \sin \theta_{i,m} & -\cos \theta_{i,m} & 0 \\ \cos \theta_{i,m} \sin \varphi_{i,m} & \sin \theta_{i,m} \sin \varphi_{i,m} & -\cos \varphi_{i,m} \end{bmatrix}. \tag{74}$$

In (74), we used the equality

$$(x^o_{m,g} - x^o_{i,s}) \cos \theta^o_{i,m} + (y^o_{m,g} - y^o_{i,s}) \sin \theta^o_{i,m} \tag{75}$$

$$= \sqrt{(x^o_{m,g} - x^o_{i,s})^2 + (y^o_{m,g} - y^o_{i,s})^2} \tag{76}$$

$$= d^o_{i,m} \cos \varphi^o_{i,m}$$

where the value $d^o_{i,m}$ is the true range of $\mathbf{g}^o_m$ to $\mathbf{s}^o_i$ and can be expressed by $d^o_{i,m} = \|\mathbf{g}^o_m - \mathbf{s}^o_i\|$.

Next, we can combine the observations into an equation

$$\mathbf{G}\mathbf{g}_m = \mathbf{y} \tag{77}$$

where

$$\mathbf{G} = \begin{bmatrix} \mathbf{G}_1 \\ \vdots \\ \mathbf{G}_{M_m} \end{bmatrix}, \quad \mathbf{y} = \begin{bmatrix} \mathbf{G}_1 \mathbf{s}_1 \\ \vdots \\ \mathbf{G}_{M_m} \mathbf{s}_{M_m} \end{bmatrix}. \tag{78}$$

It should be noted that after the data association operation, the number of samples associated with a target may not agree with the real number. For simplicity, we still use $M_m = |\mathcal{A}'_m|$ to denote the number of observations associated with the mth target.

With M_m observations, there are in total $2M_m$ equations and three unknown parameters in $\mathbf{g}_m$. Therefore, as if $M_m \geq 2$, we can use the LS method to obtain an optimal solution in the sense of the mean square error, as

$$\hat{\mathbf{g}}_m = (\mathbf{G}^\mathrm{T}\mathbf{G})^{-1}\mathbf{G}\mathbf{y}. \tag{79}$$

In practice, different observations may have different SNRs, and then, the WLS algorithm can also be used in this framework. Then, the angle measurement error and sensor positioning error are equally weighted in the process of the LS algorithm. Assume that the distribution of the angle measurement error and the positioning error of sensors are known a priori. With this information, we can impose different weights over the observations, which is the WLS algorithm.

With the following approximations

$$\sin \theta_{i,m} \approx \sin(\theta^o_{i,m}) + \Delta\boldsymbol{\theta}^\mathrm{T}_{i,m} \cos \theta^o_{i,m} \tag{80}$$

$$\cos \theta_{i,m} \approx \cos(\theta^o_{i,m}) - \Delta\boldsymbol{\theta}^\mathrm{T}_{i,m} \sin \theta^o_{i,m} \tag{81}$$

we can put (80) and (81) into (73) and then write (73) as

$$\epsilon_i = \mathbf{G}_i(\mathbf{g}_m^o - \mathbf{s}_i) \tag{82}$$

where

$$\epsilon_i = [\epsilon_{i,1}, \epsilon_{i,2}]^{\mathrm{T}} \tag{83}$$

$$\epsilon_{i,1} = \Delta\theta_{i,m} d_{i,m}^o \cos\varphi_{i,m}^o + \mathbf{a}_{i,m}^{o\mathrm{T}}\Delta\mathbf{s}_i \tag{84}$$

$$\epsilon_{i,2} = \Delta\varphi_{i,m} d_{i,m}^o + \mathbf{b}_{i,m}^{o\mathrm{T}}\Delta\mathbf{s}_i \tag{85}$$

$$\mathbf{a}_{i,m}^o = [-\sin\theta_{i,m}^o, \cos\theta_{i,m}^o, 0]^{\mathrm{T}} \tag{86}$$

$$\mathbf{b}_{i,m}^o = [-\cos\theta_{i,m}^o \sin\varphi_{i,m}^o, -\sin\theta_{i,m}^o \sin\varphi_{i,m}^o, \cos\varphi_{i,m}^o]^{\mathrm{T}}. \tag{87}$$

In (84), we have used the equality (76). In (85), we have used the equality

$$(x_{m,g}^o - x_{i,s}^o)\sin\theta_{i,m}^o - (y_{m,g}^o - y_{i,s}^o)\cos\theta_{i,m}^o = 0 \tag{88}$$

$$(\mathbf{g}_m^o - \mathbf{s}_i^o)^{\mathrm{T}}[\cos\theta_{i,m}^o \cos\varphi_{i,m}^o, \sin\theta_{i,m}^o \cos\varphi_{i,m}^o, \sin\varphi_{i,m}^o]^{\mathrm{T}} = d_{i,m}^o \tag{89}$$

which can be easily verified in the angle measurement equations.

The vector ϵ_i can be rewritten as

$$\epsilon_i = \mathbf{B}_i^o \delta_i \tag{90}$$

where

$$\mathbf{B}_i^o = \begin{bmatrix} d_{i,m}^o \cos\varphi_{i,m}^o & 0 & \mathbf{a}_{i,m}^{o,\mathrm{T}} \\ 0 & d_{i,m}^o & \mathbf{b}_{i,m}^{o,\mathrm{T}} \end{bmatrix} \tag{91}$$

$$\delta_i = [\Delta\theta_{i,m}, \Delta\varphi_{i,m}, \Delta\mathbf{s}_i^{\mathrm{T}}]^{\mathrm{T}}. \tag{92}$$

Under the assumption that the self-positioning error and the angle measurement error are decorrelated, it can be proven that the covariance matrix of ϵ_i can be written as

$$\mathbf{Q}_n = \mathbb{E}(\epsilon\epsilon^{\mathrm{T}}) = \mathrm{diag}(\mathbf{C}_{n,\theta}, \mathbf{C}_{n,s}). \tag{93}$$

Putting (90) into (82) and combining the observations into an equation yield

$$\mathbf{B}^o\delta = \mathbf{G}\mathbf{g}_m^o - \mathbf{y}. \tag{94}$$

where

$$\mathbf{B}^o = \begin{bmatrix} \mathbf{B}_1^o \\ \vdots \\ \mathbf{B}_{M_m}^o \end{bmatrix}, \quad \delta = \begin{bmatrix} \delta_1 \\ \vdots \\ \delta_{M_m} \end{bmatrix}. \tag{95}$$

In the WLS algorithm, the goal is to minimize the objective function $J(\mathbf{g}_m)$ as

$$J(\mathbf{g}_m) = (\mathbf{G}\mathbf{g}_m - \mathbf{y})^{\mathrm{T}}\mathbf{W}(\mathbf{G}\mathbf{g}_m - \mathbf{y}) \tag{96}$$

where $\mathbf{W}$ is the weighting matrix with the expression

$$\mathbf{W} = \mathbb{E}[\mathbf{B}^o\delta\delta^{\mathrm{T}}\mathbf{B}^{o\mathrm{T}}]^{-1} = (\mathbf{B}^o\mathbf{Q}\mathbf{B}^{o\mathrm{T}})^{-1} \tag{97}$$

and $\mathbf{Q}$ is the error covariance matrix with an expression $\mathbf{Q} = \mathrm{diag}(\mathbf{Q}_1, \mathbf{Q}_2, \ldots, \mathbf{Q}_N)$.

The variable $\mathbf{g}_m$ to minimize the objective function $J(\mathbf{g}_m)$ can be calculated by the least square method and the estimate of the target positioning can be expressed by

$$\hat{\mathbf{g}}_m = (\mathbf{G}^{\mathrm{T}}\mathbf{W}\mathbf{G})^{-1}\mathbf{G}^{\mathrm{T}}\mathbf{W}\mathbf{y} \tag{98}$$

which is the WLS algorithm.

In WLS, the covariance matrices of different observations should be known a priori. If the observations can be deemed to have close SNRs, then one can simply use the LS algorithm. If the covariance matrices are not known with certain accuracy, some performance loss may occur, which should be analyzed with numerical analysis.

4. Numerical Results

We first study the localization performance in the presence of a platform of self-positioning errors and angle measurement errors. Then, the data association performance will be analyzed, followed by the analysis of the impact of sensor-target geometry.

Consider a scenario where two sensors are installed on two aircraft and three targets of interest are in the scope. Assume that two aircraft move at the same speed. The parameter settings are shown in Table 1. For simplicity, we assume that two passive sensors collect their observations on the same instants, namely, $t_{k,i} = t_{k,j}$ for all k and $i,j \in \{1, \cdots, N\}$. Meanwhile, the sampling interval is 0.1 s and the simulation runs for 10 seconds. Assume that the covariance matrices of angle measurement error for all sensors and all targets are the same as $\mathbf{C}_{k,\theta} = \sigma_\theta^2 \mathbf{I}_2, \forall k$. The self-positioning error covariance is $\mathbf{C}_{n,s} = \sigma_s^2 \mathbf{I}_3, n = 1, \cdots, N$.

Table 1. Positions and velocities of sensors and targets.

	Position (m) at $t = 0$ s	**Velocity (m/s)**	**Position (m) at $t = 10$ s**
Sensor #1	$\mathbf{s}_1^o(0) = [0, 0, 0]^{\mathrm{T}}$	$[50, 100, 0]^{\mathrm{T}}$	$\mathbf{s}_1^o(10) = [500, 1000, 0]^{\mathrm{T}}$
Sensor #2	$\mathbf{s}_2^o(0) = [12{,}000, 10{,}000, -800]^{\mathrm{T}}$	$[50, 100, 0]^{\mathrm{T}}$	$\mathbf{s}_2^o(10) = [12{,}500, 11{,}000, -800]^{\mathrm{T}}$
Target #1	$\mathbf{g}_1^o(0) = [18{,}000, 12{,}000, 8000]^{\mathrm{T}}$	$[20, 30, 0]^{\mathrm{T}}$	$\mathbf{g}_1^o(10) = [18{,}200, 12{,}300, 8000]^{\mathrm{T}}$
Target #2	$\mathbf{g}_2^o(0) = [15{,}000, 13{,}000, 7000]^{\mathrm{T}}$	$[20, 30, 0]^{\mathrm{T}}$	$\mathbf{g}_2^o(10) = [15{,}200, 13{,}300, 7000]^{\mathrm{T}}$
Target #3	$\mathbf{g}_3^o(0) = [13{,}000, 12{,}000, 5000]^{\mathrm{T}}$	$[20, 30, 0]^{\mathrm{T}}$	$\mathbf{g}_3^o(10) = [13{,}200, 12{,}300, 5000]^{\mathrm{T}}$

4.1. Impact of Self-Positioning Error and Angle Measurement Error

We first study the impacts of sensor positioning error and angle measurement error in the proposed intersection method. In order to make a comparison between the intersection method, the LS algorithm [19] and the WLS algorithm [6], we first consider the positioning accuracy of target #1 with the two sensors at $t = 0$ s.

In order to analyze the positioning performance of the concerned algorithms, we measure the performance with the average root mean square error (RMSE), defined for a target, e.g., target #1, by

$$\mathrm{RMSE} = \sqrt{\frac{1}{N_s} \sum_{k=1}^{N_s} \|\hat{\mathbf{g}}_1(k) - \mathbf{g}_1^o\|^2} \tag{99}$$

where $\hat{\mathbf{g}}_1(k)$ denotes the estimate of the target position at the kth simulation run for the target #1, and N_s denotes the number of simulation runs.

With $N_s = 10{,}000$ Monte Carlo runs, the RMSE of different self-positioning errors are shown in Figure 2, under the assumption that the covariance matrix is $\mathbf{C}_{1,s} = \sigma_s^2 \mathbf{I}_3$, where we always set $\sigma_\theta = 0.03°$ in this simulation. It can be seen that as σ_s rises from 0 to 20 m, the three algorithms have close performances and the WLS performs better, under the assumption that the covariance matrices of $\mathbf{C}_{k,\theta}, k = 1, 2, 3$ and $\mathbf{C}_{n,s}, n = 1, 2$ are known exactly.

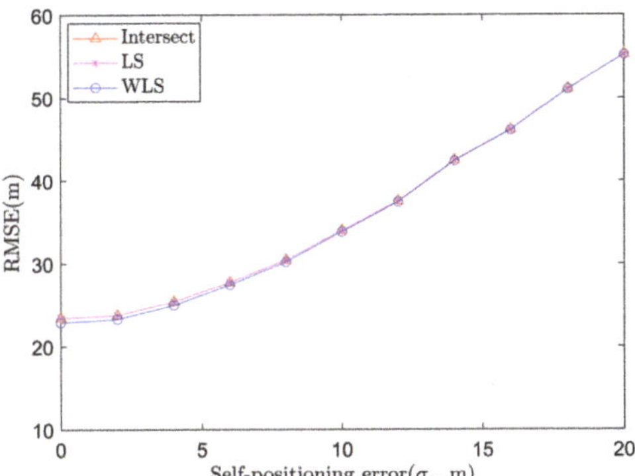

Figure 2. Average RMSE of target localization algorithms based on angle measurements of two sensors at a fixed angle measurement error.

The impacts of the angle measurement error are shown in Figure 3, where σ_θ raises from $0°$ to $1.5°$, where we set $\sigma_s = 5$ m in this simulation. It shows that the three algorithms have close performances and the intersection method and the LS method perform a little worse than the WLS method. The rise of both the self-positioning error and the angle measurement error will cause the rise of the target positioning error. However, with a better weighting, the WLS algorithm often performs better than the intersection method and the LS method.

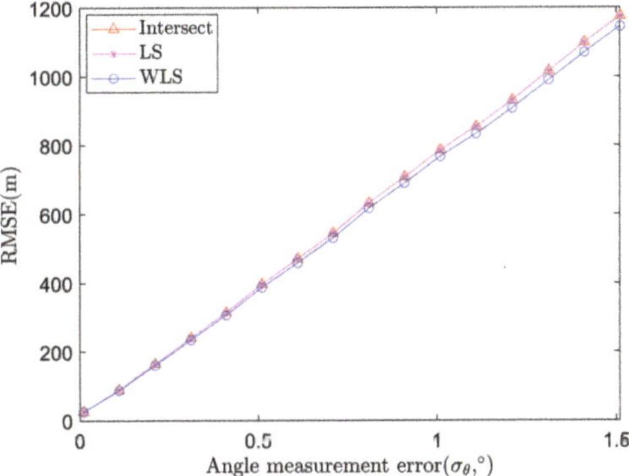

Figure 3. Average RMSE of target localization algorithms based on angle measurements of two sensors at a fixed self-positioning error.

4.2. Data Association Performance

Consider the parameters of the two sensors and the targets, as shown in Table 1. As the targets are moving in this scenario, we set the sampling interval to 0.1 s. At each sampling instant, each sensor has three measurements corresponding to the three targets. Therefore, the two sensors will totally have nine measurement combinations at each sampling instant, of which three combinations are correct. In data association, we set $p_f = 1\%$ and then the probability of correct association $p_c = 99\%$.

With $N_s = 2000$ Monte Carlo runs, the averaged correct association probability is shown in Figure 4, where the self-positioning error is $\sigma_s = 5$ m, and the angle measurement error is $\sigma_\theta = 0.03°$. It can be seen that during the whole sampling period, the correct association probability of the three targets is close to the present value 99%, namely 20 wrong combinations, on average, for each instant.

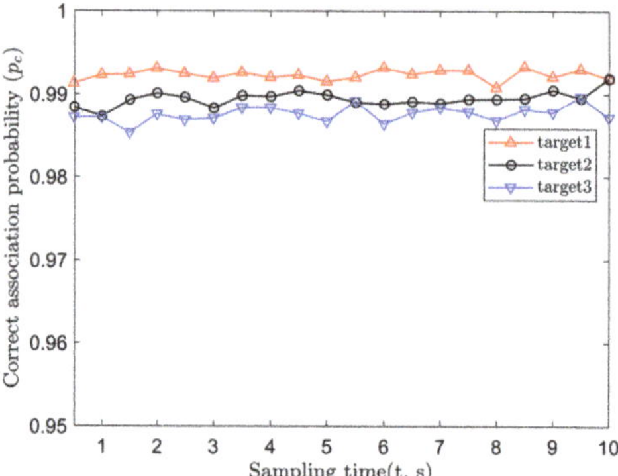

Figure 4. Probability of correct association.

Next, we consider the location performance with observations after the data association operation. The RMSEs of the algorithms over the three targets based on the proposed method are shown in Figure 5. In Figure 5, the RMSEs of the three targets are 23.43, 30.05 and 14.45 m. Note that the data association error may affect the localization performance in this case. From Figure 5, the relative position of the sensor and the target will affect the positioning performance, so the impact of the geometry of sensors and targets on the positioning performance will be studied subsequently.

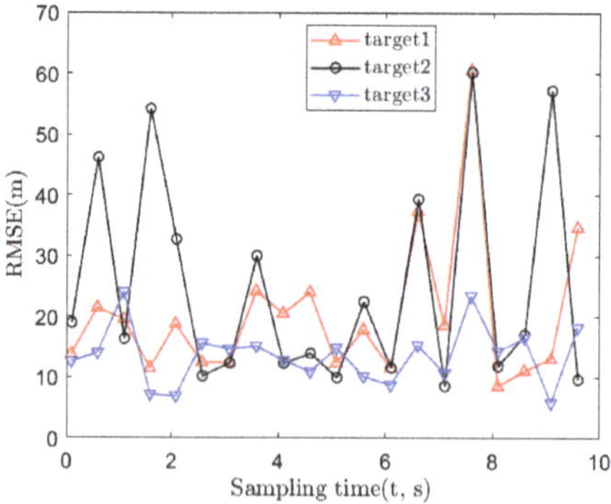

Figure 5. The localization error of the three targets.

4.3. Impact of Target-Sensor Geometry

The geometry will play an important role in the positioning performance. Consider the scenario shown in Figure 6, where a target is probed with two passive sensors and the intersection angle of two azimuth lines is denoted by ϕ. In the 3D scenario, we also define a new elevation angle η for convenience and the distance of the target is the same for both the sensors, i.e., 8000 m.

It is assumed that the positions of the two sensors are accurately measured and target #1 can be detected by both two sensors. The two sensors are symmetrically distributed on both sides of target #1. We explore the impact of geometry on the positioning accuracy by changing ϕ and η. In order to illustrate that the errors caused by geometry on the three coordinate axes are different, we define Δx, Δy and Δz to represent the RMSE on the three coordinate axes, respectively, which can be expressed by

$$\Delta x = \sqrt{\frac{1}{N_s} \sum_{k=1}^{N_s} \|\hat{x}_{1,g}(k) - x_{1,g}^o\|^2} \tag{100}$$

$$\Delta y = \sqrt{\frac{1}{N_s} \sum_{k=1}^{N_s} \|\hat{y}_{1,g}(k) - y_{1,g}^o\|^2} \tag{101}$$

$$\Delta z = \sqrt{\frac{1}{N_s} \sum_{k=1}^{N_s} \|\hat{z}_{1,g}(k) - z_{1,g}^o\|^2}. \tag{102}$$

With $N_s = 10{,}000$ Monte Carlo runs, the Δx, Δy, Δz and the spatial error *sum* denoted by $sum = \sqrt{\Delta x^2 + \Delta y^2 + \Delta z^2}$ are shown in Figure 7, under the assumption that $\eta = 15°$, and the angle measurement error $\sigma_\theta = 0.03°$. It can be seen that when the intersection angle ϕ is changed, the positioning error Δx, Δy, Δz are different. When the intersection angle ϕ is equal to $82.58°$, the spatial error *sum* is the best, about 5.58 m, and Δx, Δy and Δz are approximately equal to each other. In fact, the angle around $\phi = 90°$ will all lead to high accuracy.

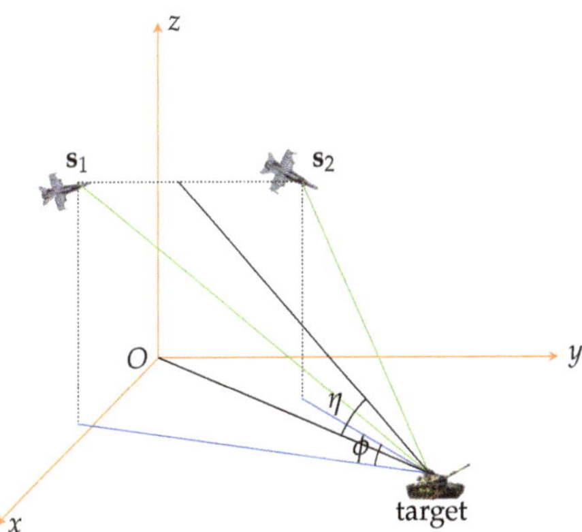

Figure 6. Definition of the intersection angle ϕ and pitch angle η of the plane.

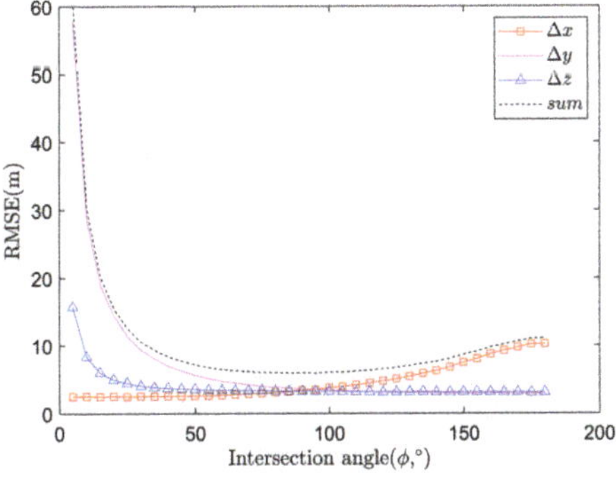

Figure 7. Relationship between localization error and intersection angle ϕ.

The impacts of intersection ϕ and η on spatial error *sum* are shown in Figure 8. For η in the range of $0°$ to $45°$ and ϕ in the range of $25°$ to $175°$, the spatial error sum is less than

20 m. Therefore, in practice, one can look for geometry with sensors and targets nearly perpendicular to each other to improve the positioning performance.

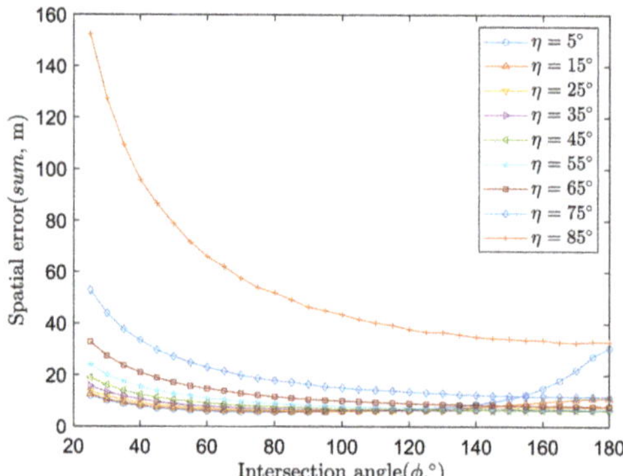

Figure 8. The relationship between spatial error *sum* and intersection angle ϕ, elevation angle η.

5. Conclusions

This paper studies the data association and signal source localization problems with distributed passive sensors with angle-only observations. A geometry-based data association method is considered, and the concept is that real targets will contribute observations with a small minimal distance. The statistical distribution of the minimal distance of two lines associated with the same target is formulated, based on which a data association method based on hypothesis testing is also developed. The decision threshold is formulated. Meanwhile, for observations that are classified into the same class, three positioning algorithms are studied, namely the intersection method, the LS method and the WLS method. Two kinds of measurements errors are considered, namely sensor self-positioning error and angle measurement error.

In numerical results, we analyze the data association performance of the concerned positioning algorithms and the signal source positioning performance in different scenarios, indicating that the data association algorithm works well and the positioning performances of the algorithms are very close to each other. Meanwhile, if the observation lines are approximately perpendicular to each other, then the localization performance is more accurate.

This algorithm can be used in laser, infrared, and other passive sensors with angle-only measurements. In practice, other information, such as range, ground surface and sea surface, may be available, which can be incorporated into the positioning and association algorithms to improve performance.

Author Contributions: Conceptualization, S.Z.; formal analysis, Y.C. and L.W.; methodology, Y.C., L.W. and S.Z.; software, Y.C. and L.W.; supervision, R.C. and S.Z.; writing—original draft, Y.C. and L.W.; revising, S.Z. All authors have read and agreed to the published version of the manuscript.

Funding: This research received no external funding.

Institutional Review Board Statement: Not applicable.

Informed Consent Statement: Not applicable.

Data Availability Statement: This study did not report any data.

Conflicts of Interest: The authors declare no conflict of interest.

Appendix A

In this section, we derive the partial derivative vectors $\frac{\partial r}{\partial \theta_{i,m}}$ and $\frac{\partial r}{\partial \theta_{j,m}}$ from (47) and (48). The elements in vector $\frac{\partial r}{\partial \theta_{i,m}}$ and $\frac{\partial r}{\partial \theta_{j,m}}$ can be expressed by

$$\frac{\partial r}{\partial \theta_{i,m}} = \left(\frac{\partial r}{\partial \mathbf{e}_i}\right)^{\mathrm{T}} \frac{\partial \mathbf{e}_i}{\partial \theta_{i,m}} = -\sin\theta_{i,m}\cos\varphi_{i,m}\frac{\partial r}{\partial e_{i,x}} + \cos\theta_{i,m}\cos\varphi_{i,m}\frac{\partial r}{\partial e_{i,y}} \tag{A1}$$

$$\frac{\partial r}{\partial \varphi_{i,m}} = \left(\frac{\partial r}{\partial \mathbf{e}_i}\right)^{\mathrm{T}} \frac{\partial \mathbf{e}_i}{\partial \varphi_{i,m}} \tag{A2}$$

$$= -\cos\theta_{i,m}\sin\varphi_{i,m}\frac{\partial r}{\partial e_{i,x}} - \cos\theta_{i,m}\sin\varphi_{i,m}\frac{\partial r}{\partial e_{i,y}} + \cos\varphi_{i,m}\frac{\partial r}{\partial e_{i,z}} \tag{A3}$$

$$\frac{\partial r}{\partial \theta_{j,m}} = \left(\frac{\partial r}{\partial \mathbf{e}_j}\right)^{\mathrm{T}} \frac{\partial \mathbf{e}_j}{\partial \theta_{j,m}} = -\sin\theta_{j,m}\cos\varphi_{j,m}\frac{\partial r}{\partial e_{j,x}} + \cos\theta_{j,m}\cos\varphi_{j,m}\frac{\partial r}{\partial e_{j,y}} \tag{A4}$$

$$\frac{\partial r}{\partial \varphi_{j,m}} = \left(\frac{\partial r}{\partial \mathbf{e}_j}\right)^{\mathrm{T}} \frac{\partial \mathbf{e}_j}{\partial \varphi_{j,m}} \tag{A5}$$

$$= -\cos\theta_{j,m}\sin\varphi_{j,m}\frac{\partial r}{\partial e_{j,x}} - \cos\theta_{j,m}\sin\varphi_{j,m}\frac{\partial r}{\partial e_{j,y}} + \cos\varphi_{j,m}\frac{\partial r}{\partial e_{j,z}} \tag{A6}$$

where

$$\frac{\partial r}{\partial e_{i,x}} = X^{-\frac{1}{2}}(e_{j,y}z_{i,j} - e_{j,z}y_{i,j}) - X^{-\frac{3}{2}}Y(e_{j,y}^2 e_{i,x} + e_{j,z}^2 e_{i,x} - e_{i,y}e_{j,y}e_{j,x} - e_{i,z}e_{j,z}e_{j,x}) \tag{A7}$$

$$\frac{\partial r}{\partial e_{i,y}} = X^{-\frac{1}{2}}(e_{j,z}x_{i,j} - e_{j,x}z_{i,j}) - X^{-\frac{3}{2}}Y(e_{j,z}^2 e_{i,y} + e_{j,x}^2 e_{i,y} - e_{i,z}e_{j,z}e_{j,y} - e_{i,x}e_{j,x}e_{j,y}) \tag{A8}$$

$$\frac{\partial r}{\partial e_{i,z}} = X^{-\frac{1}{2}}(e_{j,x}y_{i,j} - e_{j,y}x_{i,j}) - X^{-\frac{3}{2}}Y(e_{j,y}^2 e_{i,z} + e_{j,x}^2 e_{i,z} - e_{i,y}e_{j,y}e_{j,z} - e_{i,x}e_{j,x}e_{j,z}) \tag{A9}$$

$$\frac{\partial d}{\partial e_{j,x}} = X^{-\frac{1}{2}}(e_{i,z}y_{i,j} - e_{i,y}z_{i,j}) - X^{-\frac{3}{2}}Y(e_{i,y}^2 e_{j,x} + e_{i,z}^2 e_{j,x} - e_{i,y}e_{j,y}e_{i,x} - e_{i,z}e_{j,z}e_{i,x}) \tag{A10}$$

$$\frac{\partial d}{\partial e_{j,y}} = X^{-\frac{1}{2}}(e_{i,x}z_{i,j} - e_{i,z}x_{i,j}) - X^{-\frac{3}{2}}Y(e_{i,z}^2 e_{j,y} + e_{i,x}^2 e_{j,y} - e_{i,z}e_{j,z}e_{i,y} - e_{i,x}e_{j,x}e_{i,y}) \tag{A11}$$

$$\frac{\partial d}{\partial e_{j,z}} = X^{-\frac{1}{2}}(e_{i,y}x_{i,j} - e_{i,x}y_{i,j}) - X^{-\frac{3}{2}}Y(e_{i,y}^2 e_{j,z} + e_{i,x}^2 e_{j,z} - e_{i,y}e_{j,y}e_{i,z} - e_{i,x}e_{j,x}e_{i,z}) \tag{A12}$$

$$Y = \begin{vmatrix} x_{i,j} & y_{i,j} & z_{i,j} \\ e_{i,x} & e_{i,y} & e_{i,z} \\ e_{j,x} & e_{j,y} & e_{j,z} \end{vmatrix} \tag{A13}$$

$$X = \begin{vmatrix} e_{i,y} & e_{i,z} \\ e_{j,y} & e_{j,z} \end{vmatrix}^2 + \begin{vmatrix} e_{i,z} & e_{i,x} \\ e_{j,z} & e_{j,x} \end{vmatrix}^2 + \begin{vmatrix} e_{i,x} & e_{i,y} \\ e_{j,x} & e_{j,y} \end{vmatrix}^2 \tag{A14}$$

where $[x_{i,j}, y_{i,j}, z_{i,j}]^{\mathrm{T}} = \mathbf{s}_j - \mathbf{s}_i$.

References

1. Qiang, Z.; Jiaqi, C.; Rijie, Y.; Zhichao, S. Research on airborne infrared location technology based on orthogonal multi-station angle measurement method. *Infrared Phys. Technol.* **2017**, *86*, 202.
2. Sioutis, M.; Tan, Y. *User Indoor Location System with Passive Infrared Motion Sensors and Space Subdivision*; Springer International Publishing: Berlin/Heidelberg, Germany, 2014.
3. Cheng, X.; Huang, D.; Wei, H. High precision passive target localization based on airborne electro-optical payload. In Proceedings of the 2015 14th International Conference on Optical Communications and Networks (ICOCN), Nanjing, China, 3–5 July 2015.
4. Bai, G.; Liu, J.; Song, Y.; Zuo, Y. Two-UAV Intersection Localization System Based on the Airborne Optoelectronic Platform. *Sensors* **2017**, *17*, 98. [CrossRef] [PubMed]

5. Peng, S.; Zhao, Q.; Ma, Y.; Jiang, J. Research on the Technology of Cooperative Dual-Station position Based on Passive Radar System. In Proceedings of the 2020 3rd International Conference on Unmanned Systems (ICUS), Harbin, China, 27–28 November 2020.

6. Wang, Y.; Ho, V. An Asymptotically Efficient Estimator in Closed-Form for 3-D AOA Localization Using a Sensor Network. *IEEE Trans. Wirel. Commun.* **2015**, *14*, 6524–6535. [CrossRef]

7. Li, M.; Lu, Y. Angle-of-arrival estimation for localization and communication in wireless networks. In Proceedings of the 2008 16th European Signal Processing Conference, Lausanne, Switzerland, 25–29 August 2008.

8. Stolnik, M. *Radar Handbook*, 3rd ed.; McGraw-Hill Professional Publishing: New York, NY, USA, 2008.

9. Kang, S.; Kim, T.; Chung, W. Multi-Target Localization Based on Unidentified Multiple RSS/AOA Measurements in Wireless Sensor Networks. *Sensors* **2021**, *21*, 4455. [CrossRef] [PubMed]

10. Pattipati, K.R.; Deb, S. A new relaxation algorithm and passive sensor data association. *IEEE Trans. Autom. Control* **1992**, *37*, 198–213. [CrossRef]

11. Ouyang, C.; Ji, H. Modified cost function for passive sensor data association. *Electron. Lett.* **2011**, *47*, 383–385. [CrossRef]

12. Li, H.; Lu, C.; Feng, X.; Di, Z.; Wei, W. Data association algorithm for multi-infrared-sensor system. *Infrared Laser Eng.* **2014**, *79*, 511–517.

13. Wang, X.; Cai, W.; Qiu, L.; Yuan, A.; Cao, Z. Research on Data Association Algorithm Based on Line of sight Distance in Photoelectric Two-dimensional Detection System. *Air Space Def.* **2019**, *2*, 51.

14. Yin, J.; Wan, Q.; Yang, S.; Ho, K.C. A Simple and Accurate TDOA-AOA Localization Method Using Two Stations. *IEEE Signal Process. Lett.* **2015**, *23*, 144–148. [CrossRef]

15. Dogancay, K. On the bias of linear least squares algorithms for passive target localization. *Signal Process. Amst.* **2004**, *84*, 475–486. [CrossRef]

16. Doanay, K. Bearings-only target localization using total least squares. *Signal Process.* **2005**, *85*, 1695–1710.

17. Dogancay, K. Relationship Between Geometric Translations and TLS Estimation Bias in Bearings-Only Target Localization. *IEEE Trans. Signal Process.* **2008**, *56*, 1005–1017. [CrossRef]

18. Yu, Z.; Fu, Y. A passive location method based on virtual time reversal of cross antenna sensor array and Tikhonov regularized TLS. *IEEE Sens. J.* **2021**, *21*, 21931–21940 [CrossRef]

19. Li, C.; Zhuang, W. Hybrid TDOA/AOA mobile user location for wideband CDMA cellular systems. *IEEE Trans. Wirel. Commun.* **2002**, *1*, 439–447.

20. Loan, G. An Analysis of the Total Least Squares Problem. *SIAM J. Numer. Anal.* **1980**, *17*, 883–893.

21. Wu, W.; Jiang, J.; Fan, X.; Zhou, Z. Performance analysis of passive location by two airborne platforms with angle-only measurements in WGS-84. *Infrared Laser Eng.* **2015**, *44*, 654–661

22. Frew, E.W. Sensitivity of Cooperative Target Geolocalization to Orbit Coordination. *J. Guid. Control Dyn.* **2008**, *31*, 1028–1040 [CrossRef]

23. Yi, Z.; Li, Y.; Qi, G.; Sheng, A. Cooperative Target Localization and Tracking with Incomplete Measurements. *Int. J. Distrib. Sens. Netw.* **2014**, *2014*, 1–16.

24. Malick, M. A Note on Bearing Measurement Model. 2018. Available online: https://www.researchgate.net/profile/Mahendra-Mallick/publication/325214760_A_Note_on_Bearing_Measurement_Model/links/5afe2d230f7e9b98e0197b3f/A-Note-on-Bearing-Measurement-Model.pdf (accessed on 8 January 2022).

25. Mallick, M.; Nagaraju, R.M.; Duan, Z. IMM-CKF for a Highly Maneuvering Target Using Converted Measurements. In Proceedings of the 2021 International Conference on Control, Automation and Information Sciences (ICCAIS), Xi'an, China, 14–17 October 2021; pp. 15–20. [CrossRef]

26. Casella, G.; Berger, R.L. *Statistical Inference*, 2nd ed.; Duxbury Thomson Learning Press: Pacific Grove, CA, USA, 2002.

Article

Intelligence-Aware Batch Processing for TMA with Bearings-Only Measurements

Gabriele Oliva [1,*,†], Alfonso Farina [2,†] and Roberto Setola [1,†]

[1] Department of Engineering, Università Campus Bio-Medico di Roma, 00128 Rome Italy; r.setola@unicampus.it
[2] Selex-ES (retired), 00144 Rome, Italy; alfonso.farina@outlook.it
* Correspondence: g.oliva@unicampus.it; Tel.: +39-06225419636
† These authors contributed equally to this work.

Abstract: This paper develops a framework to track the trajectory of a target in 2D by considering a moving ownship able to measure bearing measurements. Notably, the framework allows one to incorporate additional information (e.g., obtained via intelligence) such as knowledge on the fact the target's trajectory is contained in the intersection of some sets or the fact it lies outside the union of other sets. The approach is formally characterized by providing a constrained maximum likelihood estimation (MLE) formulation and by extending the definition of the Cramér–Rao lower bound (CRLB) matrix to the case of MLE problems with inequality constraints, relying on the concept of generalized Jacobian matrix. Moreover, based on the additional information, the ownship motion is chosen by mimicking the Artificial Potential Fields technique that is typically used by mobile robots to aim at a goal (in this case, the region where the target is assumed to be) while avoiding obstacles (i.e., the region that is assumed not to intersect the target's trajectory). In order to show the effectiveness of the proposed approach, the paper is complemented by a simulation campaign where the MLE computations are carried out via an evolutionary ant colony optimization software, namely, mixed-integer distributed ant colony optimization solver (MIDACO-SOLVER). As a result, the proposed framework exhibits remarkably better performance, and in particular, we observe that the solution is less likely to remain stuck in unsatisfactory local minima during the MLE computation.

Keywords: target motion analysis; intelligence-aware estimation; radar; Cramér–Rao lower bound; nonlinear estimation; constrained MLE; data fusion; smart estimation; intelligence analysis; critical infrastructure protection; evolutionary ant colony optimization; MIDACO-SOLVER

Citation: Oliva, G.; Farina, A.; Setola, R. Intelligence-Aware Batch Processing for TMA with Bearings-Only Measurements. *Sensors* **2021**, *21*, 7211. https://doi.org/10.3390/s21217211

Academic Editors: Andrzej Stateczny, Mahendra Mallick and Ratnasingham Tharmarasa

Received: 20 August 2021
Accepted: 27 October 2021
Published: 29 October 2021

Publisher's Note: MDPI stays neutral with regard to jurisdictional claims in published maps and institutional affiliations.

1. Introduction

In the last decades, target motion analysis (TMA) has become an increasingly popular research field, and in the literature, several approaches have been developed, such as batch processing frameworks [1–4] and recursive ones [3,5–8]. The aim of TMA is to estimate the state of a target (usually position and velocity) from noise-corrupted measurements collected by an observer [9]. The TMA problem presents several challenges, mainly due to the nonlinear relationship between the measurements and target state. Another challenge is that the observer must outmaneuver the target in order to make the target state observable [10]. For instance, to track a target with constant velocity, the observer platform must change its speed or course. Otherwise, there exist other target trajectories that produce the same sequence of noise-free bearing angles [3].

Among other approaches, bearing-only target tracking [11–13] represents an increasingly popular topic, with application scenarios ranging from underwater tracking [14,15] to cooperative tracking for multiagent systems [15–18].

Other relevant approaches in the literature include, among other works: applications to sensor network localization [19]; algorithms based on direction-of-arrival measurements, modeled by von Mises–Fisher distributions [20]; pseudolinear estimators for 3D target motion analysis by a single moving ownship collecting azimuth and elevation angle measurements [21]; a methodology based on a bank of batch maximum a posteriori (MAP) estimators as a general estimation framework that provides the relinearization of the entire state trajectory, multihypothesis tracking and an efficient hypothesis generation scheme [22]; an approach based on Newton–Raphson methods and Particle Swarm Optimization [23]; a methodology to combine target motion compensation and track-before-detect methods within passive radar based on global navigation satellite systems (GNSS) for the detection of maritime targets [24]; TMA from cosines of conical angles acquired by a towed array [25]; and a new pseudolinear filter for bearings-only tracking without the requirement of bias compensation [26].

Notice that, in the literature, some approaches have been developed where the availability of road or traffic information is used to track a moving target. In particular, in [27], the target is assumed to move in a a road network, and the tracking is performed via an airborne sensor that exploits knowledge on the network; in [28], a similar setting is considered, and a Bayesian approach is adopted; in [29], a particle filter is developed in order to track multiple vehicles on multi-lane roads based on a microscopic traffic flow model. However, such approaches require one to rely on a large deal of fine-grained information (e.g., the structure of the road network) and can only be applied to scenarios involving roads. However, in many cases, especially considering a maritime context, only coarse-grained information is available: for instance, the environment might contain physical obstacles or deterring entities such as warships that discourage the target from passing nearby, or there might be rough evidence of the presence of a target in a given zone (e.g., due to a witness or to cheap range-free sensors able to only detect the presence of a target in a given zone). In this view, relying on such a coarse-grained information could help improve the target's trajectory estimate, also in contexts where road network information cannot be leveraged upon without requiring huge computational resources. This has been demonstrated, for instance, in [30] where such information is used in the framework of network localization to overcome localization ambiguities.

This is the aim of this paper. In particular, this paper considers a scenario where additional information is available to the ownship in charge of estimating the target's trajectory; specifically, the ownship is aware that the trajectory of the target lies in the intersection of some sets and is not contained in the union of some other sets. This additional information is exploited by developing a constrained MLE problem and an approach for the selection of the ownship's trajectory mimicking the Artificial Potential Fields technique [31,32], which is typically used by mobile robots to aim at a goal (in this case, the region where the target is assumed to be) while avoiding obstacles (i.e., the region that is assumed not to intersect the target's trajectory). Moreover, from a theoretical standpoint, the CRLB on the estimation covariance matrix is characterized in the case of MLE problems with inequality constraints; this is performed by extending the approach in [33,34], where equality constraints where discussed, via the cast of inequality constraints into nonsmooth equality ones and by the adoption of generalized Jacobian matrices [35], which are set-valued on a zero-measure set where the derivative of the resulting nonsmooth function is not defined.

The paper is complemented by an experimental analysis showing the effectiveness of the proposed approach.

To summarize, the main contributions of the paper are as follows:

- We develop a novel MLE approach to carry out batch target-tracking estimation based on noisy bearing-only measurements, which incorporates as inequality constraints additional information in terms of sets where the target's trajectory is assumed to be contained and other sets which have empty intersection with the target's trajectory;
- We characterize the CRLB associated to the constrained problem by considering a generalized set-valued Jacobian matrix of the constraints function and by resorting to nonsmooth theory;

- We provide a heuristic way to select the ownship's trajectory based on the although coarse-grained available information regarding the target's trajectory.

2. Materials and Methods

The aim of this section is to present the main algorithms, tools and derivations that are used in the paper. Specifically, the section is organized as follows: in Section 2.1, we present our problem statement; then, we discuss maximum likelihood estimation problems (in Section 2.2, we review the unconstrained case, while in Section 2.3 we develop the proposed MLE approach with additional inequality constraints); Section 2.4 is devoted to addressing the computational aspects related to the approximated solution of the above constrained and unconstrained MLE problems; Sections 2.5 and 2.6 address, respectively, the characterization of the CRLB in the unconstrained and constrained case; finally, Section 2.7 discusses a heuristic approach to choosing the ownship's direction based on the available information.

2.1. Problem Statement

Let us consider a scenario where a target moves in a linear motion on a plane; in particular, considering a discrete-time sampling, let us assume that the target moves according to the following equations:

$$
\begin{aligned}
x_t(k) &= x_{t0} + \dot{x}_{t0}kT + \frac{1}{2}\ddot{x}_t k^2 T^2 \\
y_t(k) &= y_{t0} + \dot{y}_{t0}kT + \frac{1}{2}\ddot{y}_t k^2 T^2 \\
\dot{x}_t(k) &= \dot{x}_{t0} + \ddot{x}_t kT \\
\dot{y}_t(k) &= \dot{y}_{t0} + \ddot{y}_t kT,
\end{aligned}
\tag{1}
$$

with T being the sampling time. Moreover, let us consider an ownship platform aiming to estimate the parameter vector

$$
\boldsymbol{\psi} = \begin{bmatrix} x_{t0} & y_{t0} & \dot{x}_{t0} & \dot{y}_{t0} & \ddot{x}_t & \ddot{y}_t \end{bmatrix}^T,
\tag{2}
$$

based on a batch of measurements, sampled at uniform discrete-time instants $t = kT$ during the ownship motion and evaluated over the time interval $[0, k_{\max}T]$.

In more detail, we assume that the ownship attempts to sense the following *nominal measurement* (e.g., see [36]):

$$
h(\boldsymbol{\psi}, k) = \text{atan2}(y_t(k) - y_o(k), x_t(k) - x_o(k))
\tag{3}
$$

where $x_o(k), y_o(k)$ are the coordinates of the ownship at time $t = kT$ along the x and y axes, respectively. However, we consider a scenario where the ownship is actually provided with noisy measurements with the following structure:

$$
z(k) = h(\boldsymbol{\psi}, k) + w(k),
$$

where the terms $w(k) \sim \mathcal{N}(0, \sigma^2)$ are independent and identically distributed Gaussian noises with zero-mean and variance σ^2.

Further to that, let us assume that additional information is available; specifically, let us assume that the ownship is aware that, during the considered time frame $[0, k_{\max}T]$, the position $p(k) = [x_t(k), y_t(k)]^T$ of the target is confined in a region $\mathcal{P}$ of the plane defined as follows:

$$
\mathcal{P} = \left\{ p \in \mathbb{R}^2 \mid p \in \bigcap_{i=1}^{m} \mathcal{X}_i \right\},
\tag{4}
$$

where the sets $\mathcal{X}_i \subseteq \mathbb{R}^2$ are convex, and their intersection is nonempty; moreover, the ownship is aware that the position $p(k)$ lies outside a region $\mathcal{S}$ of the plane defined as follows:

$$\mathcal{S} = \left\{ p \in \mathbb{R}^2 \mid p \in \bigcup_{i=1}^{r} \mathcal{Y}_i \right\}, \tag{5}$$

where the sets $\mathcal{Y}_i \subseteq \mathbb{R}^2$ are convex.

The aim of this paper is to investigate how this additional information influences the estimation of θ and how the ownship can leverage on this additional information to select a trajectory that improves the estimation performance.

2.2. Maximum Likelihood Estimation without Additional Information

Let us discuss how to estimate the parameter vector ψ via the *maximum* likelihood estimate (MLE) technique (see, for instance [37], p. 182). The MLE for a vector parameter ψ is defined to be the value θ^* that maximizes the likelihood function $p(z_1, z_2, \ldots, z_m, \theta)$ over the allowable domain of θ. In what follows, where understood, we abbreviate the notation by writing $p(\theta)$. When $p(\theta)$ is differentiable, we have that the MLE θ^* satisfies

$$\left. \frac{\partial \ln(p(\theta))}{\partial \theta} \right|_{\theta=\theta^*} = 0_n. \tag{6}$$

It is worthwhile to mention that the solution of the above equation is unique in this case and is theoretically asymptotically unbiased. For our case, we have the following expression for the likelihood function [3]:

$$p(\theta) = \frac{1}{2\pi\sigma} \prod_{k=1}^{k_{\max}} \exp\left(-\frac{(z(k) - h(\theta, k))^2}{2\sigma^2} \right). \tag{7}$$

Notably, the maximization problem at hand can be formulated as

$$\theta^* = \arg\max_{\theta} p(\theta) = \arg\max_{\theta} \ln(p(\theta)). \tag{8}$$

Let us define $\lambda(\theta) = -\ln(p(\theta))$. The above problem can be equivalently expressed as

$$\theta^* = \arg\min_{\theta} \lambda(\theta) \tag{9}$$

which, by simple computations is equivalent to solving [3]

$$\theta^* = \arg\min_{\theta} \sum_{k=1}^{k_{\max}} \frac{(z(k) - h(\theta, k))^2}{2\sigma^2}. \tag{10}$$

Notably, Equation (10) is recognized to be the classical least squares (LS) solution.

2.3. Maximum Likelihood Estimation with Additional Information

Let us now extend the above MLE framework in order to account for the additional intelligence available to the ownship.

In particular, we notice that, by Equations (1) and (2), it holds that

$$p(k) = Q(k)\psi, \tag{11}$$

with

$$Q(k) = \begin{bmatrix} I_2 & kT I_2 & \frac{1}{2}k^2 T^2 I_2 \end{bmatrix} \in \mathbb{R}^{2 \times 6}. \tag{12}$$

Therefore, by plugging the above expression in Equation (4), we have that $p(k) \in \mathcal{P}$ if and only if

$$Q(k)\psi \in \mathcal{P}, \quad \forall k \in \{0, \ldots, k_{\max}\}, \tag{13}$$

while $p(k) \notin \mathcal{S}$ if and only if

$$Q(k)\boldsymbol{\psi} \notin \mathcal{S}, \quad \forall k \in \{0,\ldots,k_{\max}\}.$$

In the following section, we assume that the constraints in the form

$$Q(k)\boldsymbol{\theta} \in \mathcal{P}, \quad Q(k)\boldsymbol{\theta} \notin \mathcal{S}$$

can be equivalently expressed as

$$g_k(\boldsymbol{\theta}) = g(Q(k)\boldsymbol{\theta}) \leq \mathbf{0}_q$$

for some $q \geq 0$ and for some differentiable function $g : \mathbb{R}^2 \to \mathbb{R}^q$. In this view, the above unconstrained MLE problem can be extended as follows:

$$\boldsymbol{\theta}^* = \arg\min_{\boldsymbol{\theta}} \lambda(\boldsymbol{\theta})$$

subject to

$$g_k(\boldsymbol{\theta}) \leq \mathbf{0}_q, \quad \forall k \in \{0,\ldots,k_{\max}\}. \tag{14}$$

2.4. Computational Approach to Solve MLE Problems

Notice that, as remarked in [38], the MLE for bearings-only target motion analysis does not admit a closed-from solution and must be implemented iteratively, considering an initialization close to the true solution to avoid divergence. In particular, we point out that the above MLE minimization problems (both in the unconstrained and constrained fashion) are not, in general, convex, thus calling for approximated solution schemes that typically aim to find a good local optimum. In this paper, we resort to the MIDACO-SOLVER optimization software, which implements an extension of the evolutionary ant colony optimization meta-heuristic [39] and which has been developed especially for highly nonlinear real-world applications. See [40] or [41] for a focus of the performance of MIDACO software with respect to the state of the art. Notably, MIDACO-SOLVER allows one to evaluate the satisfaction of the constraints and the objective function from an algorithmic standpoint, thus allowing one to also tackle the problems that are not easily expressed in a closed form nor easily solved by traditional solvers. Note that the suggested strategy is independent of a particular solver, but the nonconvex nature of the optimization problem suggests an evolutionary approach, such as genetic algorithms [42].

2.5. CRLB of the Estimate in the Unconstrained Case

Let us now discuss a useful metric that represents a lower bound on the covariance matrix associated to the MLE estimation process. Specifically, in this subsection, we consider the unconstrained case, while the constrained one is discussed in the next subsection.

In particular, consider the problem of evaluating the CRLB for the estimated vector $\boldsymbol{\psi}$ of target parameters (see, for instance [37], p.44). In particular, it is well known that the covariance matrix is bounded by the inverse of the Fisher information matrix (FIM) J, i.e., it holds that

$$E\{(\boldsymbol{\theta}^* - \boldsymbol{\psi})(\boldsymbol{\theta}^* - \boldsymbol{\psi})^T\} \geq J^{-1}(\overline{\boldsymbol{\psi}}) = \text{CRLB}^{\text{unconstrained}},$$

where

$$J(\overline{\boldsymbol{\psi}}) = E\{[\nabla_{\boldsymbol{\theta}}\lambda(\boldsymbol{\theta})][\nabla_{\boldsymbol{\theta}}\lambda(\boldsymbol{\theta})]^T\}\big|_{\boldsymbol{\theta}=\overline{\boldsymbol{\psi}}}.$$

Note that the expectations in the above equation are taken with respect to $p(\boldsymbol{\theta})$; moreover, the derivatives in $J(\cdot)$ are evaluated at the true value of $\boldsymbol{\psi}$ (i.e., $\overline{\boldsymbol{\psi}} = \boldsymbol{\psi}$) or, alternatively, at the estimated value $\boldsymbol{\theta}^*$ (i.e., $\overline{\boldsymbol{\psi}} = \boldsymbol{\theta}^*$) if the true value is unknown. Let us now present a more direct expression of the FIM. To this end, let us express the gradient of $\lambda(\boldsymbol{\theta})$ with respect to $\boldsymbol{\theta}$ as [3]

$$\nabla_{\boldsymbol{\theta}}\lambda(\boldsymbol{\theta}) = -\sum_{k=1}^{k_{\max}} \frac{1}{\sigma^2}(z(k) - h(\boldsymbol{\theta},k))\nabla_{\boldsymbol{\theta}}h(\boldsymbol{\theta},k); \tag{15}$$

then, we have that [3]

$$\begin{aligned} E\Big\{[\nabla_{\boldsymbol{\theta}}\lambda(\boldsymbol{\theta})][\nabla_{\boldsymbol{\theta}}\lambda(\boldsymbol{\theta})]^T\Big\}\Big|_{\boldsymbol{\psi}} &= \frac{1}{\sigma^4}\sum_{k=1}^{k_{\max}} E\{(z(k) - h(\boldsymbol{\psi},k))^2\}\nabla_x h(\boldsymbol{\psi},k)\nabla_{\boldsymbol{\psi}}h(\boldsymbol{\psi},k)^T \\ &= \frac{1}{\sigma^2}\sum_{k=1}^{k_{\max}} \nabla_{\boldsymbol{\psi}}h(\boldsymbol{\psi},k)\nabla_{\boldsymbol{\psi}}h(\boldsymbol{\psi},k)^T, \end{aligned} \tag{16}$$

where in the first equation, we used the fact that the measures are independent in order to neglect mixed terms, while the last equation follows from the consideration that

$$E\Big\{(z(k) - h(\boldsymbol{\psi},k))^2\Big\} = E\Big\{w(k)^2\Big\} = \sigma^2.$$

Notice that, in the Appendix A, we provide the analytical expression of the entries of $\nabla_{\boldsymbol{\psi}}h(\boldsymbol{\psi},k)$.

2.6. CRLB for Constrained MLE

As demonstrated in [33] (see also [34]), assuming $\boldsymbol{\theta} \in \mathbb{R}^n$, when the MLE problem has an equality constraint in the form

$$\boldsymbol{f}(\boldsymbol{\theta}) = \mathbf{0}_q, \quad \boldsymbol{f} : \mathbb{R}^n \to \mathbb{R}^q,$$

the CRLB can be computed starting from the unconstrained case, according to the following equation

$$\mathrm{CRLB}^{\mathrm{constrained}} = J^{-1}(\overline{\boldsymbol{\psi}}) - J^{-1}(\overline{\boldsymbol{\psi}})F(\overline{\boldsymbol{\psi}})\Big(F^T(\overline{\boldsymbol{\psi}})J^{-1}(\overline{\boldsymbol{\psi}})F(\overline{\boldsymbol{\psi}})\Big)^{-1}F^T(\overline{\boldsymbol{\psi}})J^{-1}(\overline{\boldsymbol{\psi}}),$$

where $F(\overline{\boldsymbol{\psi}})$ is the $n \times q$ Jacobian matrix of the constraint function $\boldsymbol{f}$, evaluated at $\overline{\boldsymbol{\psi}}$, i.e.,

$$F(\overline{\boldsymbol{\psi}}) = \nabla_{\boldsymbol{\theta}}\boldsymbol{f}^T(\boldsymbol{\theta})\Big|_{\boldsymbol{\theta}=\overline{\boldsymbol{\psi}}}.$$

Let us now extend this method to the case of inequality constraints. To this end, consider a constraint in the form of

$$\boldsymbol{g}(\boldsymbol{\theta}) \leq \mathbf{0}_q$$

where $\boldsymbol{g} : \mathbb{R}^n \to \mathbb{R}^q$ is differentiable. We point out that the inequality constraint can equivalently be expressed in the form of an equality constraint using

$$\boldsymbol{f}(\boldsymbol{\theta}) = \max\{\boldsymbol{g}(\boldsymbol{\theta}), \mathbf{0}_q\} = \mathbf{0}_q,$$

where max is intended component-wise. Notably, the max function is globally Lipschitz (e.g., see [43]); hence, if $\boldsymbol{g}$ is differentiable everywhere, we have that $\boldsymbol{f}$ is differentiable for all $\boldsymbol{\theta} \in \mathbb{R}^n \setminus \Omega$, where Ω is a zero-measure set in the form

$$\Omega = \{\boldsymbol{\theta} \in \mathbb{R}^n \,|\, g_i(\boldsymbol{\theta}) = 0, \text{ for some } i \in \{1,\ldots,q\}\}.$$

In this view, a natural extension is to resort to the *generalized Jacobian* of f. Specifically, it follows from [35] that the generalized Jacobian $F_G(\theta)$ of $f(\theta)$ is defined as

$$F_G(\theta) = \overline{co} \left\{ \lim_{i \to \infty} F(\theta_i) : \theta_i \to \theta, \, \theta_i \notin \Omega \right\}, \tag{17}$$

with $\overline{co}$ being the convex closure, $F(\theta_i) \in \mathbb{R}^{n \times m}$ the classical Jacobian whenever it exists, and Ω the set of measure zero where $F(\theta_i)$ is not defined. In other words, $F_G(\theta)$ is in general set-valued at the points where the Jacobian is not defined and $F_G(\theta) = \{F(\theta)\}$ when the Jacobian is defined. Consequently, in the case of inequality constraints, the CRLB is also, in general, set-valued, and it holds that

$$\text{CRLB}^{\text{constrained}} = \left\{ J^{-1}(\overline{\psi}) - J^{-1}(\overline{\psi}) F \left(F^T J^{-1}(\overline{\psi}) F \right)^{-1} F^T J^{-1}(\overline{\psi}) \, \middle| \, F \in F_G(\overline{\psi}) \right\}.$$

Notice that, in the event that $\overline{\psi}$ coincides with a point in the zero-measure set Ω, any metric based on the CRLB (e.g., variance for a specific parameter, norm of the matrix, etc.) is computed considering the worst case over the elements in the set $\text{CRLB}^{\text{constrained}}$.

2.7. Ownship Trajectory Selection Based on Artificial Potential Fields

The availability of additional information (i.e., knowledge on $\mathcal{P}$ and $\mathcal{S}$) can be leveraged by the ownship in order to select a trajectory that allows one to improve the accuracy of the estimate, i.e., by moving along a direction that mediates between the attempt to get closer to $\mathcal{P}$ and the will to avoid $\mathcal{S}$. In this paper, we borrow some of the key concepts of the so-called artificial potential fields (APF) technique [31,32] in order to accomplish this task. Within the APF approach, a robot has to navigate in a space toward a goal while avoiding one or more obstacles; this is achieved by associating a repulsive potential field to each obstacle and an attractive potential field to the goal so that, depending on the robot's position, the robot moves in the direction of the force that corresponds to the antigradient of the overall potential field.

For the application at hand in this paper, we consider a repulsive potential field to be associated with each set $\mathcal{Y}_i$ that the target is assumed not to cross and an attractive potential field associated with each set $\mathcal{X}_i$ where the target is assumed to be confined in. For simplicity, assume the ownship is initially in the origin, considering some fixed frame of reference.

In detail, referring to y_i and x_i as the center of mass of $\mathcal{Y}_i$ and $\mathcal{X}_i$, respectively, the potential field at a point $p \in \mathbb{R}^2$ is the superposition of a contribution

$$-\frac{1}{2} \alpha_i (x_i - p)^T (x_i - p)$$

for each set $\mathcal{X}_i$ and a contribution

$$\frac{1}{2} \beta_i (y_i - p)^T (y_i - p)$$

for each set $\mathcal{Y}_i$, i.e.,

$$U(p) = -\frac{1}{2} \sum_{i=1}^{m} \alpha_i (x_i - p)^T (x_i - p) + \frac{1}{2} \sum_{i=1}^{r} \beta_i (y_i - p)^T (y_i - p),$$

from which the corresponding force at the origin is

$$f(0_2) = -\nabla_p U(p) \Big|_{p=0_2} = \sum_{i=1}^{m} \alpha_i x_i - \sum_{i=1}^{m} \beta_i y_i,$$

and we note that the force is the composition of terms that are attracting toward the points x_i and terms that are repulsive from the points y_i.

Notice that, in this paper, we chose coefficients β_i that are proportional to the area of the corresponding set $\mathcal{Y}_i$ (i.e., the larger $\mathcal{Y}_i$ is, the more the force is repulsive); conversely, we chose coefficients α_i that are inversely proportional to the area of the corresponding set $\mathcal{Y}_i$ (i.e., the smaller $\mathcal{X}_i$ is, the more the force is attractive).

Ownship Trajectory

Based on the resulting force $f(0_2)$ at the origin, the ownship computes the angle γ between the vector $\begin{bmatrix} 0 & 1 \end{bmatrix}^T$ and $f(0_2)$ and moves according to

$$\begin{bmatrix} x_o(kT) \\ y_o(kT) \end{bmatrix} = \begin{bmatrix} \cos(\gamma) & -\sin(\gamma) \\ \sin(\gamma) & \cos(\gamma) \end{bmatrix} \begin{bmatrix} \bar{x}_o kT \\ A\sin(\omega x_o(kT)) \end{bmatrix}, \tag{18}$$

i.e., the ownship moves of linear motion (and constant velocity $\bar{x}_o$) along the direction of $f(0_2)$ while performing sinusoidal motion around such a direction. Notably, Equation (18) can be rearranged as

$$\begin{bmatrix} x_o(kT) \\ y_o(kT) \end{bmatrix} = \begin{bmatrix} \cos(\gamma)\bar{x}_o kT - A\sin(\gamma)\sin(\omega x_o(kT)) \\ \sin(\gamma)\bar{x}_o kT + A\cos(\gamma)\sin(\omega x_o(kT)) \end{bmatrix}. \tag{19}$$

Figure 1 provides an example of the above procedure for the selection of $f(0_2)$.

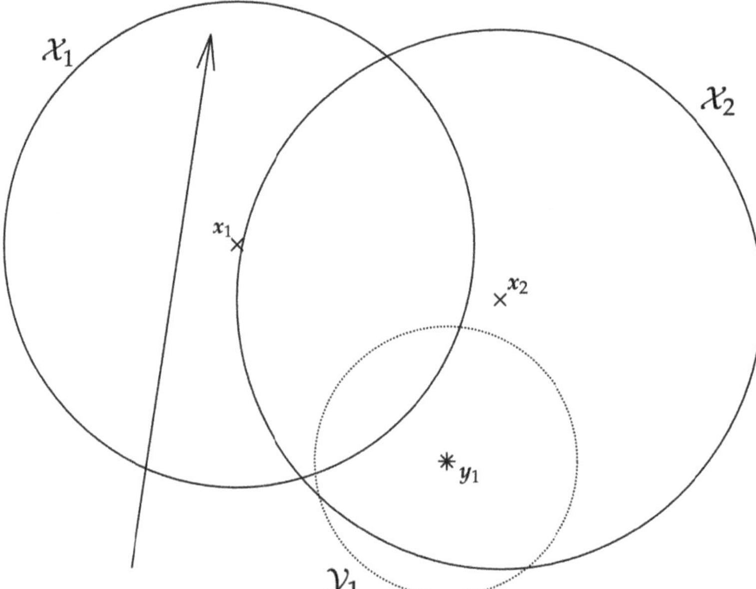

Figure 1. Example of the proposed approach for selecting the ownship's trajectory. In this example, we consider two sets, $\mathcal{X}_1$ and $\mathcal{X}_2$, and one setm $\mathcal{Y}_1$, represented by the circles shown with a solid line and by a dotted line, respectively. The points x_1 and x_2 and the point y_1 (i.e., the centers of the circles) are shown with an x mark and with a cross, respectively. In this example, we choose α_1 and α_2 equal to the area of $\mathcal{X}_1$ and $\mathcal{X}_2$, respectively, while β_1 is the reciprocal of the area of $\mathcal{Y}_1$. The resulting direction for the ownship is shown with an arrow (the initial position for the ownship is given by the starting endpoint of the arrow).

3. Experimental Analysis

In order to experimentally demonstrate the effectiveness of the proposed approach, we consider the scenario depicted in Figure 2, where the target has an initial position $x_{t0} = y_{t0} = 3 \times 10^4$ m and moves with constant velocity having components $\dot{x}_{t0} = 8.333$ m/s and $\dot{y}_{t0} = 7.778$ m/s, while the acceleration is $\ddot{x}_{t0} = \ddot{y}_{t0} = 0$ m/s^2.

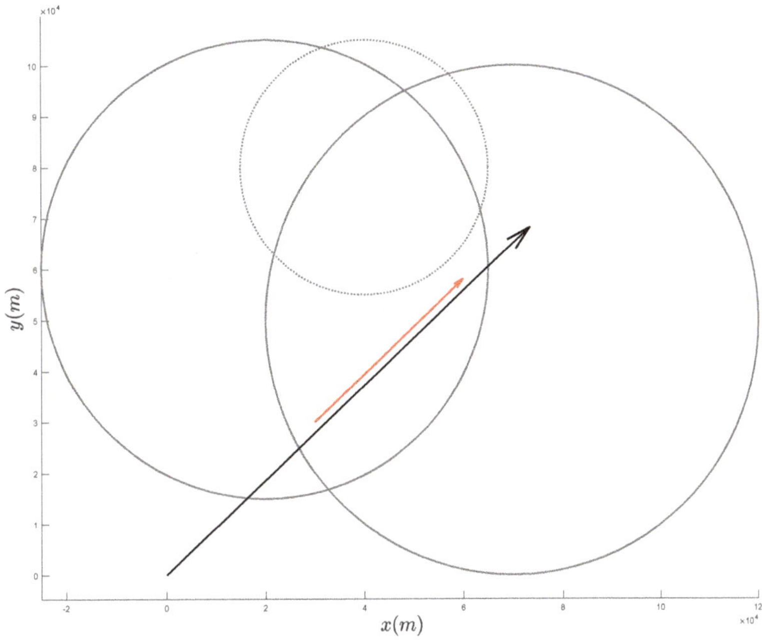

Figure 2. Scenario considered in the experimental analysis. We assume additional information is available, in that the target's trajectory is known to be confined in the intersection of the two solid circles and to lie outside the dotted circle. The red (shorter) arrow represents the target's trajectory. The APF direction is shown with a black (longer) arrow.

Notice that we assume the additional information is available to the ownship regarding the target's trajectory. Specifically, we assume the target's trajectory lies in the intersection of the circles $\mathcal{X}_1$ and $\mathcal{X}_2$ and outside of the circle $\mathcal{Y}_1$; the centers x_1, x_2, y_1 and the radii $\rho_{\mathcal{X}_1}, \rho_{\mathcal{X}_2}$ and $\rho_{\mathcal{Y}_1}$ of such circles are reported in Table 1.

Table 1. Parameters describing the sets $\mathcal{X}_1, \mathcal{X}_2$ and $\mathcal{Y}_1$ considered in the experimental analysis.

Set	Centroid	Radius
$\mathcal{X}_1$	$x_1 = [10^4, 5 \times 10^4]^T$ m	$\rho_{\mathcal{X}_1} = 5 \times 10^4$ m
$\mathcal{X}_2$	$x_2 = [2 \times 10^4, 6 \times 10^4]^T$ m	$\rho_{\mathcal{X}_2} = 4.5 \times 10^4$ m
$\mathcal{Y}_1$	$y_1 = [4 \times 10^4, 8 \times 10^4]^T$ m	$\rho_{\mathcal{Y}_1} = 2.5 \times 10^4$ m

In our simulations, we consider a sampling time $T = 2$ s, a time horizon of 3600 s and noise variance set to $\sigma = 0.5°$.

As for the ownship, we consider four operational scenarios:

- No information: the ownship does not rely on the additional information and selects the trajectory in Equation (19) with $\gamma = 0$ rad.

- Unconstrained, APF direction: the ownship does not rely on the additional information for the computations but selects the trajectory according to the proposed APF approach; in other words, it selects the trajectory in Equation (19) with $\gamma = 0.749$ rad.
- Constrained, no APF direction: the ownship relies on the additional information for the computations but selects the trajectory in Equation (19) with $\gamma = 0$ rad.
- Proposed Approach: the ownship follows the trajectory in Equation (19) with $\gamma = 0.749$ rad and, further to that, actively relies on the additional information during the computation of the MLE solution, of the experimental covariance matrix and CRLB.

In all cases, we select the following parameters for the ownship: $x_o(0) = y_o(0) = 0$ m, $\dot{x}_o = 7.778$ m/s, $\omega = 2.5 \times 10^{-4}$ rad/s and $A = 5 \times 10^3$ m.

The computation of the MLE solution with MIDACO-SOLVER was conducted on an Intel® i7 quad-core @ 2.27 GHz. For each execution of MIDACO-SOLVER, we set the number of evaluated solutions to 10^6. All other MIDACO-SOLVER parameters were used by their default values, which especially means that a feasibility accuracy of 0.001 was used for all individual constraints. In all cases, we compute the MLE solution with MIDACO-SOLVER, and we consider 100 MLE solutions for each operational scenario.

Table 2 reports the results obtained for the four operational scenarios, considering both the best solution found (in terms of the objective function of the MLE) over the 100 runs and the average of the solutions found. For each solution reported in the table, we use, as a measure of quality of the estimate the relative position and relative velocity, defined as follows:

$$\text{rel. pos. err.} = \left\| \begin{bmatrix} \dfrac{x_{t0} - x_{t0}^*}{x_{t0}} \\ \dfrac{y_{t0} - y_{t0}^*}{y_{t0}} \end{bmatrix} \right\|_2 \tag{20}$$

$$\text{rel. vel. err.} = \left\| \begin{bmatrix} \dfrac{\dot{x}_{t0} - \dot{x}_{t0}^*}{\dot{x}_{t0}} \\ \dfrac{\dot{y}_{t0} - \ddot{y}_{t0}^*}{\dot{y}_{t0}} \end{bmatrix} \right\|_2 \tag{21}$$

while, since the target moves of with zero acceleration, we consider the absolute acceleration error

$$\text{abs. acc. err.} = \left\| \begin{bmatrix} \ddot{x}_t^* \\ \ddot{y}_t^* \end{bmatrix} \right\|_2 \tag{22}$$

where the asterisk superscript is used for the estimated parameters. According to the table, the best solution found in the first and proposed operative scenarios is comparable and exhibits small error, while the error is larger for the third operational scenario. Conversely, within the second operational scenario, the best solution also found is not satisfactory, due to large relative velocity error. The situation is radically different considering the average of the found solutions; in fact, we observe that while the proposed approach shows limited error, the first and second operational conditions are characterized by erroneous average solutions (especially for what concerns the estimate of the target's initial velocity), while the third one exhibits a limited but significant degradation. This suggests that, while the proposed approach consistently finds a good solution over the different trials, the other methods may become trapped by worst local minima.

This intuition is supported by Figure 3, where we show the distribution of the position and velocity error values over the 100 trials. According to the figure, it can be noted that the first two scenarios have, overall, worse performance than the second and third scenario. Moreover, while outliers in the first two operative scenarios assume remarkably large values, in the latter two scenarios the outliers exhibit only a limited increase.

Table 2. MLE parameter estimation via MIDACO-SOLVER for the different operational scenarios.

	x_{t0} m	y_{t0} m	$\dot{x}_{t0}$ m/s	$\dot{y}_{t0}$ m/s	$\ddot{x}_t$ m/s^2	$\ddot{y}_t$ m/s^2	rel. pos. err.	rel. vel. err.	abs. acc. err.
Ground truth	3.000×10^4	3.000×10^4	8.333	7.778	0.000	0.000	-	-	-
No information, average	3.933×10^4	3.848×10^4	-5.048	-4.123	3.437×10^{-3}	2.303×10^{-3}	4.202×10^{-1}	2.218	4.137×10^{-3}
Unconstrained, APF direction, average	3.796×10^4	3.733×10^3	-4.714	-4.143	4.780×10^{-3}	3.958×10^{-3}	3.605×10^{-1}	2.191	6.206×10^{-3}
Constrained, no APF direction, average	3.204×10^4	3.193×10^4	7.161	6.884	2.543×10^{-4}	4.987×10^{-5}	9.384×10^{-2}	1.815×10^{-1}	2.591×10^{-4}
Proposed Approach, average	3.040×10^4	3.035×10^4	7.912	7.382	2.652×10^{-4}	2.598×10^{-4}	1.767×10^{-2}	7.169×10^{-2}	3.713×10^{-4}
No information, best	3.028×10^4	3.024×10^4	8.045	7.507	1.989×10^{-4}	1.969×10^{-4}	1.243×10^{-2}	4.902×10^{-2}	2.789×10^{-4}
Unconstrained, APF direction, best	3.704×10^4	3.640×10^4	0.686	0.618	3.754×10^{-3}	3.560×10^{-3}	3.171×10^{-1}	1.299	5.173×10^{-3}
Constrained, no APF direction, best	2.811×10^4	2.806×10^4	10.081	7.382	-1.460×10^{-3}	-2.202×10^{-3}	9.023×10^{-2}	3.412×10^{-1}	2.642×10^{-3}
Proposed Approach, best	3.026×10^4	3.023×10^4	8.070	7.531	1.835×10^{-4}	1.821×10^{-4}	1.179×10^{-2}	4.471×10^{-2}	2.586×10^{-4}

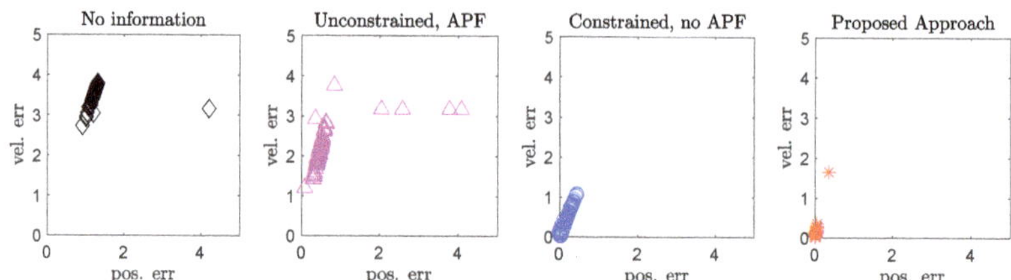

Figure 3. Ensemble view of the (relative) position and velocity error values over 100 runs (i.e., see Equations (20) and (21)), considering the four operative scenarios.

In order to further analyze the different operational scenarios, Table 3 reports the norm of the covariance matrix obtained experimentally from the 100 trials, the experimental covariance limited to the solutions with errors within the 50th percentile, and the norm of the CRLB. In other words, we experimentally evaluate the covariance by executing 100 instances of MIDACO-SOLVER with random initial choice for the parameters and then taking the covariance of the 100 (or less, when only the 50th percentile is considered), possibly different, solutions obtained via MIDACO-SOLVER.

Notably, for the proposed approach, due to the presence of inequality constraints, the CRLB is in general set-valued and, following a worst-case philosophy, when the set is not a singleton, we consider

$$\max_{C \in \text{CRLB}^{\text{constrained}}} \|C\|_2. \tag{23}$$

According to the table, the magnitude of the norm of the experimental covariance matrices associated with the proposed approach is three orders of magnitude smaller than those corresponding to the first and second operational scenarios, while it is two orders of magnitude smaller than the covariance associated to the third operational scenario. Moreover, the norm of the CRLB covariance matrices is two orders of magnitude smaller than the one obtained for the first and second operational scenarios, while it is comparable to the one associated with the third operational scenario. Moreover, we observe that, while in the other cases the experimental covariance is between three and four orders of magnitude larger than the CRLB, for the proposed approach, the experimental covariance is just two orders of magnitude larger than the CRLB. Notably, the discrepancy experienced between the experimental covariance computed over 100 trials and the CRLB one is due to the structure of the problem at hand. In fact, the MLE problem being solved amounts to a nonconvex, nonlinear programming problem and is thus characterized by the presence of local minima. Since we are adopting an approximated solver for finding a solution, the large experimental covariance is explained by the reach of different local minima across the different trials. In fact, while analyzing the covariance restricted to solutions with errors within the 50th percentile, we observe a significant drop with respect to the covariance over all trials; in particular, we observe a reduction of two orders of magnitude for the first two operational scenarios and one order of magnitude for the other two; moreover, also in this case, the proposed approach shows a two orders of magnitude reduction of the covariance with respect to the others.

Let us now discuss the computation times, which are collected in Table 4. According to the table, we observe that the main differences arise between the unconstrained and the constrained cases, the latter requiring a computational time that is, on average, about 46% larger than the unconstrained case, while the standard deviation is sensibly larger, being about 6.92 times the one obtained in the unconstrained case.

Table 3. Euclidean norm of the parameter estimation via MIDACO-SOLVER for the different operational scenarios.

	Experimental Covariance $\|\cdot\|_2$	Experimental Covariance $\|\cdot\|_2$ (Solutions with Errors $\leq$ 50th Percentile)	CRLB $\|\cdot\|_2$
No information	3.851×10^8	3.743×10^6	7.713×10^5
Unconstrained, APF direction	5.666×10^8	4.144×10^6	7.713×10^5
Constrained, no APF direction	3.696×10^7	5.410×10^6	1.454×10^3
Proposed Approach	4.784×10^5	6.600×10^4	1.873×10^3

Table 4. Average and standard deviation over 100 trials of the computational time (in seconds) for the computation of the MLE solution via MIDACO-SOLVER for the different operational scenarios.

	Time (Average) s	Time (Standard Deviation) s
No information	58.790	3.235
Unconstrained, APF direction	59.124	3.049
Constrained, no APF direction	85.926	21.231
Proposed Approach	86.230	22.395

Overall, the above results suggest that, while the implementation of the constrained MLE computation without considering the APF direction has a positive effect on the estimate, the APF direction alone has no particular benefit without constraining the MLE based on the available information. Instead, when such innovations are combined, the MLE error, the experimental covariance and the CRLB are greatly reduced. Notably, since the proposed approach amounts to a constrained problem, the price to pay is a noticeable but limited increase in the computation times.

Figures 4–9 provide an at-a-glance view of the performance gap when the additional information is actively relied upon. Specifically, Figures 4 and 5 show the results of the proposed operational scenario considering the average MLE solution over the 100 trials (blue dashed line), along with 100 trajectories (in cyan) obtained by sampling the parameters from a Gaussian distribution with mean given by the average MLE result and covariance corresponding to the experimental covariance matrix or the CRLB, respectively (the ownship's trajectory is shown via the blue sinusoidal dashed curve). Conversely, Figure 6 shows the results obtained considering the case where the APF direction is used but the additional information is not relied upon during the MLE computation; specifically, the figure considers the best solution found and the CRLB matrix. Notably, in the latter case, the large covariance yields sampled trajectories that are characterized by remarkably large error, while the proposed approach (both considering the experimental and CRLB covariance matrices) yields significantly better results. Figure 7 shows a zoomed version of Figure 6; according to the figure, we observe that, while the APF approach without using the additional information in the MLE computations is characterized by an initial position that is relatively accurate, the velocities and accelerations are characterized by highly variable and erroneous values, resulting in the inaccurate trajectories. Finally, Figures 8 and 9 show the results obtained considering the case where the APF direction is not used but the additional information is relied upon during the MLE computation; specifically, Figure 8 consider the best solution found and the experimental covariance matrix, while Figure 8 considers the best solution found and the CRLB matrix; in this case, the trajectory is characterized by an intermediate error and is outperformed by the proposed approach.

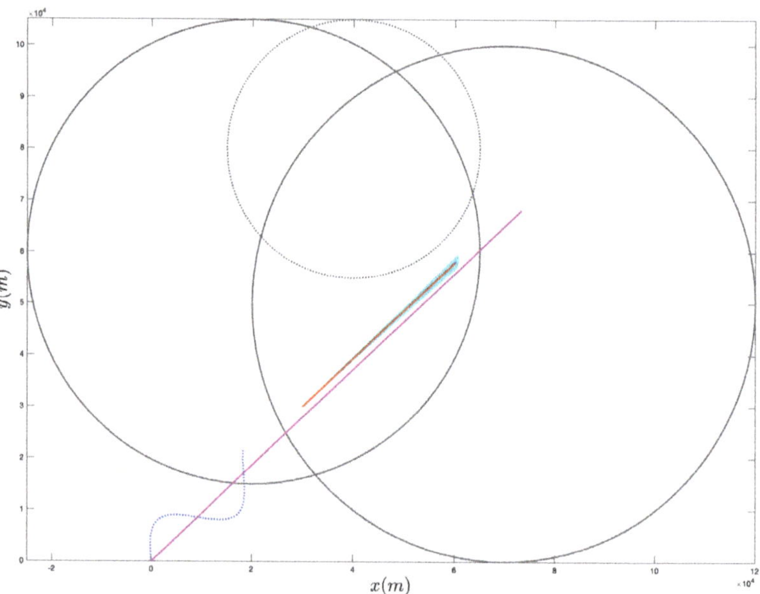

Figure 4. Sampling of 100 trajectories (in cyan) based on the average solution found via the proposed approach and on the experimental covariance matrix.

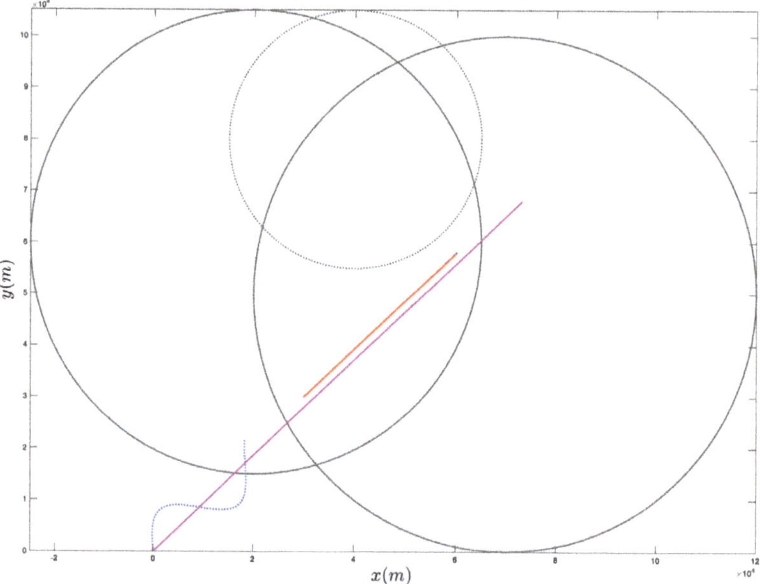

Figure 5. Sampling of 100 trajectories (cyan) based on the average solution found via the proposed approach and on the CRLB covariance matrix.

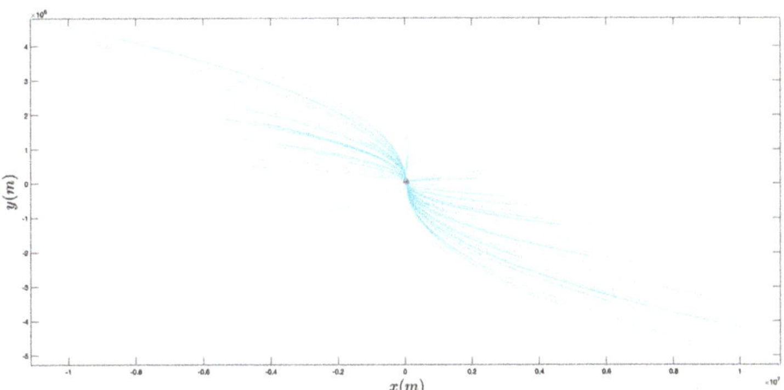

Figure 6. Sampling of 100 trajectories (cyan) based on the best solution found via the unconstrained, APF direction approach and on the CRLB covariance matrix.

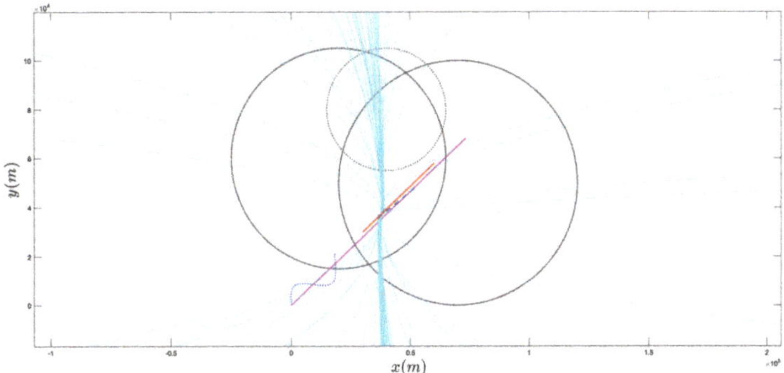

Figure 7. Zoomed version of a portion of Figure 6.

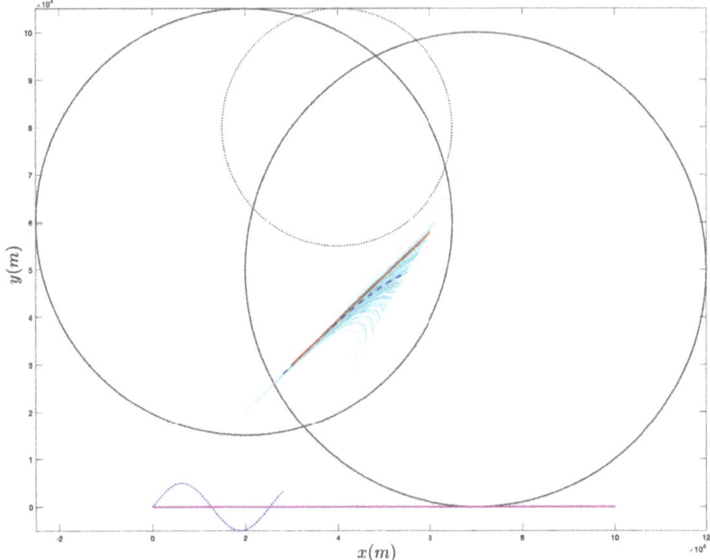

Figure 8. Sampling of 100 trajectories (cyan) based on the best solution found via for the constrained case, but without relying on the APF, and on the experimental covariance matrix.

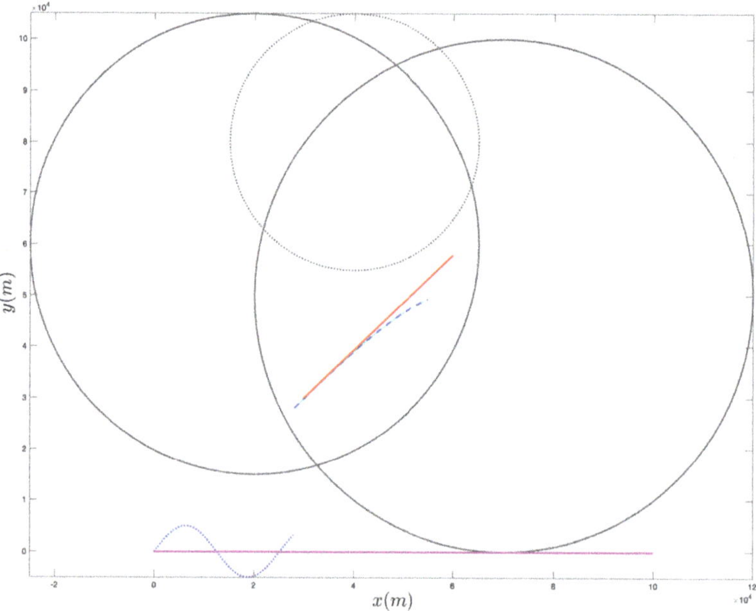

Figure 9. Sampling of 100 trajectories (cyan) based on the best solution found via for the constrained case, but without relying on the APF, and on the CRLB covariance matrix.

To conclude the section, we compare the proposed approach against other methods in the literature. Specifically, our proposed comparison is based on two macro-steps: (i) the estimation of the target's trajectory for $k \in \{0, \ldots k_{\max}\}$ via other approaches in the literature and (ii) the comparison with the trajectory obtained based on the estimation of the parameter vector ψ via the proposed approach. In particular, we estimate the target's trajectory by resorting to the following four algorithms:

- a standard extended Kalman filter (EKF) (e.g., see [44] and references therein), where we approximate the nonlinear output function $h(\cdot)$ by its Jacobian matrix at each time instant;

- the pseudolinear Kalman filter (PL-KF) [45], where the nonlinear and noisy output $z(k) = \mathrm{atan2}(y_t(k) - y_o(k), x_t(k) - x_o(k)) + w(k)$ is approximated by

$$\tilde{z}(k) = \begin{bmatrix} \sin(z(k)) \\ -\cos(z(k)) \end{bmatrix} \underbrace{\begin{bmatrix} 1 & 0 & 0 & 0 & 0 & 0 \\ 0 & 1 & 0 & 0 & 0 & 0 \end{bmatrix}}_{M} \widehat{x}_{k|k} + \eta(k),$$

where $\widehat{x}_{k|k}$ is the vector collecting the filtered states (i.e., the stack of the positions, velocities and accelerations) of the target at time k and $\eta(k)$ is a pseudolinear noise in the form

$$\eta(k) \sim \mathcal{N}(0, R_k),$$

with

$$R_k \approx \|\widehat{d}_{k|k-1}\|^2 \sigma^2$$

and

$$\widehat{d}_{k|k-1} = M\widehat{x}_{k|k-1} - \begin{bmatrix} x_o(k) \\ y_o(k) \end{bmatrix},$$

$\widehat{x}_{k|k-1}$ being the vector collecting the prediction of the target's states at time k;

- a statistical linearization extended Kalman filter (SL-EKF) [46] where, instead of the Jacobian of the output function $h(\cdot)$, we approximate the nonlinear measurement function $y(k) = h(x, k) + w(k)$ via the linear approximation H^*x, with

$$H^* = \underset{H \in \mathbb{R}^{2 \times 6}}{\arg\min} \|y(k) - H\hat{x}_{k|k}\|^2.$$

- the shifted Rayleigh filter (SRF) [47], where $z(k)$ is approximated by

$$z(k) \approx \Pi(Mx(k) + u(k) + \omega(k)),$$

where $\Pi(r) = r/\|r\|$, $u(k) = \begin{bmatrix} x_o(k) & y_o(k) \end{bmatrix}$ and

$$\omega(k) \sim \mathcal{N}\left(0_2, \sigma^2 E\left[\|Mx(k) + u(k)\|^2 \,\big|\, z(1), \ldots, z(k)\right] I_2\right).$$

Conversely, within the proposed approach, we estimate the parameter vector

$$\psi = \begin{bmatrix} x_{t0} & y_{t0} & \dot{x}_{t0} & \dot{y}_{t0} & \ddot{x}_t & \ddot{y}_t \end{bmatrix}^T$$

via the proposed constrained MLE formulation, and we compute the trajectory of the target as follows

$$x(k) = Q(k)\psi,$$

where

$$x(k) = \begin{bmatrix} x_t(k) & y_t(k) & \dot{x}_t(k) & \dot{y}_t(k) & \ddot{x}_t(k) & \ddot{y}_t(k) \end{bmatrix}^T,$$

and

$$Q(k) = \begin{bmatrix} I_2 & kTI_2 & \frac{1}{2}k^2T^2I_2 \\ 0_{2\times2} & I_2 & kTI_2 \\ 0_{2\times2} & 0_{2\times2} & I_2 \end{bmatrix}.$$

Regarding the initial condition for the algorithms compared against the proposed approach, we consider three different scenarios, with a decreasing degree of uncertainty:

1. a scenario where the average initial condition $\hat{x}_{0|0}$ is drawn from a zero-mean Gaussian variable with standard deviation equal to ψ, while the initial covariance $\Sigma_{0|0}$ is equal to the square of ψ, i.e.,

$$\hat{x}_{0|0} \sim \mathcal{N}\left(0_6, \mathrm{diag}(\psi)^2\right), \quad \Sigma_{0|0} = \mathrm{diag}(\psi)^2;$$

2. a scenario where the average initial condition is drawn from a Gaussian variable with a mean equal to ψ and standard deviation equal to ψ, while the initial covariance is equal to the square of ψ, i.e.,

$$\hat{x}_{0|0} \sim \mathcal{N}\left(\psi, \mathrm{diag}(\psi)^2\right), \quad \Sigma_{0|0} = \mathrm{diag}(\psi)^2$$

3. a scenario where the average initial condition is exactly ψ, while the initial covariance is equal to the square of ψ, i.e.,

$$\hat{x}_{0|0} = \psi, \quad \Sigma_{0|0} = \mathrm{diag}(\psi)^2.$$

Figures 10–12 report the results of the comparison, where the four aforementioned approaches are compared against the average MLE solution over 100 trials obtained via the proposed methodology, considering the same simulation setting as in the fourth operational scenario described above. Specifically, we report the relative position and velocity errors between the real target's trajectory at time k and the estimated one, i.e.,

$$\frac{x_t(k) - \widehat{x}(k)}{x_t(k)}, \quad \frac{y_t(k) - \widehat{y}(k)}{y_t(k)}, \quad \frac{\dot{x}_t(k) - \widehat{\dot{x}}(k)}{\dot{x}_t(k)}, \quad \frac{\dot{y}_t(k) - \widehat{\dot{y}}(k)}{\dot{y}_t(k)},$$

where we denote by $\widehat{x}(k), \widehat{y}(k), \widehat{\dot{x}}(k)$ and $\widehat{\dot{y}}(k)$, the estimate of the target's positions and velocities obtained according to the generic technique being compared. According to Figure 10, the proposed approach achieves an error that is at least two orders of magnitude less than the other approaches. Notably, while moving to a scenario where more information on the initial condition is available for the four approaches used in the comparison, the gap with PLKF and SL-EKF reduces to about one order of magnitude. Finally, when we compare the initial condition for the methods against the proposed one, which is assumed to have a mean that corresponds to the actual vector of parameter being estimated, we observe that the proposed approach is comparable with SL-EKF, while being in general better than the others.

Overall, the proposed comparison contributes to experimentally demonstrating the effectiveness of the proposed approach with respect to the literature.

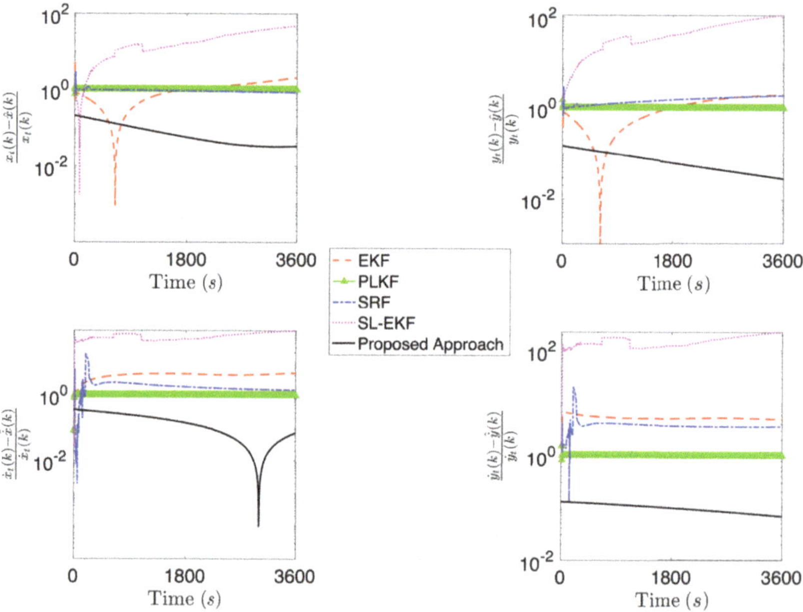

Figure 10. Comparison of the proposed approach against EKF, PLKF, SRF and SL-EKF, considering a scenario where the methods compared with the proposed one are initialized with $\widehat{x}_{0|0} \sim \mathcal{N}(0_6, \mathrm{diag}(\boldsymbol{\psi})^2)$ and $\Sigma_{0|0} = \mathrm{diag}(\boldsymbol{\psi})^2$.

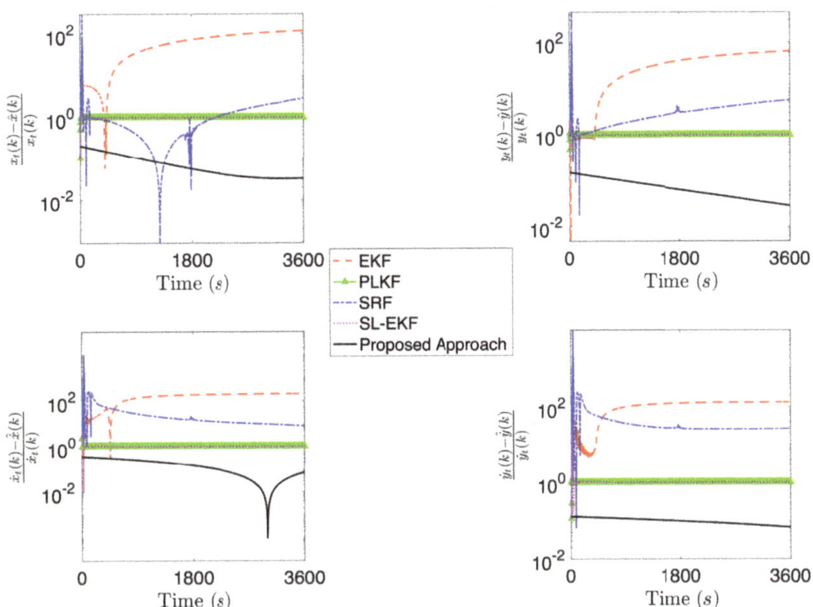

Figure 11. Comparison of the proposed approach against EKF, PLKF, SRF and SL-EKF, considering a scenario where the methods compared with the proposed one are initialized with $\widehat{x}_{0|0} \sim \mathcal{N}(\boldsymbol{\psi}, \mathrm{diag}(\boldsymbol{\psi})^2)$ and $\Sigma_{0|0} = \mathrm{diag}(\boldsymbol{\psi})^2$.

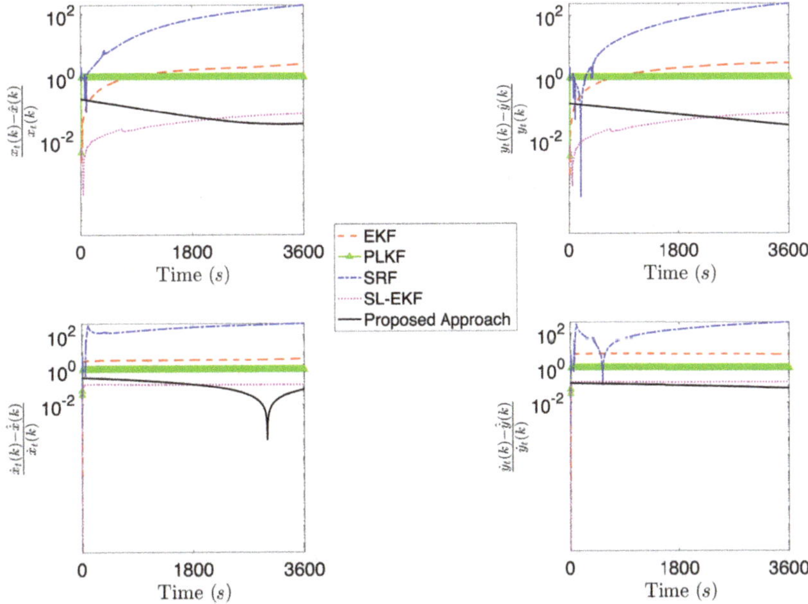

Figure 12. Comparison of the proposed approach against EKF, PLKF, SRF and SL-EKF, considering a scenario where the methods compared with the proposed one are initialized with $\widehat{x}_{0|0} = \boldsymbol{\psi}$ and $\Sigma_{0|0} = \mathrm{diag}(\boldsymbol{\psi})^2$.

4. Conclusions and Future Work

This paper presents a batch strategy to estimate the parameters describing the trajectory of a target, based on a moving ownship able to measure bearings. In particular, the proposed methodology allows one to incorporate additional information (e.g., obtained via intelligence) such as knowledge of the fact that the target's trajectory is contained in the intersection of some sets or the fact it lies outside the union of other sets. The approach is formally characterized by providing a constrained MLE formulation and by extending the definition of the CRLB matrix to the case of MLE problems with inequality constraints, relying on the concept of generalized Jacobian matrix. Moreover, based on the additional information, the ownship motion is chosen by mimicking the Artificial Potential Fields technique that is typically used by mobile robots to aim to a goal (in this case, the region where the target is assumed to be) while avoiding obstacles (i.e., the region that is assumed not to intersect with the target's trajectory). As a result, the proposed framework exhibits remarkably better performance, and in particular, we observe that the solution is less likely to remain stuck in unsatisfactory local minima during the MLE computation and is characterized by smaller covariance, (both considering the experimental and the CRLB ones).

Future work will be mainly devoted to extending the framework along the following research directions: (i) consider more sophisticated models for the target's motion (e.g., nearly constant acceleration); (ii) consider dynamically changing constraints on the target's trajectory; (iii) provide adaptive strategies for the ownship trajectory based on the, although partial, initial estimates (e.g., in order to avoid crossing the space where the target's trajectory is contained); (iv) filter possible outliers (e.g., resorting to the approach in [3], Section 2.7); and (v) consider multiple targets.

Author Contributions: Conceptualization, G.O., R.S. and A.F.; methodology, A.F.; software, G.O.; validation, R.S.; formal analysis, R.S.; investigation, R.S.; resources, G.O. and A.F.; data curation, G.O. and A.F.; writing—original draft preparation, G.O.; writing—review and editing, R.S. and A.F.; visualization, G.O.; supervision, A.F. All authors have read and agreed to the published version of the manuscript.

Funding: This work was supported by POR FESR Lazio Region Project RESIM A0375-2020-36673 (CUP: F89J21004860008).

Institutional Review Board Statement: Not applicable.

Informed Consent Statement: Not applicable.

Data Availability Statement: The data presented in this study are available on reasonable request from the corresponding author.

Acknowledgments: The authors would like to express their sincere gratitude for the useful comments and suggestions by the guest editors and the anonymous reviewers during the review process.

Conflicts of Interest: The authors declare no conflict of interest.

Abbreviations

The following abbreviations are used in this manuscript:

APF	Artificial potential fields
CRLB	Cramér–Rao lower bound
DOAJ	Directory of open access journals
EKF	Extended Kalman filter
FIM	Fisher information matrix
MAP	Maximum a posteriori
MDPI	Multidisciplinary Digital Publishing Institute
MIDACO-SOLVER	Mixed integer distributed ant colony optimization solver
MLE	Maximum likelihood estimation

PL-KF	Pseudo-linear Kalman filter
SL-EKF	Statistical linearization extended Kalman filter
SRF	Shifted Rayleigh filter
TMA	Target motion analysis

Appendix A

In this appendix we provide the analytical expression of the partial derivatives of the function $h(\boldsymbol{\psi}, k)$ with respect to the target's motion parameters. For the sake of readability let us define

$$
\begin{aligned}
\Delta_x(k) &= x_t(k) - x(k), \\
\Delta_y(k) &= y_t(k) - y(k).
\end{aligned}
\tag{A1}
$$

The partial derivatives of $h(\boldsymbol{\psi}, k)$ with respect to the target's motion parameters are as follows

$$
\begin{aligned}
\frac{\partial h(\boldsymbol{\psi}, k)}{\partial x_{t0}} &= -\frac{\Delta_y(k)}{\Delta_x(k)^2 + \Delta_y(k)^2} \\[2mm]
\frac{\partial h(\boldsymbol{\psi}, k)}{\partial y_{t0}} &= \frac{\Delta_x(k)}{\Delta_x(k)^2 + \Delta_y(k)^2} \\[2mm]
\frac{\partial h(\boldsymbol{\psi}, k)}{\partial \dot{x}_{t0}} &= -kT\frac{\Delta_y(k)}{\Delta_x(k)^2 + \Delta_y(k)^2} \\[2mm]
\frac{\partial h(\boldsymbol{\psi}, k)}{\partial \dot{y}_{t0}} &= kT\frac{\Delta_x(k)}{\Delta_x(k)^2 + \Delta_y(k)^2} \\[2mm]
\frac{\partial h(\boldsymbol{\psi}, k)}{\partial \ddot{x}_{t0}} &= -\frac{1}{2}k^2 T^2 \frac{\Delta_y(k)}{\Delta_x(k)^2 + \Delta_y(k)^2} \\[2mm]
\frac{\partial h(\boldsymbol{\psi}, k)}{\partial \ddot{y}_{t0}} &= \frac{1}{2}k^2 T^2 \frac{\Delta_x(k)}{\Delta_x(k)^2 + \Delta_y(k)^2}.
\end{aligned}
\tag{A2}
$$

References

1. Nardone, S.C.; Graham, M.L. A closed-form solution to bearings-only target motion analysis. *IEEE J. Ocean. Eng.* **1997**, *22*, 168–178. [CrossRef]
2. Song, T.L.; Um, T.Y. Practical guidance for homing missiles with bearings-only measurements. *IEEE Trans. Aerosp. Electron. Syst.* **1996**, *32*, 434–443. [CrossRef]
3. Farina, A. Target tracking with bearings–only measurements. *Signal Process.* **1999**, *78*, 61–78. [CrossRef]
4. Oh, R.; Song, T.L.; Choi, J.W. Batch Processing through Particle Swarm Optimization for Target Motion Analysis with Bottom Bounce Underwater Acoustic Signals. *Sensors* **2020**, *20*, 1234. [CrossRef]
5. Kronhamn, T. Bearings-only target motion analysis based on a multihypothesis Kalman filter and adaptive ownship motion control. *IEE Proc.-Radar Sonar Navig.* **1998**, *145*, 247–252. [CrossRef]
6. Mehrjouyan, A.; Alfi, A. Robust adaptive unscented Kalman filter for bearings-only tracking in three dimensional case. *Appl. Ocean. Res.* **2019**, *87*, 223–232. [CrossRef]
7. Daowang, F.; Teng, L.; Tao, H. Square-root second-order extended Kalman filter and its application in target motion analysis. *IET Radar Sonar Navig.* **2010**, *4*, 329–335. [CrossRef]
8. Liu, J.; Guo, G. A Recursive Estimator for Pseudolinear Target Motion Analysis Using Multiple Hybrid Sensors. *IEEE Trans. Instrum. Meas.* **2021**, *70*. [CrossRef]
9. Nardone, S.; Lindgren, A.; Gong, K. Fundamental properties and performance of conventional bearings-only target motion analysis. *IEEE Trans. Autom. Control* **1984**, *29*, 775–787. [CrossRef]
10. Blackman, S.; Popoli, R. *Design and Analysis of Modern Tracking Systems(Book)*; Artech House: Norwood, MA, USA, 1999.
11. Wang, Y.; Bai, Y.; Wang, X.; Shan, Y.; Shui, Y.; Cui, N.; Li, Y. Event-based distributed bias compensation pseudomeasurement information filter for 3D bearing-only target tracking. *Aerosp. Sci. Technol.* **2021**, *117*, 106956. [CrossRef]
12. Voronina, N.G.; Shafranyuk, A.V. Algorithm for constructing trajectories of maneuvering object based on bearing-only information using the Basis Pursuit method. *J. Phys. Conf. Ser.* **2021**, *1864*, 012139. [CrossRef]

13. Shalev, H.; Klein, I. BOTNet: Deep Learning-Based Bearings-Only Tracking Using Multiple Passive Sensors. *Sensors* **2021**, *21*, 4457. [CrossRef]

14. Miller, A.B.; Miller, B.M. Underwater target tracking using bearing-only measurements. *J. Commun. Technol. Electron.* **2018**, *63*, 643–649. [CrossRef]

15. Hou, X.; Zhou, J.; Yang, Y.; Yang, L.; Qiao, G. Adaptive Two-Step Bearing-Only Underwater Uncooperative Target Tracking with Uncertain Underwater Disturbances. *Entropy* **2021**, *23*, 907. [CrossRef]

16. Han, X.; Liu, M.; Zhang, S.; Zhang, Q. A multi-node cooperative bearing-only target passive tracking algorithm via UWSNs. *IEEE Sens. J.* **2019**, *19*, 10609–10623. [CrossRef]

17. Wu, K.; Hu, J.; Lennox, B.; Arvin, F. Finite-time bearing-only formation tracking of heterogeneous mobile robots with collision avoidance. *IEEE Trans. Circuits Syst. II Express Briefs* **2021**, *68*, 3316–3320. [CrossRef]

18. Zhao, S.; Zhenhong, L.; Zhengtao, D. Bearing-only formation tracking control of multiagent systems. *IEEE Trans. Autom. Control* **2019**, *64*, 4541–4554. [CrossRef]

19. Hejazi, F.; Joneidi, M.; Rahnavard, N. A tensor-based localization framework exploiting phase interferometry measurements. In Proceedings of the 2020 IEEE International Radar Conference (RADAR), Washington, DC, USA, 28–30 April 2020; pp. 554–559.

20. García-Fernández, Á.F.; Tronarp, F.; Särkkä, S. Gaussian target tracking with direction-of-arrival von Mises–Fisher measurements. *IEEE Trans. Signal Process.* **2019**, *67*, 2960–2972. [CrossRef]

21. Doğançay, K. 3D pseudolinear target motion analysis from angle measurements. *IEEE Trans. Signal Process.* **2015**, *63*, 1570–1580. [CrossRef]

22. Huang, G.; Zhou, K.; Trawny, N.; Roumeliotis, S.I. A bank of maximum a posteriori (MAP) estimators for target tracking. *IEEE Trans. Robot.* **2015**, *31*, 85–103. [CrossRef]

23. Oh, R.; Shi, Y.; Choi, J.W. A Hybrid Newton–Raphson and Particle Swarm Optimization Method for Target Motion Analysis by Batch Processing. *Sensors* **2021**, *21*, 2033. [CrossRef]

24. Santi, F.; Pastina, D.; Bucciarelli, M. Experimental demonstration of ship target detection in GNSS-based passive radar combining target motion compensation and track-before-detect strategies. *Sensors* **2020**, *20*, 599. [CrossRef]

25. Lebon, A.; Perez, A.C.; Jauffret, C.; Laneuville, D. TMA from Cosines of Conical Angles Acquired by a Towed Array. *Sensors* **2021**, *21*, 4797. [CrossRef]

26. Bu, S.; Meng, A.; Zhou, G. A New Pseudolinear Filter for Bearings-Only Tracking without Requirement of Bias Compensation. *Sensors* **2021**, *21*, 5444. [CrossRef]

27. Kirubarajan, T.; Bar-Shalom, Y.; Pattipati, K.R.; Kadar, I. Ground target tracking with variable structure IMM estimator. *IEEE Trans. Aerosp. Electron. Syst.* **2000**, *36*, 26–46. [CrossRef]

28. Ulmke, M.; Koch, W. Road-map assisted ground moving target tracking. *IEEE Trans. Aerosp. Electron. Syst.* **2006**, *42*, 1264–1274. [CrossRef]

29. Song, D.; Tharmarasa, R.; Florea, M.C.; Duclos-Hindie, N.; Fernando, X.N.; Kirubarajan, T. Multi-vehicle tracking with microscopic traffic flow model-based particle filtering. *Automatica* **2019**, *105*, 28–35. [CrossRef]

30. Oliva, G.; Panzieri, S.; Pascucci, F.; Setola, R. Sensor networks localization: Extending trilateration via shadow edges. *IEEE Trans. Autom. Control* **2015**, *60*, 2752–2755. [CrossRef]

31. Khatib, O. Real-time obstacle avoidance for manipulators and mobile robots. In *Autonomous Robot Vehicles*; Springer: Berlin/Heidelberg, Germany, 1986; pp. 396–404. [CrossRef]

32. Vadakkepat, P.; Tan, K.C.; Ming-Liang, W. Evolutionary artificial potential fields and their application in real time robot path planning. In Proceedings of the 2000 Congress on Evolutionary Computation. CEC00 (Cat. No. 00TH8512), La Jolla, CA, USA, 16–19 July 2000; Volume 1, pp. 256–263. [CrossRef]

33. Marzetta, T.L. A simple derivation of the constrained multiple parameter Cramér-Rao bound. *IEEE Trans. Signal Process.* **1993**, *41*, 2247–2249. [CrossRef]

34. Benavoli, A.; Farina, A.; Ortenzi, L. MLE in presence of equality and inequality nonlinear constraints for the ballistic target problem. In Proceedings of the 2008 IEEE Radar Conference, Rome, Italy, 26–30 May 2008; pp. 1–6.

35. Clarke, F.H. *Optimization and Nonsmooth Analysis*; SIAM: Philadelphia, PA, USA, 1990.

36. Mallick, M. A note on bearing measurement model. *Researchgate* **2018**. [CrossRef]

37. Kay, S.M. *Fundamentals of Statistical Signal Processing*; Prentice Hall PTR: Upper Saddle River, NJ, USA, 1993. [CrossRef]

38. Doğançay, K. On the efficiency of a bearings-only instrumental variable estimator for target motion analysis. *Signal Process.* **2005**, *85*, 481–490.

39. Dorigo, M.; Birattari, M.; Stutzle, T. Ant colony optimization. *IEEE Comput. Intell. Mag.* **2006**, *1*, 28–39.

40. Schlueuter, M.; Gerdts, M.; Rückmann, J.J. A numerical study of MIDACO on 100 MINLP benchmarks. *Optimization* **2012**, *61*, 873–900. [CrossRef]

41. Schlueter, M.; Erb, S.O.; Gerdts, M.; Kemble, S.; Rückmann, J.J. MIDACO on MINLP space applications. *Adv. Space Res.* **2013**, *51*, 1116–1131.

42. Deb, K.; Pratap, A.; Agarwal, S.; Meyarivan, T. A fast and elitist multiobjective genetic algorithm: NSGA-II. *IEEE Trans. Evol. Comput.* **2002**, *6*, 182–197. [CrossRef]

43. Oliva, G.; Rikos, A.I.; Hadjicostis, C.N.; Gasparri, A. Distributed flow network balancing with minimal effort. *IEEE Trans. Autom. Control* **2019**, *64*, 3529–3543. [CrossRef]

44. Aidala, V.J. Kalman filter behavior in bearings-only tracking applications. *IEEE Trans. Aerosp. Electron. Syst.* **1979**, *AES-15*, 29–39.[CrossRef]
45. Lingren, A.G.; Gong, K.F. Position and velocity estimation via bearing observations. *IEEE Trans. Aerosp. Electron. Syst.* **1978**, *AES-14*, 564–577. [CrossRef]
46. Farina, A.; Benvenuti, D.; Ristic, B. A comparative study of the Benes filtering problem. *Signal Process.* **2002**, *82*, 133–147. [CrossRef]
47. Clark, J.; Vinter, R.; Yaqoob, M. Shifted Rayleigh filter: A new algorithm for bearings-only tracking. *IEEE Trans. Aerosp. Electron. Syst.* **2007**, *43*, 1373–1384. [CrossRef]

sensors

Article

Signal Source Localization with Long-Term Observations in Distributed Angle-Only Sensors

Shenghua Zhou [1,*], Linhai Wang [1], Ran Liu [1], Yidi Chen [2], Xiaojun Peng [1], Xiaoyang Xie [2], Jian Yang [2], Shibo Gao [3] and Xuehui Shao [3]

[1]   National Laboratory of Radar Signal Processing, Xidian University, Xi'an 710071, China
[2]   China Academy of Launch Vehicle Technology, Beijing 100076, China
[3]   Beijing Aerospace Automatic Control Institute, Beijing 100854, China
*   Correspondence: shzhou@mail.xidian.edu.cn; Tel.: +86-29-88201220

Abstract: Angle-only sensors cannot provide range information of targets and in order to determine accurate position of a signal source, one can connect distributed passive sensors with communication links and implement a fusion algorithm to estimate target position. To measure moving targets with sensors on moving platforms, most of existing algorithms resort to the filtering method. In this paper, we present two fusion algorithms to estimate both the position and velocity of moving target with distributed angle-only sensors in motion. The first algorithm is termed as the gross least square (LS) algorithm, which takes all observations from distributed sensors together to form an estimate of the position and velocity and thus needs a huge communication cost and a huge computation cost. The second algorithm is termed as the linear LS algorithm, which approximates locations of sensors, locations of targets, and angle-only measures for each sensor by linear models and thus does not need each local sensors to transmit raw data of angle-only observations, resulting in a lower communication cost between sensors and then a lower computation cost at the fusion center. Based on the second algorithm, a truncated LS algorithm, which estimates the target velocity through an average operation, is also presented. Numerical results indicate that the gross LS algorithm, without linear approximation operation, often benefits from more observations, whereas the linear LS algorithm and the truncated LS algorithm, both bear lower communication and computation costs, may endure performance loss if the observations are collected in a long period such that the linear approximation model becomes mismatch.

Keywords: passive sensor networks; signal localization; angle-only observations; accuracy analysis

Citation: Zhou, S.; Wang, L.; Liu, R.; Chen, Y.; Peng, X.; Xie, X.; Yang, J.; Gao, S.; Shao, X. Signal Source Localization with Long-Term Observations in Distributed Angle-Only Sensors. *Sensors* **2022**, *22*, 9655. https://doi.org/10.3390/s22249655

Academic Editors: Mahendra Mallick and Ratnasingham Tharmarasa

Received: 8 November 2022
Accepted: 5 December 2022
Published: 9 December 2022

1. Introduction

For some passive sensors, such as infrared sensors, photoelectric sensors and cameras, they can detect targets by receiving electromagnetic signals. As they do not emit signals, they can probe targets in a stealth manner [1]. However, such sensors can measure only angles of signals and thus are termed as angle-only sensors subsequently. The signal position information, which is of great concern in many situations, cannot be obtained with a sensor. To determine the position of signal sources, one can connect distributed sensors with communication links and then estimate the position through a fusion algorithm. This is a hot topic in recent years and gains wide attentions of scholars in different fields [2–5].

In the 3-dimensional (3D) scenario, the angle information measured by each passive sensor includes the azimuth and elevation of the signal. From a mathematical perspective, each angle observation can be represented by a straight line passing the sensor and a target in space. If no error occurs in this process, all the lines will intersect in a point in space, which is the location of a signal source. In practice, both sensor location measures and angle measures are inevitably contaminated by measurement noises and then the lines may not intersect a point in space. However, as if the signals are from the same target, the lines will intersect in a small volume, whose center can be deemed as the location of a target. Following this concept, an angle-only positioning algorithm is presented in [6] and a closed-form solution is derived.

In real applications, if the targets of interest are static, or if the sampling frequency to the signals is too high in contrast to the velocities of possible targets, the algorithms can be developed under an assumption that the velocity of the target is static. The least squares (LS) algorithm is applied in the target position estimation based on angle-only measurements by linearizing the angle observation equations [7–10]. The intersection localization algorithm is obtained by considering that the straight lines formed by the angle observations will intersect in a small volume in space [11–13]. Real sensor often makes observations in an asynchronous manner, namely the observations are not obtained at the same instants. The stationary target assumption will also make the sensor synchronization problem easier, because we can totally drop the timing information of the observations. If the target is stationary, even if the sensors are moving, the straight lines formed by the angle measurements of multiple sensors at different times will converge to a small area near the target location. Therefore, in this scenario, one just needs to solve a target positioning algorithm in an asynchronous manner.

Once the target motion should be considered at different observation instants, target location estimation will face greater biases and then one has to take the target motion issue into consideration. Meanwhile, for moving targets, the observation instants should be taken into account and then as the distributed fusion algorithm should take instant information into account, the fusion algorithm becomes more complicated. There are mainly two strategies available so far. The first strategy is to use the filtering algorithms, such as the Kalman filter that can estimate target velocity through observations from different instants. In the target tracking theoretical framework, the angle-only observations can be described by a measurement equation, although it is heavily nonlinear. Therefore, a nonlinear filtering algorithm should be used [14,15]. For instance, the extended Kalman filtering (EKF) algorithm linearizes the angle measurement function through the first-order Taylor approximation, and then uses the standard Kalman filtering algorithm for the angle-only target tracking problem [16]. The cubature Kalman filter (CKF) [17], the unscented Kalman filter (UKF) [18], the pseudo linear Kalman filter (PLKF) [19,20], the particle filter [21] and a series of sigma-point based algorithms can also be used in the target tracking problem with distributed angle-only sensors.

Although the tracking algorithms have been widely used to estimate moving target positions, it requires the noise distribution parameters known a priori. It also faces the convergence problem if the initial state is set improperly [22,23]. In distributed sensor networks, if each observation undergoes a tracking process, the computation cost will also be high since the data amount of observations are often intensive in practice. Therefore, a good positioning algorithm should be implemented before filtering. For instance, in [15], short-term angle-only observations are fused by a distributed positioning algorithm, whose outputs are then processed by a tracking algorithm.

In the other strategy, the target position and velocity can be estimated together and then the result is valid in a longer period. In this case, the tracking operation can be performed in a longer period, so that the computation cost can be further reduced. However, if the velocity is estimated, more optimization variables are involved and then the optimization problem is more complicated. Meanwhile, in a distributed sensor configuration, the communication cost between the sensors may be high if all the observations are transmitted to a fusion center. In this paper, we study the distributed positioning of moving targets with distributed asynchronous angle-only sensors. We consider the scenario where multiple asynchronous passive sensors are linked with the fusion center through communication links. First, we formulate an algorithm, termed as gross LS algorithm, that takes all angle observations of multiple sensors together with their positions in certain period to estimate the position and velocity of the target. Different observations contribute different lines and with many lines available, both the target position and its velocity can be estimated. The classical LS algorithm is formulated such that the computation cost is reduced a lot.

Due to the huge amount of data, this algorithm still has high computational complexity and high communication cost. In order to reduce the communication and computation cost, we further present a distributed positioning algorithm, termed as linear LS algorithm, that can implement the fusion algorithm in a parallel computation manner. In detail, both positions and angle observations of local sensors are processed by LS operations, whose outputs are zero and first orders of the Taylor series of corresponding parameters. The outputs are then transmitted to a fusion center for which we derive a fusion algorithm to efficiently combine position and velocity estimates for a higher parameter estimation accuracy. The later algorithm can greatly compress the data rate from local sensors to the fusion center, such that the communication cost is greatly reduced. Meanwhile, local observations are represented by a few parameters and thus the fusion algorithm also needs a lower computation cost. The sensor location can be recorded asynchronously with the angle observations and thus can make the algorithm easier in applications. Meanwhile, a truncated LS algorithm, which replaces the velocity estimation of the linear LS algorithm by a simply average operation, is also presented.

Numerical results are obtained with distributed asynchronous angle-only sensors measuring a moving target with certain velocity. The convergence performance of both the algorithms are presented first, in order to examine the impact of the number of observations on the positioning performance. Then the impact of the linear approximation of position and angle measures on the estimation accuracy is analyzed. It will be found that the gross LS algorithm often benefits from more observations. However, although the linear LS algorithm and the truncated LS algorithm will perform good if the number of observations is small, as the number of observations increase, their performances will degrade, as a resulting of the linear model mismatching. The truncated LS algorithm will perform better in a short period than the linear LS algorithm but worse in a longer time. To an extreme, the estimation performance of the linear LS algorithm may deteriorate with more observations if the model mismatching is severe. We also verify that the linear LS algorithm has a lower communication cost in most situations and examine the performance loss due to inaccurate platform velocity estimates. Numerical angle distortion errors under the linear approximations are also analyzed.

2. Localization with Angle-Only Passive Sensors

2.1. Signal Model of Passive Observations

Consider a passive sensor network with N widely separated sensors and M targets in the surveillance volume. All the N passive sensors can measure only direction of arrival (DOA) of signals, based on which real position of a signal emitter can be estimated. Assume that all the sensors operate in the same coordinate system through some inherent position and attitude measurement devices, such as the Global Positioning System (GPS) and inertial sensors. A typical coordinate system is the earth-centered earth-fixed (ECEF) of the World Geodetic System 84 (WGS84). Both the targets and the sensors are in motion by assumption. The real position of the nth sensor at instant t is denoted by $\mathbf{p}_n^o(t) = [x_{n,s}^o(t), y_{n,s}^o(t), z_{n,s}^o(t)]^{\mathrm{T}}$, $n = 1, 2, \ldots, N$, where $(\cdot)^{\mathrm{T}}$ denotes the transpose operation, and $x_{n,s}^o(t), y_{n,s}^o(t), z_{n,s}^o(t)$ denote the x, y, z coordinates of the nth sensor in the common coordinate system at instant t, respectively. The real position of the mth target at instant t is denoted by $\mathbf{g}_m^o(t) = [x_{m,g}^o(t), y_{m,g}^o(t), z_{m,g}^o(t)]^{\mathrm{T}}$, $m = 1, \cdots, M$, where $x_{m,g}^o(t), y_{m,g}^o(t), z_{m,g}^o(t)$ denote the x, y, z coordinates of the mth target at instant t, respectively. The topology of the passive sensors and targets are shown in Figure 1.

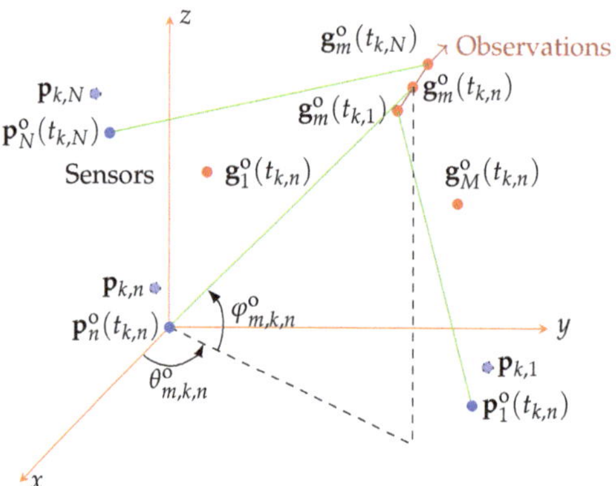

Figure 1. Measurement scenario of the passive sensors.

For the nth sensor, signals are detected and their DOAs are measured at instants denoted by $t_{k,n}, k = 1, \cdots, N_n$, where N_n denotes the number of observations of the nth sensor. At the instant $t_{k,n}$, assume that the position of the nth sensors is measured as

$$\mathbf{p}_{k,n}(t_{k,n}) = \mathbf{p}^o_n(t_{k,n}) + \Delta\mathbf{p}_n(t_{k,n}) = [x_{n,s}(t_{k,n}), y_{n,s}(t_{k,n}), z_{n,s}(t_{k,n})]^T, k = 1, \cdots, N_n \quad (1)$$

where $\Delta\mathbf{p}_n(t_{k,n})$ denotes the sensor self-positioning error. For simplicity, we assume that the sensor self-positioning error follows zero mean Gaussian distributions with covariance matrices $\mathbf{C}_s(k, n) = \mathbb{E}(\Delta\mathbf{p}_n(t_{k,n})\Delta\mathbf{p}^T_n(t_{k,n}))$, where $\mathbb{E}$ denotes the expectation operation.

At instant $t_{k,n}$, the real position of the mth signal source is denoted by

$$\mathbf{g}^o_m(t_{k,n}) = [x^o_{m,g}(t_{k,n}), y^o_{m,g}(t_{k,n}), z^o_{m,g}(t_{k,n})]^T, m = 1, \cdots, M. \quad (2)$$

Assume that all the observations regarding the same target are obtained in a short period $\mathcal{T} = [T_1, T_2]$. In this period, assume that the location of the mth target can be expressed by

$$\mathbf{g}^o_m(t) \approx \mathbf{g}^o_m(t_0) + \mathbf{v}^o_m(t - t_0) \quad (3)$$

where $\mathbf{g}^o_m(t_0)$ denotes the location of the target at the reference instant t_0, and $\mathbf{v}^o_m$ denotes the velocity over $\mathcal{T}$. The signal model in use depends on the velocity of the target and the period of observations. If all the observations are collected in a short period and the velocity is small, then one can simply assume $\mathbf{g}^o_m(t) \approx \mathbf{g}^o_m(t_0)$ as [6]. Under the signal model (3), more observations can be used to make an estimation of the target space locations. If the observations are obtained in a long period and the velocity is huge, then this model may also mismatch and higher order approximations may be used.

For the nth angle-only passive sensors, the lth observation at $t_{k,n}$ is indexed by a triple $(l, k, n), l = 1, \cdots, L_{k,n}$. For simplicity, we also encode all the triples available, corresponding to all the observations available, with a one-to-one function $\Omega(l, k, n) \rightarrow i$. Then we define a set $\mathcal{L}_{k,n}$ by

$$\mathcal{L}_{k,n} = \{i | i = (l, k, n), l = 1, \cdots, L_{k,n}\}, k = 1, \cdots, N_n, n = 1, \cdots, N \quad (4)$$

denotes a set of signal indices detected at the instant $t_{k,n}$ by the nth sensor. Therefore, $L_{k,n} = |\mathcal{L}_{k,n}|$, where $|\cdot|$ over a set denotes the cardinality of the set. As the possibility of miss detection, false alarms and overlapping of signal sources, $L_{k,n}$ may not be equal to M. Denote

$$\mathcal{L}_n = \cup^{N_n}_{k=1}\mathcal{L}_{k,n}, \mathcal{L} = \cup^{N}_{n=1}\mathcal{L}_n \quad (5)$$

where $\cup$ denotes the union operation. The total number of observations by N sensors is denoted by

$$N_s = |\mathcal{L}| = \sum_{n=1}^{N} \sum_{k=1}^{N_n} L_{k,n}. \tag{6}$$

Each observation is associated with one of M targets or the false alarm indexed by 0, represented by a set $\mathcal{M} = \{0, 1, \cdots, M\}$. It can be considered as a mapping $\psi : \mathcal{L} \to \mathcal{M}$, which is a correct mapping and is thus typically unknown in practice. According to our setting, the index set $\mathcal{L}$ can be partitioned into $M + 1$ disjoint sets $\mathcal{A}_0, \mathcal{A}_1, \cdots, \mathcal{A}_M$, and $\mathcal{A}_m$ is defined by

$$\mathcal{A}_m = \{i | \psi(i) = m, i \in \mathcal{L}\}, \tag{7}$$

where $\mathcal{A}_0$ denotes the index of observations corresponding to false alarms, and $\mathcal{A}_m$ denotes the index set of observations from the mth signal source. As a partition of $\mathcal{A}$, we have $\mathcal{A}_i \cap \mathcal{A}_j = \varnothing, i, j \in \mathcal{M}, i \neq j$, and $\mathcal{A} = \cup_{i=0}^{M} \mathcal{A}_i$, where $\cap$ denotes the intersection operation of sets. Assume that $|\mathcal{A}_m| = L_\mathrm{m}$ and there are totally $L = \sum_{m=0}^{M} L_\mathrm{m}$ observations available.

The signal indices in $\mathcal{A}_m$ are composed of signal indices from all the sensors and the sub set for the nth sensor is denoted by

$$\mathcal{A}_{n,m} = \mathcal{L}_n \cap \mathcal{A}_m \tag{8}$$

which indicates the observations from the nth sensor probing the mth target. Denote $K_{n,m} = |\mathcal{A}_{n,m}|$ and then we have $L_\mathrm{m} = \sum_{n=1}^{N} K_{n,m}$ and $\mathcal{A}_m = \cup \mathcal{A}_{n,m}$.

For simplicity, we first assume that the mapping ψ is exactly known and then observations associated with $\mathcal{A}_m$ is exactly known. For observation $i \in \mathcal{A}_m$, real azimuth angle and elevation angle, regarding the nth sensor at $t_{k,n}$, can be expressed by

$$\theta_i^\mathrm{o} = \tan^{-1}\left(y_{m,g}^\mathrm{o}(t_{k,n}) - y_{n,s}^\mathrm{o}(t_{k,n}), x_{m,g}^\mathrm{o}(t_{k,n}) - x_{n,s}^\mathrm{o}(t_{k,n}) \right) \tag{9}$$

$$\varphi_i^\mathrm{o} = \arctan \frac{z_{m,g}^\mathrm{o}(t_{k,n}) - z_{n,s}^\mathrm{o}(t_{k,n})}{\sqrt{(x_{m,g}^\mathrm{o}(t_{k,n}) - x_{n,s}^\mathrm{o}(t_{k,n}))^2 + (y_{m,g}^\mathrm{o}(t_{k,n}) - y_{n,s}^\mathrm{o}(t_{k,n}))^2}}$$

respectively, where $\theta_i^\mathrm{o} \in (-\pi, \pi)$, $\varphi_i^\mathrm{o} \in (-\frac{\pi}{2}, \frac{\pi}{2})$, $\tan^{-1}(*)$ is called the two-argument inverse tangent function [24,25] and $\arctan(*)$ is the inverse tangent function. Denote $\eta_i^\mathrm{o} = [\theta_i^\mathrm{o}, \varphi_i^\mathrm{o}]^\mathrm{T}$. The azimuth angle and elevation angle measures can be written as

$$\eta_i = [\theta_i, \varphi_i]^\mathrm{T} = \eta_i^\mathrm{o} + \Delta\eta_i \tag{10}$$
$$\theta_i = \theta_i^\mathrm{o} + \Delta\theta_i \tag{11}$$
$$\varphi_i = \varphi_i^\mathrm{o} + \Delta\varphi_i \tag{12}$$
$$\Delta\eta_i = [\Delta\theta_i, \Delta\varphi_i]^\mathrm{T} \tag{13}$$

where $\Delta\theta_i$ and $\Delta\varphi_i$ represent the measurement noise of the azimuth angle and elevation angle, respectively.

For simplicity, we assume that observation noises $\Delta\theta_i$ and $\Delta\varphi_i$ are statistically independent and follow zero-mean Gaussian distribution. The covariance matrices of $\Delta\eta$ are denoted by $\mathbf{C}_\eta(i) = \mathbb{E}(\Delta\eta\Delta\eta^\mathrm{T}) \in \mathbb{C}^{2\times2}$, namely $\Delta\eta \sim \mathcal{N}(\mathbf{0}, \mathbf{C}_\eta(i))$, which is typically affected by the SNR of the signal, where $\mathcal{N}(\mathbf{0}, \mathbf{C}_\eta(i))$ denotes the zero-mean Gaussian distribution with mean $\mathbf{0}$ and covariance matrix $\mathbf{C}_\eta(i)$.

2.2. Estimation of Target Track

Each angle-only observation contributes a line in 3D space and without measurement error, a target will be present at the line. With many angle-only observations, real position of the target can be determined. The line associated with the ith observation can be expressed by

$$L_i : \mathbf{x}(t_i) = \mathbf{p}_i + \alpha_i \mathbf{e}_i, \alpha_i \in \mathbb{R} \tag{14}$$

where $\mathbf{p}_i$ denotes the sensor location regarding the ith observation, α_i is a parameter indicating the distance to the origin $\mathbf{p}_i$, $\mathbf{e}_i = \mathbf{e}(\boldsymbol{\eta}_i) = [e_{i,x}, e_{i,y}, e_{i,z}]^{\mathrm{T}} \in \mathbb{R}^{3\times 1}$ is the normalized direction vector associated with the angle observation $\boldsymbol{\eta}_i$, namely $\|\mathbf{e}_i\| = 1$, $\|\cdot\|$ over a vector denotes the ℓ_2-norm, and

$$e_{i,x} = \cos(\theta_i)\cos(\varphi_i) \tag{15}$$

$$e_{i,y} = \sin(\theta_i)\cos(\varphi_i) \tag{16}$$

$$e_{i,y} = \sin(\varphi_i). \tag{17}$$

In what follows, we consider the observations in $\mathcal{A}_m, m \neq 0$. From (3), we can rewrite

$$\mathbf{x}(t_i) = \mathbf{p}_i + \alpha_i \mathbf{e}_i = \mathbf{g}_{0,m} + \mathbf{v}_{g,m}(t_i - t_0) + \boldsymbol{\epsilon}_e(i) = \mathbf{A}_e(i)\mathbf{q}_m + \boldsymbol{\epsilon}_e(i) \tag{18}$$

where $\mathbf{q}_m = [\mathbf{g}_{0,m}^{\mathrm{T}}, \mathbf{v}_{g,m}^{\mathrm{T}}]^{\mathrm{T}}, \mathbf{g}_{0,m} = \mathbf{g}_m^{\mathrm{o}}(t_0), \mathbf{v}_{g,m} = \mathbf{v}_m^{\mathrm{o}}(t_0), \boldsymbol{\epsilon}_e(i)$ denotes the bias term,

$$\mathbf{A}_e(i) = [\mathbf{I}, (t_i - t_0)\mathbf{I}] \in \mathbb{R}^{3\times 6} \tag{19}$$

and $\mathbf{I}$ denotes the identity matrix. In (18), there are totally 7 unknown parameters and an observation can provide 3 equations. In addition to an observation, one can obtain another 3 equations and the number of unknown parameters will increase by 1. Unless specified, we always refer to the mth target and drop the subscripts m in situations without ambiguity subsequently, e.g., denote $\mathbf{g}_{0,m} \rightarrow \mathbf{g}_0, \mathbf{v}_{g,m} \rightarrow \mathbf{v}_g, \mathbf{q}_m \rightarrow \mathbf{q}$.

In order to determine the location and velocity of the target, the optimization problem can be formulated as

$$\min_{\boldsymbol{\alpha},\mathbf{q}} \|\mathbf{P} + \mathbf{E}\,\mathrm{diag}(\boldsymbol{\alpha}) - \mathbf{A}_{\mathrm{h}}(\mathbf{1}_6 \otimes \mathbf{q})\| \tag{20}$$

where $\|\cdot\|$ refers to the ℓ_2-norm subsequently unless explicitly specified, $\boldsymbol{\alpha} = [\alpha_1, \cdots, \alpha_{L_{\mathrm{m}}}]^{\mathrm{T}}$, $\mathbf{P}$ is a matrix whose columns are $\mathbf{p}_i, i \in \mathcal{A}_m, \mathbf{P} = [\mathbf{p}, \cdots, \mathbf{p}_{L_{\mathrm{m}}}], \mathbf{1}_6$ denotes a 6×1 all-one vector of length $L_{\mathrm{m}}, \mathbf{E}$ is a matrix whose columns are $\mathbf{e}_i, i \in \mathcal{A}_m, \otimes$ denotes the Kronecker product operation,

$$\mathbf{A}_{\mathrm{h}} = [\mathbf{A}_e(i), \cdots] \in \mathbb{R}^{3\times 6L_{\mathrm{m}}}, i \in \mathcal{A}_m, \tag{21}$$

and $\mathrm{diag}(\cdot)$ with a vector entry denotes a diagonal matrix with the vector as diagonal elements.

2.3. The Gross LS Algorithm

In practice, the observations are generally contaminated by measurement noise and then the lines often do not intersect into a point in space. With L_{m} observations, there are totally $L_{\mathrm{m}} + 6$ unknown parameters and $3L_{\mathrm{m}}$ equations. Therefore, if we have at least 3 observations from angle-only sensors, we can find 9 unknown parameters together. For that purpose, let $\bar{\mathbf{q}} = [\mathbf{q}^{\mathrm{T}}, \boldsymbol{\alpha}^{\mathrm{T}}]^{\mathrm{T}}$ and then we can reformulate (18). In order to minimize the mismatch, the optimization problem can be rewritten as

$$\min_{\boldsymbol{\alpha},\mathbf{q}} \|\mathbf{p} + \mathbf{D}\boldsymbol{\alpha} - \mathbf{A}_{\mathrm{d}}\mathbf{q}\|. \tag{22}$$

The combination of equations for all observation indices in $\mathcal{A}_m$ can be formulated as

$$\mathbf{A}_a\bar{\mathbf{q}} = \mathbf{p} + \boldsymbol{\epsilon}_e \tag{23}$$

where $\mathbf{p} = \mathrm{vec}(\mathbf{P})$, $\mathrm{vec}(\cdot)$ denotes the vectorization operation.

$$\boldsymbol{\epsilon}_e = [\boldsymbol{\epsilon}^{\mathrm{T}}(1), \cdots, \boldsymbol{\epsilon}^{\mathrm{T}}(L_m)]^{\mathrm{T}} \tag{24}$$

$$\mathbf{A}_a = [\mathbf{A}_d, -\mathbf{D}] \in \mathbb{R}^{3L_m \times (6+L_m)} \tag{25}$$

$$\mathbf{A}_d = [\mathbf{A}_e^{\mathrm{T}}(1), \cdots, \mathbf{A}_e^{\mathrm{T}}(L_m)]^{\mathrm{T}} \tag{26}$$

$$= [\mathbf{1}, \mathbf{t}] \otimes \mathbf{I} \in \mathbb{R}^{3L_m \times 3} \tag{27}$$

$$\mathbf{D} = \mathrm{diag}(\mathbf{e}_1, \cdots, \mathbf{e}_{L_m}) \in \mathbb{R}^{3L_m \times L_m} \tag{28}$$

where $\mathbf{t} = [t_1 - t_0, t_2 - t_0, \ldots, t_{L_m} - t_0]^{\mathrm{T}}$, and $\mathrm{diag}(\cdot)$ with some matrices inputs denotes a block diagonal matrix with the input matrices as block diagonal elements.

For this optimization problem, we can find the classical LS solution as

$$\hat{\mathbf{q}}_a = (\mathbf{A}_a^{\mathrm{T}} \mathbf{A}_a)^{-1} \mathbf{A}_a^{\mathrm{T}} \mathbf{p}. \tag{29}$$

It can be proved that for symmetric matrices $\mathbf{A}$ and $\mathbf{B}$, and $\mathbf{C}$, all of appropriate sizes,

$$\begin{bmatrix} \mathbf{A} & \mathbf{C}^{\mathrm{T}} \\ \mathbf{C} & \mathbf{B} \end{bmatrix}^{-1} = \begin{bmatrix} (\mathbf{A} - \mathbf{C}^{\mathrm{T}}\mathbf{B}^{-1}\mathbf{C})^{-1} & -(\mathbf{A} - \mathbf{C}^{\mathrm{T}}\mathbf{B}^{-1}\mathbf{C})^{-1}\mathbf{C}^{\mathrm{T}}\mathbf{B} \\ -\mathbf{B}^{-1}\mathbf{C}(\mathbf{A} - \mathbf{C}^{\mathrm{T}}\mathbf{B}^{-1}\mathbf{C})^{-1} & (\mathbf{B} - \mathbf{C}\mathbf{A}^{-1}\mathbf{C}^{\mathrm{T}})^{-1} \end{bmatrix} \tag{30}$$

Consequently, with a fact that $\mathbf{D}^{\mathrm{T}}\mathbf{D} = \mathbf{I}$, we have

$$(\mathbf{A}_a^{\mathrm{T}} \mathbf{A}_a)^{-1} = \begin{bmatrix} \mathbf{A}_d^{\mathrm{T}}\mathbf{A}_d & -\mathbf{A}_d^{\mathrm{T}}\mathbf{D} \\ -\mathbf{D}^{\mathrm{T}}\mathbf{A}_d & \mathbf{D}^{\mathrm{T}}\mathbf{D} \end{bmatrix}^{-1} = \begin{bmatrix} \mathbf{X} & \mathbf{Z}^{\mathrm{T}} \\ \mathbf{Z} & \mathbf{Y} \end{bmatrix} \tag{31}$$

where

$$\mathbf{X} = (\mathbf{A}_d^{\mathrm{T}}\mathbf{A}_d - \mathbf{A}_d^{\mathrm{T}}\mathbf{D}(\mathbf{D}^{\mathrm{T}}\mathbf{D})^{-1}\mathbf{D}^{\mathrm{T}}\mathbf{A}_d)^{-1} \tag{32}$$

$$= (\mathbf{A}_d^{\mathrm{T}}\mathbf{A}_d - \mathbf{A}_d^{\mathrm{T}}\mathbf{D}\mathbf{D}^{\mathrm{T}}\mathbf{A}_d)^{-1} \in \mathbb{R}^{6 \times 6} \tag{33}$$

$$\mathbf{Y} = (\mathbf{D}^{\mathrm{T}}\mathbf{D} - \mathbf{D}^{\mathrm{T}}\mathbf{A}_d(\mathbf{A}_d^{\mathrm{T}}\mathbf{A}_d)^{-1}\mathbf{A}_d^{\mathrm{T}}\mathbf{D})^{-1} \tag{34}$$

$$= (\mathbf{I} - \mathbf{D}^{\mathrm{T}}\mathbf{A}_d(\mathbf{A}_d^{\mathrm{T}}\mathbf{A}_d)^{-1}\mathbf{A}_d^{\mathrm{T}}\mathbf{D})^{-1} \in \mathbb{R}^{L_m \times L_m} \tag{35}$$

$$\mathbf{Z} = (\mathbf{D}^{\mathrm{T}}\mathbf{D})^{-1}\mathbf{D}^{\mathrm{T}}\mathbf{A}_d\mathbf{X} \tag{36}$$

$$= \mathbf{D}^{\mathrm{T}}\mathbf{A}_d(\mathbf{A}_d^{\mathrm{T}}\mathbf{A}_d - \mathbf{A}_d^{\mathrm{T}}\mathbf{D}\mathbf{D}^{\mathrm{T}}\mathbf{A}_d)^{-1} \in \mathbb{R}^{L_m \times 3} \tag{37}$$

and the following equation is used in above formulations.

It can also be proved that

$$\mathbf{A}_d^{\mathrm{T}}\mathbf{A}_d = \begin{bmatrix} L\mathbf{I} & T_d\mathbf{I} \\ T_d\mathbf{I} & T_D\mathbf{I} \end{bmatrix} = \mathbf{M} \otimes \mathbf{I} \tag{38}$$

$$\mathbf{A}_d^{\mathrm{T}}\mathbf{D} = \begin{bmatrix} \mathbf{E} \\ \mathbf{E}_t \end{bmatrix}, \tag{39}$$

where $T_d = \sum_{i=1}^{L}(t_i - t_0)$, $T_D = \sum_{i=1}^{L}(t_i - t_0)^2$, $\mathbf{E}_t = \mathbf{E}\,\mathrm{diag}(\mathbf{t})$

$$\mathbf{M} = \begin{bmatrix} L & T_d \\ T_d & T_D \end{bmatrix} \Leftrightarrow \mathbf{M}^{-1} = \frac{1}{LT_D - T_d^2} \begin{bmatrix} T_D & -T_d \\ -T_d & L \end{bmatrix} \tag{40}$$

and $\otimes$ denotes the Kronecker product operation.

Still evoke (30) and we have

$$(\mathbf{A}_d^{\mathrm{T}}\mathbf{A}_d)^{-1} = \mathbf{M}^{-1} \otimes \mathbf{I} \tag{41}$$

and thus

$$\mathbf{Y}^{-1} = \mathbf{I} - \frac{1}{LT_\mathrm{D} - T_\mathrm{d}^2}[\mathbf{E}^\mathrm{T}, \mathbf{E}_\mathrm{t}^\mathrm{T}]\begin{bmatrix} T_\mathrm{D}\mathbf{E} - T_\mathrm{d}\mathbf{E}_\mathrm{t} \\ -T_\mathrm{d}\mathbf{E} + L\mathbf{E}_\mathrm{t} \end{bmatrix} \tag{42}$$

$$= \mathbf{I} - \frac{1}{LT_\mathrm{D} - T_\mathrm{d}^2}(T_\mathrm{D}\mathbf{E}^\mathrm{T}\mathbf{E} - T_\mathrm{d}\mathbf{E}_\mathrm{t}^\mathrm{T}\mathbf{E} - T_\mathrm{d}\mathbf{E}_\mathrm{t}^\mathrm{T}\mathbf{E} + L\mathbf{E}_\mathrm{t}^\mathrm{T}\mathbf{E}_\mathrm{t}). \tag{43}$$

Meanwhile,

$$\mathbf{A}_\mathrm{a}^\mathrm{T}\mathbf{p} = \begin{bmatrix} \mathbf{A}_\mathrm{d}^\mathrm{T}\mathbf{p} \\ -\mathbf{D}^\mathrm{T}\mathbf{p} \end{bmatrix} = \begin{bmatrix} \mathbf{q}_\mathrm{t} \\ -\mathbf{p}_\mathrm{e} \end{bmatrix} \tag{44}$$

where

$$\mathbf{q}_\mathrm{t} = \mathbf{P}[\mathbf{1}, \mathbf{t}^\mathrm{T}]^\mathrm{T} \tag{45}$$

$$\mathbf{p}_\mathrm{e} = \mathrm{diag}(\mathbf{E}^\mathrm{T}\mathbf{P}). \tag{46}$$

Therefore, the estimates for $\mathbf{q}_m$ and α can be written separately as

$$\hat{\mathbf{q}}_m = \begin{bmatrix} \hat{\mathbf{g}}_0 \\ \hat{\mathbf{v}}_\mathrm{g} \end{bmatrix} = \mathbf{X}\mathbf{q}_\mathrm{t} - \mathbf{X}\begin{bmatrix} \mathbf{E} \\ \mathbf{E}_\mathrm{t} \end{bmatrix}\mathbf{p}_\mathrm{e} \tag{47}$$

$$\hat{\alpha}_m = \mathbf{A}_\mathrm{d}^\mathrm{T}\mathbf{D}\mathbf{X}\mathbf{q}_\mathrm{t} - \mathbf{Y}\mathbf{p}_\mathrm{e} \tag{48}$$

where $\mathbf{X}$ can be written in a concise form as

$$\mathbf{X}^{-1} = \begin{bmatrix} L\mathbf{I} - \mathbf{E}\mathbf{E}^\mathrm{T} & T_\mathrm{d}\mathbf{I} - \mathbf{E}\mathbf{E}_\mathrm{t}^\mathrm{T} \\ T_\mathrm{d}\mathbf{I} - \mathbf{E}_\mathrm{t}\mathbf{E}^\mathrm{T} & T_\mathrm{D}\mathbf{I} - \mathbf{E}_\mathrm{t}\mathbf{E}_\mathrm{t}^\mathrm{T} \end{bmatrix}. \tag{49}$$

Under the assumption that all the sample under consideration is from the same target, the track, parameterized by $\mathbf{q}$, is identical for all observations hereafter.

2.4. The Linear LS Algorithm

In a long period, one can obtain a sequence of $\eta = (\theta, \phi)$ observations. If we take all the observations into consideration for optimization, a huge computation cost may be required. In some cases, it is also unnecessary at all. We can extract information from local observations and then transmit estimated parameters to the fusion center for target location estimation. The target location has been approximated by a linear model. Next, we express the DOA and sensor location by a linear model as well.

For DOA measures from a sensor, we can approximate a series of angle measures by

$$\eta = \begin{cases} \theta \approx \theta_0 + \dot{\theta}_0(t - t_0) \\ \varphi \approx \varphi_0 + \dot{\varphi}_0(t - t_0). \end{cases} \tag{50}$$

Now consider the nth sensor and let

$$\begin{cases} \theta_0 \to \theta_{0,n} \\ \dot{\theta}_0 \to \dot{\theta}_{0,n} \end{cases} \Rightarrow \bar{\boldsymbol{\theta}}_n = \begin{bmatrix} \theta_{0,n} \\ \dot{\theta}_{0,n} \end{bmatrix}, \quad \begin{cases} \varphi_0 \to \varphi_{0,n} \\ \dot{\varphi}_0 \to \dot{\varphi}_{0,n} \end{cases} \Rightarrow \bar{\boldsymbol{\varphi}}_n = \begin{bmatrix} \varphi_{0,n} \\ \dot{\varphi}_{0,n} \end{bmatrix}. \tag{51}$$

For all observations in $\mathcal{A}_{n,m}$, we can write a polynomial regression problem as

$$\begin{cases} \boldsymbol{\theta}_n = \mathbf{A}_n\bar{\boldsymbol{\theta}}_n + \boldsymbol{\epsilon}_\theta \\ \boldsymbol{\varphi}_n = \mathbf{A}_n\bar{\boldsymbol{\varphi}}_n + \boldsymbol{\epsilon}_\varphi \end{cases} \tag{52}$$

where $\boldsymbol{\theta}_n = [\theta_i]_{i \in \mathcal{A}_{n,m}}, \boldsymbol{\varphi}_n = [\varphi_i]_{i \in \mathcal{A}_{n,m}}$, ϵ_θ denotes bias for $\boldsymbol{\theta}_n$, ϵ_ϕ denotes bias for $\boldsymbol{\varphi}_n$, and

$$\mathbf{A}_{\mathrm{a}}(n) = \begin{bmatrix} 1, t_i - t_0 \\ \vdots \end{bmatrix} \in \mathbb{R}^{K_{n,m} \times 2}, i \in \mathcal{A}_{n,m}. \tag{53}$$

It can be proved that the LS estimate of the directions for the nth sensor can be directly written as

$$\hat{\boldsymbol{\theta}}_n = (\mathbf{A}_{\mathrm{a}}(n)^{\mathrm{T}} \mathbf{A}_{\mathrm{a}}(n))^{-1} \mathbf{A}_{\mathrm{a}}(n)^{\mathrm{T}} \boldsymbol{\theta}_n = \frac{1}{LT_{\mathrm{D}} - T_{\mathrm{d}}^2} \begin{bmatrix} T_{\mathrm{D}} \boldsymbol{\theta}_n^{\mathrm{T}} \mathbf{1} - T_{\mathrm{d}} \boldsymbol{\theta}_n^{\mathrm{T}} \mathbf{t} \\ -T_{\mathrm{d}} \boldsymbol{\theta}_n^{\mathrm{T}} \mathbf{1} + L \boldsymbol{\theta}_n^{\mathrm{T}} \mathbf{t} \end{bmatrix} \tag{54}$$

$$\hat{\boldsymbol{\varphi}}_n = (\mathbf{A}_{\mathrm{a}}(n)^{\mathrm{T}} \mathbf{A}_{\mathrm{a}}(n))^{-1} \mathbf{A}_{\mathrm{a}}(n)^{\mathrm{T}} \boldsymbol{\varphi}_n = \frac{1}{LT_{\mathrm{D}} - T_{\mathrm{d}}^2} \begin{bmatrix} T_{\mathrm{D}} \boldsymbol{\varphi}_n^{\mathrm{T}} \mathbf{1} - T_{\mathrm{d}} \boldsymbol{\varphi}_n^{\mathrm{T}} \mathbf{t} \\ -T_{\mathrm{d}} \boldsymbol{\varphi}_n^{\mathrm{T}} \mathbf{1} + L \boldsymbol{\varphi}_n^{\mathrm{T}} \mathbf{t} \end{bmatrix}. \tag{55}$$

In this case, $\hat{\boldsymbol{\theta}}_n$ and $\hat{\boldsymbol{\varphi}}_n$, instead of $\boldsymbol{\theta}_n$ and $\boldsymbol{\varphi}_n$, will be transmitted to the fusion center, such that the communication cost will be greatly reduced.

In (18), $\hat{\boldsymbol{\theta}}_n$ and $\hat{\boldsymbol{\varphi}}_n$ affects the positioning accuracy through the normalized direction vector $\mathbf{e}_i, i \in \mathcal{A}_{n,m}$. With a linear approximation model, $\mathbf{e}$ can be rewritten as

$$\mathbf{e}(t) \approx \mathbf{e}_0 + \dot{\mathbf{E}}_0^{\mathrm{T}} \dot{\boldsymbol{\eta}}_0 (t - t_0) \tag{56}$$

where $\mathbf{e}_0 = \mathbf{e}(t_0)$,

$$\dot{\theta} = \frac{\partial \theta}{\partial t}, \dot{\varphi} = \frac{\partial \varphi}{\partial t} \tag{57}$$

$$\boldsymbol{\eta}_0 = [\theta_0, \varphi_0]^{\mathrm{T}}, \boldsymbol{\eta} = [\theta, \varphi]^{\mathrm{T}} \tag{58}$$

$$\dot{\boldsymbol{\eta}}_0 = \dot{\boldsymbol{\eta}}|_{t=t_0} = [\dot{\theta}_0, \dot{\varphi}_0]^{\mathrm{T}}, \dot{\boldsymbol{\eta}} = [\dot{\theta}, \dot{\varphi}]^{\mathrm{T}} \tag{59}$$

$$\dot{\mathbf{E}}_0 = \dot{\mathbf{E}}(\boldsymbol{\eta})|_{t=t_0} = \dot{\mathbf{E}}(\boldsymbol{\eta}_0) \tag{60}$$

and $\dot{\mathbf{E}}(\boldsymbol{\eta})$ denotes the Jacobi matrix defined by

$$\dot{\mathbf{E}}^{\mathrm{T}}(\boldsymbol{\eta}) = \frac{\partial \mathbf{e}}{\partial \boldsymbol{\eta}^{\mathrm{T}}} = \begin{bmatrix} -\sin(\theta)\cos(\varphi) & -\cos(\theta)\sin(\varphi) \\ \cos(\theta)\cos(\varphi) & -\sin(\theta)\sin(\varphi) \\ 0 & \cos(\varphi). \end{bmatrix} \tag{61}$$

For the nth sensor, denote $\mathbf{e}_0 \to \mathbf{e}_{0,n}$, $\dot{\mathbf{E}}_0 \to \dot{\mathbf{E}}_{0,n}$, $\dot{\boldsymbol{\eta}}_0 \to \dot{\boldsymbol{\eta}}_{0,n}$, $\boldsymbol{\eta}_0 \to \boldsymbol{\eta}_{0,n}$ and so on.

In practice, the position of the platform is also measured by a device, such as an inertial system or a positioning system. In either case, if the nth sensor is moving with speed $\mathbf{v}_{\mathrm{s}}$ at $t = t_0$, the position can be expressed by

$$\mathbf{p}_n(t) = \mathbf{p}_{0,n} + \mathbf{v}_{\mathrm{s},n}(t - t_0) + \epsilon_{\mathrm{s}}(n) \tag{62}$$

where $\mathbf{p}_{0,n}$ denotes the position of the nth sensor, $\mathbf{v}_{\mathrm{s},n}$ denotes the velocity, both at $t = t_0$, and ϵ_{s} denotes the bias of position estimation error.

For simplicity, we assume that the sensor location is measured at $t = \bar{t}_{k,n}, k = 1, \cdots, \bar{N}_n$. In practice, in this configuration, the sensor location can be measured at instants other than $t_{k,n}, k = 1, \cdots, N_n$. Now we can construct equations as

$$\mathbf{p}_{k,n} = \mathbf{p}_{0,n} + \mathbf{v}_{\mathrm{s},n}(\bar{t}_{k,n} - t_0) + \epsilon_{\mathrm{s}}(k,n), k = 1, \cdots, \bar{N}_n, n = 1, \cdots, N \tag{63}$$

or in another form as

$$\mathbf{p}_{\mathrm{s},n} = \mathbf{A}_{\mathrm{s}} \bar{\mathbf{p}}_n + \epsilon_{\mathrm{s}}(n) \tag{64}$$

where $\bar{\mathbf{p}}_n = [\mathbf{p}_{0,n}^{\mathrm{T}}, \mathbf{v}_{\mathrm{s},n}^{\mathrm{T}}]^{\mathrm{T}}$, $\mathbf{p}_{\mathrm{s},n} = [\mathbf{p}_{1,n}^{\mathrm{T}}, \cdots, \mathbf{p}_{\tilde{N}_n,n}^{\mathrm{T}}]^{\mathrm{T}}$, and

$$\mathbf{A}_{\mathrm{s}}(n) = \begin{bmatrix} \mathbf{I}, (\bar{t}_{k,n} - t_0)\mathbf{I} \\ \vdots \end{bmatrix}_{k=1,\cdots,\tilde{N}_n} = [\mathbf{1}, \mathbf{t}_{\mathrm{s}}] \otimes \mathbf{I} \tag{65}$$

for which

$$(\mathbf{A}_{\mathrm{s}}^{\mathrm{T}}(n)\mathbf{A}_{\mathrm{s}}(n))^{-1} = \mathbf{M}_{\mathrm{s}}^{-1} \otimes \mathbf{I} \tag{66}$$

where $\mathbf{M}_{\mathrm{s}}^{-1}$ is similar to $\mathbf{M}$ in (40).

The LS estimate of $\bar{\mathbf{p}}$ is

$$\hat{\mathbf{p}}_n = \begin{bmatrix} \hat{\mathbf{p}}_{0,n}^{\mathrm{T}} \\ \hat{\mathbf{v}}_{\mathrm{s},n} \end{bmatrix} = (\mathbf{A}_{\mathrm{s}}^{\mathrm{T}}(n)\mathbf{A}_{\mathrm{s}}(n))^{-1}\mathbf{A}_{\mathrm{s}}^{\mathrm{T}}(n)\mathbf{p}_{\mathrm{s},n} \tag{67}$$

$$= (\mathbf{M}_{\mathrm{s}}^{-1} \otimes \mathbf{I}) \begin{bmatrix} \mathbf{P1} \\ \mathbf{P}_{\mathrm{t}}\mathbf{t} \end{bmatrix} \tag{68}$$

$$\hat{\mathbf{p}}_{0,n} = \frac{1}{LT_{\mathrm{D}} - T_{\mathrm{d}}^2}(T_{\mathrm{D}}\mathbf{P1} - T_{\mathrm{d}}\mathbf{P}_{\mathrm{t}}\mathbf{t}) \tag{69}$$

and

$$\hat{\mathbf{v}}_{\mathrm{s},n} = \frac{1}{LT_{\mathrm{D}} - T_{\mathrm{d}}^2}(-T_{\mathrm{d}}\mathbf{P1} + L\mathbf{P}_{\mathrm{t}}\mathbf{t}). \tag{70}$$

With above operations, we can obtain a linear sensor location parameter $\hat{\mathbf{p}}_n$, linear DOA parameters $\hat{\theta}_n$, $\hat{\phi}_n$, and linear target location parameters $\mathbf{q}_m$. The accuracy depends on the interval of observations, the speed of the target, and the speed of the sensors. A series of observations can now be approximated by two parameters and it is now unnecessary to transmit all the local observation to a fusion center anymore.

With only one observation available, a fusion center can now estimate the target position and velocity with the following equation

$$\begin{aligned} \mathbf{x}(t) &= \mathbf{g}_0 + \mathbf{v}_g(t - t_0) + \boldsymbol{\epsilon}_{\mathrm{a}}(n) = \mathbf{p}_{0,n} + \mathbf{v}_{\mathrm{s},n}(t - t_0) + \alpha_n(t)\mathbf{e}_n(t) \\ &\approx \mathbf{p}_{0,n} + \mathbf{v}_{\mathrm{s},n}(t - t_0) + (\alpha_{0,n} + \dot{\alpha}_{0,n}(t - t_0))(\mathbf{e}_{0,n} + \dot{\mathbf{E}}_{0,n}^{\mathrm{T}}\boldsymbol{\eta}_{0,n}(t - t_0)) \end{aligned} \tag{71}$$

where $\dot{\alpha}_{0,n}$ represents the change rate of $\alpha_{0,n}$. By expanding equation (71) and ignoring the second-order term, $\boldsymbol{\epsilon}_{\mathrm{a}}(n)$ can be reformulated as

$$\boldsymbol{\epsilon}_{\mathrm{a}}(n) = \mathbf{p}_{0,n} + \alpha_{0,n}\mathbf{e}_{0,n} - \mathbf{g}_0 + (\alpha_{0,n}\dot{\mathbf{E}}_{0,n}^{\mathrm{T}}\boldsymbol{\eta}_{0,n} + \mathbf{v}_{\mathrm{s},n} + \dot{\alpha}_{0,n}\mathbf{e}_{0,n} - \mathbf{v}_g)(t - t_0) \tag{72}$$

where $n = 1, 2\ldots, N$ and $\boldsymbol{\epsilon}_{\mathrm{a}}(n)$ denotes the bias term in approximation target position and direction of arrival by the linear models.

The following equation can be obtained using (72) for N sensors,

$$\boldsymbol{\epsilon}_{\mathrm{a}} = \mathbf{p}_0 + \mathbf{D}_0\boldsymbol{\alpha}_0 - \check{\mathbf{I}}\mathbf{g}_0 + (\mathbf{X}_{\mathrm{d}}\boldsymbol{\alpha}_0 + \mathbf{v}_s + \mathbf{D}_0\dot{\boldsymbol{\alpha}}_0 - \check{\mathbf{I}}\mathbf{v}_g)(t - t_0) \tag{73}$$

where

$$\boldsymbol{\alpha}_0 = [\alpha_{0,1}, \alpha_{0,2}, \cdots, \alpha_{0,N}]^{\mathrm{T}} \tag{74}$$

$$\dot{\boldsymbol{\alpha}}_0 = [\dot{\alpha}_{0,1}, \dot{\alpha}_{0,2}, \cdots, \dot{\alpha}_{0,N}]^{\mathrm{T}} \tag{75}$$

$$\mathbf{p}_0 = [\mathbf{p}_{0,1}^{\mathrm{T}}, \mathbf{p}_{0,2}^{\mathrm{T}}, \ldots, \mathbf{p}_{0,N}^{\mathrm{T}}]^{\mathrm{T}} \tag{76}$$

$$\mathbf{D}_0 = \mathrm{diag}(\mathbf{e}_{0,1}, \mathbf{e}_{0,2}, \ldots, \mathbf{e}_{0,N}) \tag{77}$$

$$\mathbf{X}_{\mathrm{d}} = \mathrm{diag}(\dot{\mathbf{E}}_{0,1}^{\mathrm{T}} \dot{\boldsymbol{\eta}}_{0,1}, \dot{\mathbf{E}}_{0,2}^{\mathrm{T}} \dot{\boldsymbol{\eta}}_{0,2}, \ldots, \dot{\mathbf{E}}_{0,N}^{\mathrm{T}} \dot{\boldsymbol{\eta}}_{0,N}) \tag{78}$$

$$\mathbf{v}_{\mathrm{s}} = [\mathbf{v}_{s,1}^{\mathrm{T}}, \mathbf{v}_{s,2}^{\mathrm{T}}, \ldots, \mathbf{v}_{s,N}^{\mathrm{T}}]^{\mathrm{T}} \tag{79}$$

$$\bar{\mathbf{I}} = \mathbf{1}_N \otimes \mathbf{I}_3 \in \mathbb{R}^{3N \times 3} \tag{80}$$

and

$$\boldsymbol{\epsilon}_{\mathrm{a}} = [\boldsymbol{\epsilon}_{\mathrm{a}}^{\mathrm{T}}(1), \boldsymbol{\epsilon}_{\mathrm{a}}^{\mathrm{T}}(2), \cdots, \boldsymbol{\epsilon}_{\mathrm{a}}^{\mathrm{T}}(N)]^{\mathrm{T}}. \tag{81}$$

In order to minimize the total bias ϵ_a, we can minimize

$$\min_{\mathbf{q}_c, \mathbf{q}_v} \| \mathbf{p}_0 + \mathbf{D}_0 \boldsymbol{\alpha}_0 - \bar{\mathbf{I}} \mathbf{g}_0 + (\mathbf{X}_{\mathrm{d}} \boldsymbol{\alpha}_0 + \mathbf{v}_s + \mathbf{D}_0 \dot{\boldsymbol{\alpha}}_0 - \bar{\mathbf{I}} \mathbf{v}_g)(t - t_0) \|, t \in \mathcal{T}. \tag{82}$$

where $\mathbf{q}_c = [\mathbf{g}_0^{\mathrm{T}}, \boldsymbol{\alpha}_0^{\mathrm{T}}]^{\mathrm{T}}$ and $\mathbf{q}_v = [\mathbf{v}_g^{\mathrm{T}}, \dot{\boldsymbol{\alpha}}_0^{\mathrm{T}}]^{\mathrm{T}}$.

To ensure the bias is minimized for $t \in \mathcal{T}$, both the initial position bias, $\mathbf{p}_0 + \mathbf{D}_0 \boldsymbol{\alpha}_0 - \bar{\mathbf{I}} \mathbf{g}_0$, and the speed bias, $\mathbf{X}_{\mathrm{d}} \boldsymbol{\alpha}_0 + \mathbf{v}_s + \mathbf{D}_0 \dot{\boldsymbol{\alpha}}_0 - \bar{\mathbf{I}} \mathbf{v}_g$ should be minimized. Therefore, we can solve the optimization problem through solving the following two optimization problems of smaller scale,

$$\min_{\mathbf{q}_c} \| \mathbf{p}_0 + \mathbf{D}_0 \boldsymbol{\alpha}_0 - \bar{\mathbf{I}} \mathbf{g}_0 \| \tag{83}$$

$$\min_{\mathbf{q}_v} \| \mathbf{X}_{\mathrm{d}} \boldsymbol{\alpha}_0 + \mathbf{v}_s + \mathbf{D}_0 \dot{\boldsymbol{\alpha}}_0 - \bar{\mathbf{I}} \mathbf{v}_g \|. \tag{84}$$

The solutions to the problems can be found directly through the LS algorithm as

$$\hat{\mathbf{q}}_c = [\hat{\mathbf{g}}_0^{\mathrm{T}}, \hat{\boldsymbol{\alpha}}_0^{\mathrm{T}}]^{\mathrm{T}} = (\mathbf{A}_c^{\mathrm{T}} \mathbf{A}_c)^{-1} \mathbf{A}_c^{\mathrm{T}} \mathbf{p}_0 \tag{85}$$

$$\hat{\mathbf{q}}_v = [\hat{\mathbf{v}}_g^{\mathrm{T}}, \hat{\boldsymbol{\alpha}}_0^{\mathrm{T}}]^{\mathrm{T}} = (\mathbf{A}_c^{\mathrm{T}} \mathbf{A}_c)^{-1} \mathbf{A}_c^{\mathrm{T}} (\mathbf{X}_{\mathrm{d}} \boldsymbol{\alpha}_0 + \mathbf{v}_s) \tag{86}$$

where

$$\mathbf{A}_c = [\mathbf{1}_N \otimes \mathbf{I}_3, -\mathbf{D}_0]. \tag{87}$$

It can be proved that

$$\mathbf{A}_c^{\mathrm{T}} \mathbf{A}_c = \begin{bmatrix} \bar{\mathbf{I}}^{\mathrm{T}} \bar{\mathbf{I}} & -\bar{\mathbf{I}}^{\mathrm{T}} \mathbf{D}_0 \\ -\mathbf{D}_0^{\mathrm{T}} \bar{\mathbf{I}} & \mathbf{D}_0^{\mathrm{T}} \mathbf{D}_0 \end{bmatrix} = \begin{bmatrix} N\mathbf{I}_3 & -\mathbf{E}_0 \\ -\mathbf{E}_0^{\mathrm{T}} & \mathbf{I} \end{bmatrix} \tag{88}$$

and then

$$(\mathbf{A}_c^{\mathrm{T}} \mathbf{A}_c)^{-1} = \begin{bmatrix} (N\mathbf{I} - \mathbf{E}_0 \mathbf{E}_0^{\mathrm{T}})^{-1} & (N\mathbf{I} - \mathbf{E}_0 \mathbf{E}_0^{\mathrm{T}})^{-1} \mathbf{E}_0 \\ \mathbf{E}_0^{\mathrm{T}} (N\mathbf{I} - \mathbf{E}_0 \mathbf{E}_0^{\mathrm{T}})^{-1} & \mathbf{I} + \mathbf{E}_0^{\mathrm{T}} (N\mathbf{I} - \mathbf{E}_0 \mathbf{E}_0^{\mathrm{T}})^{-1} \mathbf{E}_0 \end{bmatrix}. \tag{89}$$

where $\mathbf{E}_0 = [\mathbf{e}_{0,1}, \mathbf{e}_{0,2}, \ldots, \mathbf{e}_{0,N}]$ and $\mathbf{P}_0 = [\mathbf{p}_{0,1}, \mathbf{p}_{0,2}, \ldots, \mathbf{p}_{0,N}]$.

Meanwhile,

$$\mathbf{A}_c^{\mathrm{T}} \mathbf{p}_0 = \begin{bmatrix} \mathbf{P}_0 \mathbf{1} \\ -\mathbf{D}_0^{\mathrm{T}} \mathbf{p}_0 \end{bmatrix} \tag{90}$$

and thus,

$$\hat{\mathbf{q}}_0 = [\hat{\mathbf{g}}_0^\mathrm{T}, \hat{\boldsymbol{\alpha}}_0^\mathrm{T}]^\mathrm{T} = (\mathbf{A}_c^\mathrm{T}\mathbf{A}_c)^{-1}\mathbf{A}_c^\mathrm{T}\mathbf{p}_0 \tag{91}$$

$$\hat{\mathbf{g}}_0 = (N\mathbf{I} - \mathbf{E}_0\mathbf{E}_0^\mathrm{T})^{-1}\mathbf{P}_0\mathbf{1} - (N\mathbf{I} - \mathbf{E}_0\mathbf{E}_0^\mathrm{T})^{-1}\mathbf{E}_0\mathbf{D}_0^\mathrm{T}\mathbf{p}_0$$
$$= (N\mathbf{I} - \mathbf{E}_0\mathbf{E}_0^\mathrm{T})^{-1}(\mathbf{P}_0\mathbf{1} - \mathbf{E}_0\mathbf{D}_0^\mathrm{T}\mathbf{p}_0) \tag{92}$$

$$\hat{\boldsymbol{\alpha}}_0 = \mathbf{D}_0^\mathrm{T}\bar{\mathbf{I}}\hat{\mathbf{g}}_0 - \mathbf{D}_0^\mathrm{T}\mathbf{p}_0$$
$$= \mathbf{D}_0^\mathrm{T}\bar{\mathbf{I}}(N\mathbf{I} - \mathbf{E}_0\mathbf{E}_0^\mathrm{T})^{-1}(\mathbf{P}_0\mathbf{1} - \mathbf{E}_0\mathbf{D}_0^\mathrm{T}\mathbf{p}_0) - \mathbf{D}_0^\mathrm{T}\mathbf{p}_0 \tag{93}$$

which is identical to the solution of (82). A minor difference is that the matrix inverse operation is over an $N \times N$ matrix $N\mathbf{I} - \mathbf{E}_0\mathbf{E}_0^\mathrm{T}$.

The solution to $\mathbf{q}_v$ can be expressed by

$$\mathbf{A}_c^\mathrm{T}(\mathbf{X}_d\boldsymbol{\alpha}_0 + \mathbf{v}_s) = \begin{bmatrix} \mathbf{1}_N^\mathrm{T} \otimes \mathbf{I}_3 \\ -\mathbf{D}_0^\mathrm{T} \end{bmatrix}(\mathrm{vec}(\mathbf{X}\,\mathrm{diag}(\boldsymbol{\alpha}_0)) + \mathrm{vec}(\mathbf{V}_s))$$
$$= \begin{bmatrix} \mathbf{X}\boldsymbol{\alpha}_0 + \mathbf{V}_s\mathbf{1} \\ -\mathbf{v}_a - \mathbf{v}_e \end{bmatrix} \tag{94}$$

where

$$\mathbf{X} = [\dot{\mathbf{E}}_{0,1}^\mathrm{T}\dot{\boldsymbol{\eta}}_{0,1}, \dot{\mathbf{E}}_{0,2}^\mathrm{T}\dot{\boldsymbol{\eta}}_{0,2}, \ldots, \dot{\mathbf{E}}_{0,N}^\mathrm{T}\dot{\boldsymbol{\eta}}_{0,N}] \tag{95}$$

$$\mathbf{V}_s = [\mathbf{v}_{s,1}, \mathbf{v}_{s,2}, \ldots, \mathbf{v}_{s,N}] \tag{96}$$

$$\mathbf{v}_a = \mathrm{diag}(\boldsymbol{\alpha}_0)\mathbf{E}_0^\mathrm{T}\mathbf{X}$$
$$= [\alpha_{0,1}\mathbf{e}_{0,1}^\mathrm{T}\dot{\mathbf{E}}_{0,1}^\mathrm{T}\dot{\boldsymbol{\eta}}_{0,1}, \cdots, \alpha_{0,N}\mathbf{e}_{0,N}^\mathrm{T}\dot{\mathbf{E}}_{0,N}^\mathrm{T}\dot{\boldsymbol{\eta}}_{0,N}]^\mathrm{T} \tag{97}$$

and

$$\mathbf{v}_e = [\mathbf{e}_{0,1}^\mathrm{T}\mathbf{v}_{s,1}, \cdots, \mathbf{e}_{0,N}^\mathrm{T}\mathbf{v}_{s,N}]^\mathrm{T}. \tag{98}$$

Consequently, we can obtain

$$\hat{\mathbf{v}}_g = (N\mathbf{I} - \mathbf{E}_0\mathbf{E}_0^\mathrm{T})^{-1}(\mathbf{X}\boldsymbol{\alpha}_0 + \mathbf{V}_s\mathbf{1} - \mathbf{E}_0\mathbf{v}_a - \mathbf{E}_0\mathbf{v}_e) \tag{99}$$

$$\hat{\boldsymbol{\alpha}}_0 = \mathbf{E}_0^\mathrm{T}(N\mathbf{I} - \mathbf{E}_0\mathbf{E}_0^\mathrm{T})^{-1}(\mathbf{X}\boldsymbol{\alpha}_0 + \mathbf{V}_s\mathbf{1}) - \mathbf{v}_a - \mathbf{v}_e - \mathbf{E}_0^\mathrm{T}(N\mathbf{I} - \mathbf{E}_0\mathbf{E}_0^\mathrm{T})^{-1}\mathbf{E}_0(\mathbf{v}_a + \mathbf{v}_e) \tag{100}$$

$$= \mathbf{E}_0^\mathrm{T}(N\mathbf{I} - \mathbf{E}_0\mathbf{E}_0^\mathrm{T})^{-1}(\mathbf{X}\boldsymbol{\alpha}_0 + \mathbf{V}_s\mathbf{1} - \mathbf{E}_0\mathbf{v}_a - \mathbf{E}_0\mathbf{v}_e) - \mathbf{v}_a - \mathbf{v}_e \tag{101}$$

$$= \mathbf{E}_0^\mathrm{T}\hat{\mathbf{v}}_g - \mathbf{v}_a - \mathbf{v}_e. \tag{102}$$

One can also derive in another way. According to (93), the change rate $\dot{\boldsymbol{\alpha}}_0$ of $\boldsymbol{\alpha}_0$ can be expressed as

$$\dot{\boldsymbol{\alpha}}_0 = \mathbf{X}_d^\mathrm{T}\bar{\mathbf{I}}\hat{\mathbf{g}}_0 + \mathbf{D}_0^\mathrm{T}\bar{\mathbf{I}}\mathbf{v}_g - \mathbf{X}_d^\mathrm{T}\mathbf{p}_0 - \mathbf{D}_0^\mathrm{T}\mathbf{v}_s. \tag{103}$$

Take (103) into (84), which can be rewritten as

$$\min_{\mathbf{v}_g} \|\mathbf{X}_d\hat{\boldsymbol{\alpha}}_0 + \mathbf{v}_s + \mathbf{D}_0\mathbf{X}_d^\mathrm{T}\bar{\mathbf{I}}\hat{\mathbf{g}}_0 - \mathbf{D}_0\mathbf{D}_0^\mathrm{T}\mathbf{v}_s - \mathbf{D}_0\mathbf{X}_d^\mathrm{T}\mathbf{p}_0 + (\mathbf{D}_0\mathbf{D}_0^\mathrm{T} - \mathbf{I})\bar{\mathbf{I}}\mathbf{v}_g\|. \tag{104}$$

The solution of (104) can also be found through the LS algorithm, which can be expressed as

$$\hat{\mathbf{v}}_g = (\mathbf{A}_v^\mathrm{T}\mathbf{A}_v)^{-1}\mathbf{A}_v^\mathrm{T}\mathbf{b}_v \tag{105}$$

where

$$\mathbf{A}_v = (\mathbf{I} - \mathbf{D}_0\mathbf{D}_0^\mathrm{T})\bar{\mathbf{I}} \tag{106}$$

$$\mathbf{b}_v = \mathbf{X}_d\hat{\boldsymbol{\alpha}}_0 + \mathbf{v}_s + \mathbf{D}_0\mathbf{X}_d^\mathrm{T}\bar{\mathbf{I}}\hat{\mathbf{g}}_0 - \mathbf{D}_0\mathbf{D}_0^\mathrm{T}\mathbf{v}_s - \mathbf{D}_0\mathbf{X}_d^\mathrm{T}\mathbf{p}_0. \tag{107}$$

2.5. The Truncated LS Algorithm

For a better performance, it is necessary to estimate the change rate $\dot{\alpha}$ and if we ignore this term, the optimization problem becomes

$$\min_{\mathbf{v}_g} \|\mathbf{X}_d \boldsymbol{\alpha}_0 + \mathbf{v}_s - \bar{\mathbf{I}} \mathbf{v}_g\| \tag{108}$$

whose solution, termed as truncated LS algorithm subsequently, is

$$\hat{\mathbf{v}}_g = \frac{1}{N}(\mathbf{V}_s \mathbf{1} + \mathbf{X}\hat{\boldsymbol{\alpha}}_0) \tag{109}$$

which is an average operation. Note that the truncated LS algorithm shares the same position estimate with the linear LS algorithm.

The estimate of the target location at $t \in \mathcal{T}$ can be written as

$$\hat{\mathbf{g}}_m(t) = \hat{\mathbf{g}}_0 + \hat{\mathbf{v}}_g(t - t_0), t \in \mathcal{T}. \tag{110}$$

In (99) and (109), the velocity terms $\mathbf{v}_{s,k}, k = 1, \cdots, N$ are unknown and should be replaced by their estimates, typically $\hat{\mathbf{v}}_s$ estimated in the LS algorithm as in (70). In practice, besides the linear regression method performed at local sensors, there may be other methods that can output more accurate velocity and angle difference information. For instance, some inertial devices can measure the velocity more accurately than the LS algorithm in use. With a more accurate velocity estimate, it is possible to obtain a better positioning performance.

Both the linear LS algorithm and the truncated LS algorithm estimate the velocity of the target and thus can make the time-consuming nonlinear filtering operation update in a longer time interval. Subsequently, the performances of these algorithms will be analyzed in numerical results.

3. Numerical Results

In order to evaluate the performance of the concerned positioning algorithms, we first consider a scenario where four angle-only sensors are estimating the position of a target with their angle-only observations. Both the sensors and the target are moving with a constant speed during the period of observation by assumption. The initial position and the constant speed of the sensors and the target are shown in Table 1. The scenario is illustrated in Figure 2.

All the sensors output observations at a frequency of 50 Hz, i.e., with a period of 20 ms. But they operate on an asynchronous manner, namely the sensors record the observations at independent instants. The differences of the sampling instants are randomly generated within 20ms. This assumption is important in real situations because it allows distributed sensors to operate asynchronously. We also assume that there is no error in recording the instants of the observations and for all the sensors, no signal is missed in detection during the observation period.

Table 1. Positions and velocities of sensors and the target.

	Position (m) at $t = 0$ s	Velocity (m/s)
Sensor #1	$\mathbf{p}_1^o(0) = [1000, 1000, 0]^\mathrm{T}$	$[-100, 0, 0]^\mathrm{T}$
Sensor #2	$\mathbf{p}_2^o(0) = [1000, 2000, 0]^\mathrm{T}$	$[-100, -80, 0]^\mathrm{T}$
Sensor #3	$\mathbf{p}_3^o(0) = [2000, 1000, 0]^\mathrm{T}$	$[-100, -50, 0]^\mathrm{T}$
Sensor #4	$\mathbf{p}_4^o(0) = [1500, 1500, 0]^\mathrm{T}$	$[-100, -60, 0]^\mathrm{T}$
Target #1	$\mathbf{g}_1^o(0) = [0, 100, 1000]^\mathrm{T}$	$[200, 100, 0]^\mathrm{T}$

Assume that the self-positioning error is distributed with zero-mean normal distribution, whose variance is 1m for all the sensors, namely

$$\mathbf{C}_s(k,n) = \mathbb{E}(\Delta\mathbf{p}_n(t_{k,n})\Delta\mathbf{p}_n^{\mathrm{T}}(t_{k,n})) = \mathbf{I}, n = 1, \cdots, N. \tag{111}$$

The angle measurement error also follows zero-mean normal distribution with variance of 0.5 degree for all the observations, namely

$$\mathbf{C}_\eta(i) = \mathbb{E}(\Delta\boldsymbol{\eta}_i\Delta\boldsymbol{\eta}_i^{\mathrm{T}}) = 0.5\mathbf{I}, i \in \mathcal{A}_m. \tag{112}$$

At the current stage, we do not consider the measurement errors from the gyroscopes installed on the platforms along with the sensors. Therefore, the angle measurement error is caused by the sensors only.

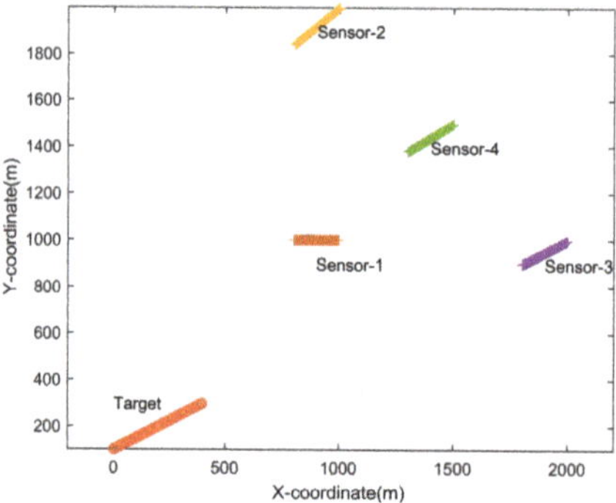

Figure 2. The topology of the sensors and the target.

In order to evaluate the performance of the algorithms, we run $N_e = 20$ random experiments and take the root mean square error (RMSE) as the resulting performance metric. The RMSE of position, RMSE of velocity and the gross RMSE at instant t in scale and in dB are defined by

$$\mathrm{RMSE}_\mathrm{p}(t) = \sqrt{\frac{1}{N_e}\sum_{k=1}^{N_e}|\hat{\mathbf{g}}_0(t;k) - \mathbf{g}_1^\mathrm{o}(0)|^2}, \quad \mathrm{RMSE}_\mathrm{p}(t) : \mathrm{dB} = 20\log_{10}\mathrm{RMSE}_\mathrm{p}(t) \tag{113}$$

$$\mathrm{RMSE}_\mathrm{v}(t) = \sqrt{\frac{1}{N_e}\sum_{k=1}^{N_e}|\hat{\mathbf{v}}_\mathrm{g}(t;k) - \mathbf{v}_{\mathrm{g},1}^\mathrm{o}(0)|^2}, \quad \mathrm{RMSE}_\mathrm{v}(t) : \mathrm{dB} = 20\log_{10}\mathrm{RMSE}_\mathrm{v}(t) \tag{114}$$

$$\mathrm{RMSE}_\mathrm{g}(t) = \sqrt{\mathrm{RMSE}_\mathrm{p}^2(t) + \mathrm{RMSE}_\mathrm{v}^2(t)}, \quad \mathrm{RMSE}_\mathrm{g}(t) : \mathrm{dB} = 20\log_{10}\mathrm{RMSE}_\mathrm{g}(t) \tag{115}$$

respectively, where $\hat{\mathbf{g}}_0(t;k)$ denotes the initial position of the target at the kth experiment, $\mathbf{g}_1^\mathrm{o}(0)$ and $\mathbf{v}_1^\mathrm{o}(0)$ are constants during experiments, and $\hat{\mathbf{v}}_\mathrm{g}(t;k)$ denotes the estimate of the target speed at the kth experiment. At each random experiment, the position error and the angle measurement error are generated randomly.

3.1. The Convergence Curves

As the number of observations increase, the localization performance will improve. Figure 3 shows the mean RMSE of position and velocity and the gross RMSE of the gross LS algorithm, the linear LS algorithm and the truncated LS algorithm.

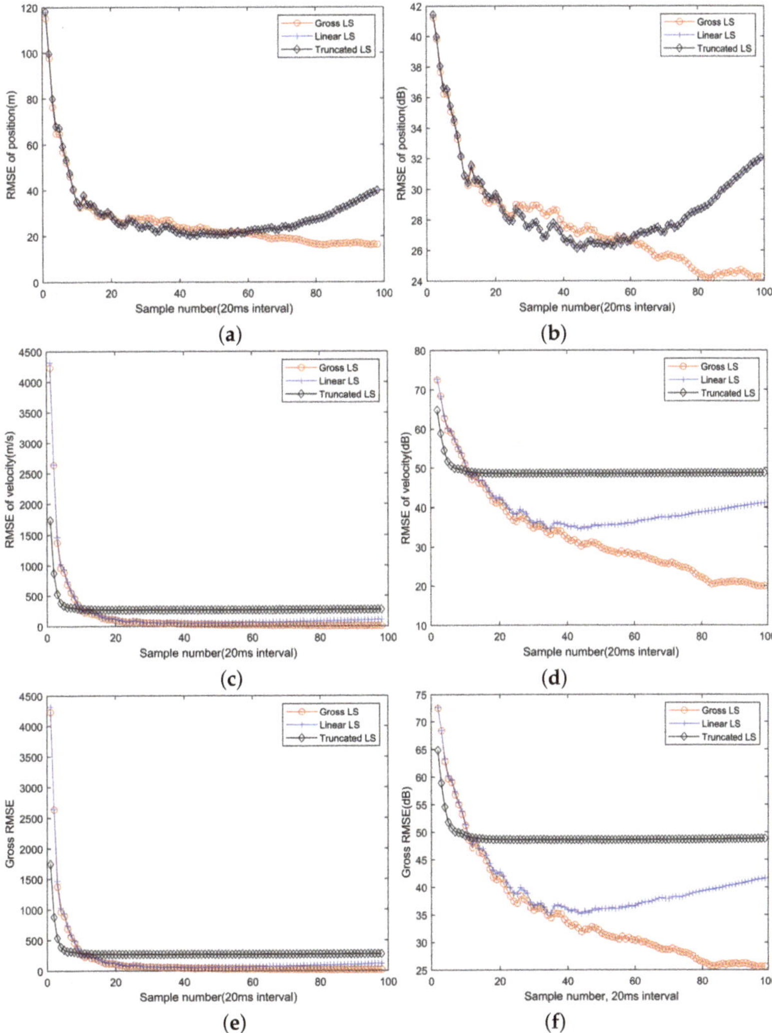

Figure 3. The convergence curves of the positioning errors with different numbers of observations. (a) The RMSE of position estimation and its dB form (b), (c) The RMSE of velocity estimation and its dB form (d); (e) The gross RMSE and its dB form (f).

From Figure 3a,b, it can be seen that with observations in a short while, roughly in about 0.8 s corresponding to 40 observations, all the algorithms have close gross RMSE curves. However, as more observations are available, the linear LS algorithm and the truncated LS algorithm will perform worse and the position RMSE even increase with sample number. It is a predictable result, because the linear approximations of the target and platform motion will be inaccurate gradually, resulting a deteriorated positioning performance. The gross LS algorithm will always benefit from the increase of the observations, because it does not rely on the linear approximation, and more observations will contribute more information of the target position.

From Figure 3c,d, with some initial observations, the truncated LS algorithm performs the best and the linear LS algorithm performs the worst and close to the gross LS algorithm. As more observations are involved, the truncated LS algorithm converges to a level much higher than that can be achieved by the gross LS algorithm and the linear LS algorithm. Therefore, ignoring the term $\dot{\alpha}_0$ will cause performance loss for long term observations. The gross LS algorithm is still benefitting from the increase of observations and it is slightly better than the linear LS algorithm for short term observations. The linear LS algorithm can reach a lower RMSE level but will still suffer performance degradation due to the linear model mismatch. With about 40 snapshots of observations, corresponding to 160 observations and 0.8 s period, the velocity estimation performances of two algorithms will depart.

The gross RMSE of all the algorithms are shown in Figure 3e,f, which have very close appearances to Figure 3c,d. That is because the velocity estimation errors are much greater than the positioning errors. Therefore, although the algorithm can estimate the velocity of targets, the accuracy is low due to a short observation period. In order to estimate the velocity in a higher accuracy, one needs to use observations from a longer period, which can be achieved through a filtering operation.

In order to show the way in which the algorithms converge to the real value, Figure 4a,b are presented to show estimated target positions and velocities at the first 20 snapshots and the latest 20 snapshots, respectively. It can be seen that with a few observations, the linear LS algorithm will converge to the real position of the target to a high accuracy. However, as more observations are available, the gross LS algorithm is closer to the real target position and the linear LS algorithm converges to other locations. Therefore, the gross LS algorithm is more robust in real applications.

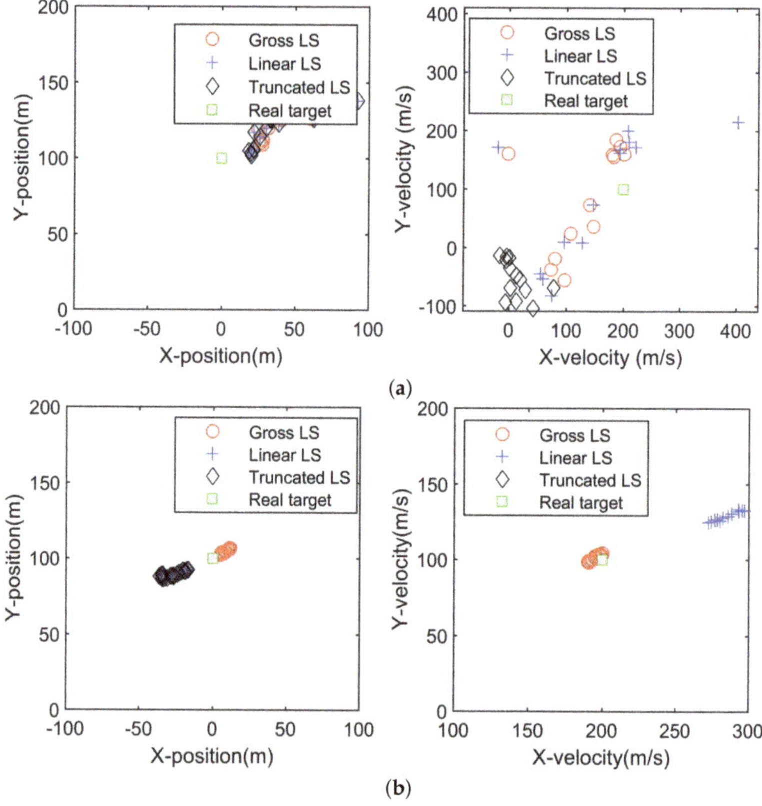

Figure 4. The first (**a**) and last (**b**) 20 estimates of the position and velocity of the gross LS, the linear LS and the truncated LS algorithms.

3.2. Computation Cost

The advantage of the linear LS algorithm and the truncated LS algorithm lies in its computation cost and communication cost. In applications of the LS algorithms, the locations of the sensor will be approximated by a linear model, which is described by an initial position and a velocity term, with totally 6 parameters. Therefore, it is unnecessary to transmit all observations to the fusion center anymore and thus the communication cost will be reduced. If 100 position estimates are described by 6 parameters, the data to transmit will be reduced to 2%. Of course, there is a limit to which the data can be reduced and the limit depends on the platform speed of the sensors, the positions of sensors, the position of the target, and the periods of the observations.

The computation cost reduction, for both the linear LS algorithm and the truncated LS algorithm, stems from reduced number of multiplication and summation operations at the fusion center. The linear LS algorithm and the truncated LS algorithm can be implemented in a structure like parallel computation, namely, the linear regression of the platform position and the local DOA measures are performed at local sensors, and the fusion center just operates on the results of local sensors. To illustrate this fact, we record the computation times of the 20 random experiments for both the algorithms and show the computation times in Figure 5a,b, in scale mode and dB mode respectively. It can be seen that as the number of observations increases, the gross LS algorithm requires a longer computation time, but the linear LS algorithm and the truncated LS algorithm have much plain slopes. Meanwhile, the linear LS algorithm needs more computation cost, as a result of estimating $\mathbf{v}_a$ and $\mathbf{v}_e$. In fact, the computation cost of the linear LS algorithm does not vary with the sample number too much because it always computes with the same number of parameters, namely the number of sensors N. The computation cost increase due to more observations is imposed over local sensors now.

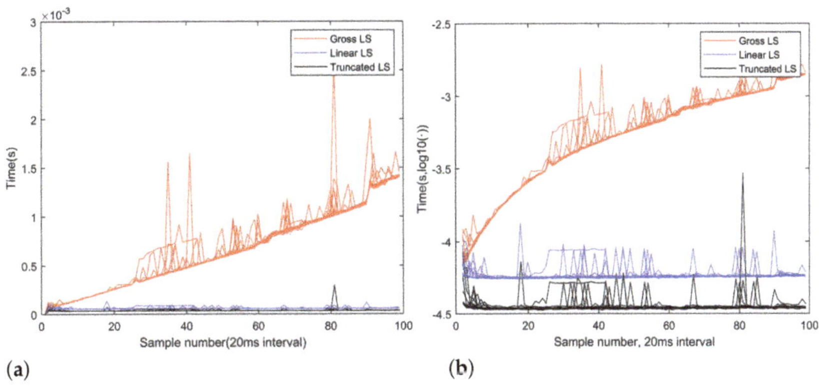

Figure 5. The computation time in 10 random simulations of the gross LS, linear LS and truncated LS algorithms, in scale (**a**) and dB (**b**). The program is run on a computer with an Intel™i7-10700 CPU and 16 GB memory.

3.3. The Impact of Velocity Estimation Error

In theoretical derivations, we assumed that the velocities of sensors are estimated by position measures from a device on a platform. In practice, the platform may provide other means to measure the velocity in a higher accuracy. Meanwhile, in order to check whether the performance degradation of the linear LS algorithm in a long period is a result of inaccurate estimation of the platform velocity, we perform a simulation in a way that the estimated velocity $\hat{\mathbf{v}}_g$ is replaced by its real value $\mathbf{v}_g$. In this case, there is no velocity error and the only measurement bias is from position measurement. Note subsequently that the linear LS algorithm shares the same position estimate with the truncated LS algorithm.

The RMSE of position estimation is shown in Figure 6. In Figure 6a,b, the sensor location uncertainty is zero-mean normal distributed with $\mathbf{C}_s(k,n) = \mathbf{I}$. It can be clearly seen that for the linear LS algorithm, the RMSE curve with real sensor velocity is very close to the RMSE curve using estimated sensor velocity. In order to examine whether a higher position estimation error will make a difference, we make another simulation with $\mathbf{C}_s(k,n) = 10\mathbf{I}$ and the results are shown in Figure 6c,d. Two RMSE curves are still very close. After some experiments with other position measurement errors, we find that the platform location and velocity regression algorithms can reach a high accuracy and thus will not cause too much performance degradation. This conclusion depends heavily on a fact that the numbers of observations under consideration is often huge according to our configurations.

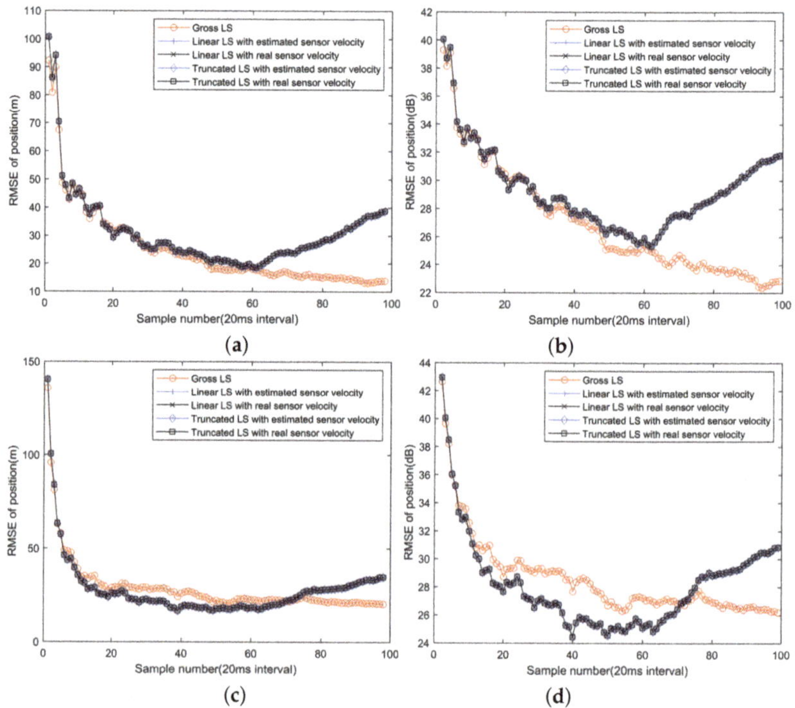

Figure 6. The RMSE of position with velocity estimated replaced by real velocity in scale (**a**) and dB (**b**) for $\mathbf{C}_s(k,n) = \mathbf{I}$, and in scale (**c**) and dB (**d**) for $\mathbf{C}_s(k,n) = 10\mathbf{I}$.

In fact, from (92), the position estimate of the target does not depend on the velocity estimation of the sensor platform too much. However, there still an insignificant impact, because in our simulation configuration, the sensor location at $t = 0$ is obtained by an interpolation operation and if the platform velocity is exactly known *a priori*, the position estimation will be more accurate.

From (105), the target velocity estimation depends on the sensor velocity more. In order the examine the impact of the sensor velocity estimation on the target velocity estimation performance, we run a simulation with $\mathbf{C}_s(k,n) = 10\mathbf{I}$ and the results are shown in Figure 7a,b, in scale and dB respectively. It can be seen that it makes a little difference to use real sensor velocity instead of estimated velocity, especially in few earlier observations. As more observations are taken into account, it makes a minor different to replace by real platform velocities. That is because more observations make the velocity estimation more accurate. However, accurate sensor velocity information does not make the target velocity estimation better necessarily, and sometimes, its impact is a bit negative. As the target also moves in a constant velocity, it is reasonable to infer that the linear approximation of the signal DOA has a great impact on the target position and velocity estimation accuracy, as will be analyzed in the subsequent results.

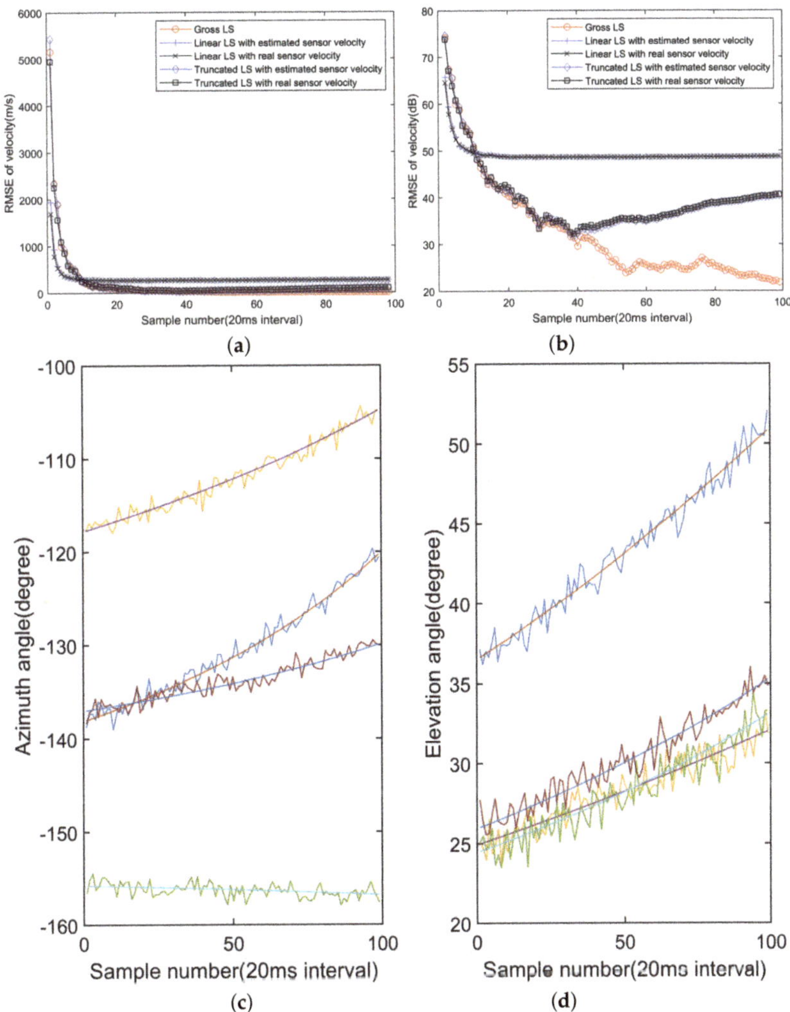

Figure 7. The target velocity estimation results with real and estimated sensor velocity are shown in (**a**) and (**b**) for $\mathbf{C}_s(k,n) = 10\mathbf{I}$. The azimuth (**c**) and elevation (**d**) angles of the target in the four sensors.

3.4. Nonlinearity of the DOA Approximation

In order to examine the impact of DOA nonlinearity on the final performance, Figure 7c,d show the azimuth angles and elevation angles of the target in the four sensors. The azimuth angle and elevation angle change by about 15° at most during 100 snapshots. Over 100 observations, corresponding to 2 s, the nonlinearity of both DOA angles becomes obvious. One should refer to explicit numerical quantities to evaluate the nonlinearity acceptable.

4. Conclusions

This paper studies the target position and velocity estimation problem with distributed passive sensors. The problem is formulated with distributed asynchronous sensors connected to a fusion center with communication links. We first present a gross LS algorithm that takes all angle observations from distributed sensors into account to make a LS estimation. The algorithm is simplified after some matrix manipulations, but as it needs local sensors transmitting all local observations to a fusion center, the communication cost is high. Meanwhile, the computation cost at the fusion center is also high. The communication cost is mainly a result of high-dimensional received data. In order to reduce the communication cost and computation cost, we present a linear LS algorithm that approximates local sensor locations and angle observations with linear models and then estimate target position and velocity with the parameters of the linear models. In order to simplify the velocity estimation, we also present a truncated LS algorithm that just take an average operation to estimate target velocity. In this manner, both the communication cost and the computation cost at the fusion center are reduced significantly. However, the linear LS algorithm and the truncated LS algorithm faces the model mismatching problem, namely, if the linear approximation is not accurate anymore, the performance may degrade greatly. That is a difference from the gross LS algorithm, which always benefits from more observations, as if the linear target position model holds.

The performance of the concerned algorithms is verified with numerical results. It is found that with less observations, the truncated LS algorithm performs the best. As the number of observations increase, the linear LS algorithm and the gross LS algorithm perform better. With more observations available, the linear model mismatch and then the gross LS algorithm perform the best. The gross LS algorithm always benefits from more local observations, which is a difference from the other two algorithms. The cost is a higher communication cost and a higher computation cost at fusion center. We also examined the angle distortion problem that is the only nonlinear term in the simulation configurations. Our matrix operations often make the estimation need less computation costs.

Compared to localization and tracking framework, the algorithms with velocity estimation needs a much lower rate of tracking operations, whose matrix inverse operation often need huge computation cost. Meanwhile, it can provide more accurate measures of target states and the tracking algorithm will also benefit from that. In our simulations, the sensor location error is not taken into account. In practice, this is inevitable. If the self-positioning error is non zero-mean Gaussian distributed, one may incorporate this goal in a distributed angle-only based positioning algorithm, which will be considered in our future works.

Author Contributions: Conceptualization, S.Z. and Y.C.; derivation, S.Z.; numerical simulation, R.L., L.W.; simulation configuration, Y.C., X.P. and X.X.; supervision, X.P.; writing, S.Z. and L.W.; proof reading, S.G., X.S. and J.Y. All authors have read and agreed to the published version of the manuscript.

Funding: This research received no external funding.

Institutional Review Board Statement: Not applicable.

Informed Consent Statement: Not applicable.

Data Availability Statement: This study did not report any data.

Conflicts of Interest: The authors declare no conflict of interest.

References

1. Wang, C.L.; Yang, R.J. Research on Airborne Infrared Passive Location Method Based on Orthogonal Multi-station Triangulation. *Laser Infrared* **2007**, *11*, 1184–1187+1191.
2. Wang, Y.; Ho, V. An Asymptotically Efficient Estimator in Closed-Form for 3-D AOA Localization Using a Sensor Network. *IEEE Trans. Wirel. Commun.* **2015**, *14*, 6524–6535. [CrossRef]
3. Bai, G.; Liu, J.; Song, Y.; Zuo, Y. Two-UAV Intersection Localization System Based on the Airborne Optoelectronic Platform. *Sensors* **2017**, *17*, 98. [CrossRef] [PubMed]

4. Peng, S.; Zhao, Q.; Ma, Y.; Jiang, J. Research on the Technology of Cooperative Dual-Station position Based on Passive Radar System. In Proceedings of the 2020 3rd International Conference on Unmanned Systems (ICUS), Harbin, China, 27–28 November 2020.

5. Zhu, Y.; Liang, S.; Gong, M.; Yan, J. Decomposed POMDP Optimization-Based Sensor Management for Multi-Target Tracking in Passive Multi-Sensor Systems. *IEEE Sens. J.* **2022**, *22*, 3565–3578. [CrossRef]

6. Chen, Y.; Wang, L.; Zhou, S.; Chen, R. Signal Source Positioning Based on Angle-Only Measurements in Passive Sensor Networks. *Sensors* **2022**, *22*, 1554. [CrossRef] [PubMed]

7. Yin, J.; Wan, Q.; Yang, S.; Ho, K.C. A Simple and Accurate TDOA-AOA Localization Method Using Two Stations. *IEEE Signal Process. Lett.* **2015**, *23*, 144–148. [CrossRef]

8. Dogancay, K. On the bias of linear least squares algorithms for passive target localization. *Signal Process.* **2004**, *84*, 475–486. [CrossRef]

9. Dogancay, K. Bearings-only target localization using total least squares. *Signal Process.* **2005**, *85*, 1695–1710. [CrossRef]

10. Dogancay, K. Relationship Between Geometric Translations and TLS Estimation Bias in Bearings-Only Target Localization. *IEEE Trans. Signal Process.* **2008**, *56*, 1005–1017. [CrossRef]

11. Wu, W.; Jiang, J.; Fan, X.; Zhou, Z. Performance analysis of passive location by two airborne platforms with angle-only measurements in WGS-84. *Infrared Laser Eng.* **2015**, *44*, 654–661.

12. Frew, E.W. Sensitivity of Cooperative Target Geolocalization to Orbit Coordination. *J. Guid. Control. Dyn.* **2008**, *31*, 1028–1040. [CrossRef]

13. Yi, Z.; Li, Y.; Qi, G.; Sheng, A. Cooperative Target Localization and Tracking with Incomplete Measurements. *Int. J. Distrib. Sens. Networks* **2014**, *2014*, 1–16.

14. Dogancay, K. 3D Pseudolinear Target Motion Analysis From Angle Measurements. *IEEE Trans. Signal Process.* **2015**, *63*, 1570–1580. [CrossRef]

15. Wang, L.; Zhou, S.; Chen, Y.; Xie, X. Bias Compensation Kalman Filter for 3D Angle-only Measurements Target Tracking. In Proceedings of the 2022 International Conference on Radar Systems, Edinburgh, UK, 24–27 October 2022.

16. Mallick, M.; Krishnamurthy, V.; Vo, B.N. Angle-Only Filtering in Three Dimensions. In *Integrated Tracking, Classification, and Sensor Management: Theory and Applications*; Wiley: Hoboken, NJ, USA, 2012; pp. 1–42. [CrossRef]

17. Arasaratnam, I.; Haykin, S. Cubature Kalman Filters. *IEEE Trans. Autom. Control.* **2009**, *54*, 1254–1269. [CrossRef]

18. Julier, S.; Uhlmann, J. Unscented filtering and nonlinear estimation. *Proc. IEEE* **2004**, *92*, 401–422. [CrossRef]

19. Rao, S.K. Pseudo-linear estimator for bearings-only passive target tracking. *IEE Proc. Radar Sonar Navig.* **2001**, *148*, 16–22. [CrossRef]

20. He, S.; Wang, J.; Lin, D. Three-Dimensional Bias-Compensation Pseudomeasurement Kalman Filter for Bearing-Only Measurement. *J. Guid. Control. Dyn.* **2018**, *41*, 2678–2686. [CrossRef]

21. Yu, J.Y.; Coates, M.J.; Rabbat, M.G.; Blouin, S. A Distributed Particle Filter for Bearings-Only Tracking on Spherical Surfaces. *IEEE Signal Process. Lett.* **2016**, *23*, 326–330. [CrossRef]

22. Pang, F.; Dogancay, K.; Nguyen, N.H.; Zhang, Q. AOA Pseudolinear Target Motion Analysis in the Presence of Sensor Location Errors. *IEEE Trans. Signal Process.* **2020**, *68*, 3385–3399. [CrossRef]

23. Kolawole, M.O. Estimation and tracking. *Radar Syst. Peak Detect. Track.* **2002**, *287*. [CrossRef]

24. Mallick, M. A Note on Bearing Measurement Model. 2018 . Available online: https://www.researchgate.net/publication/325214760_A_Note_on_Bearing_Measurement_Model (accessed on 7 November 2022).

25. Mallick, M.; Nagaraju, R.M.; Duan, Z. IMM-CKF for a Highly Maneuvering Target Using Converted Measurements. In Proceedings of the 2021 International Conference on Control, Automation and Information Sciences (ICCAIS), Xi'an, China, 14–17 October 2021; pp. 15–20. [CrossRef]

MDPI

St. Alban-Anlage 66

4052 Basel

Switzerland

Tel. +41 61 683 77 34

Fax +41 61 302 89 18

www.mdpi.com

Sensors Editorial Office

E-mail: sensors@mdpi.com

www.mdpi.com/journal/sensors